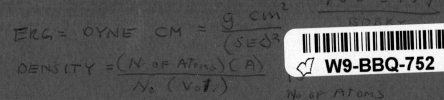
$ERG = DYNE \cdot CM = \dfrac{g \cdot cm^2}{(SEC)^2}$

$DENSITY = \dfrac{(N\ OF\ ATOMS)(A)}{N_0\ (VOL.)}$

$PLANAR\ ATOMIC\ DENSITY = \dfrac{No\ OF\ ATOMS}{AREA\ OF\ PLANE}$

$PEARLITE = \alpha + Fe_3C = FERRITE + CEMENTITE$
$\qquad A\ EUTECTOID\ MIXTURE$

$LEDEBURITE = \gamma + Fe_3C = AUSTENITE + CEMENTITE$
$\qquad A\ EUTECTIC\ MIXTURE$

$TRANSFORMED\ LEDEBURITE = PEARLITE + Fe_3C$

$2\% \ C\ OR\ LESS = \underline{STEELS}\ ;\ GREATER = \underline{CAST\ IRON}$

$ARRHENIUS\ EQN = R = Ae^{-Q/RT}$

$DIFFUSION = D = D_0 e^{-Q/RT}$

7.3

$\tau = G\gamma$

$2(1+r) = 2(1+.35) = 2.7$

$\tau = F/s = \dfrac{Y \times 10^4}{16}$

$\quad = 2.68° = .0446° = 7.8 \times 10^{-4} \text{ rad.}$

$\quad = 5 \times 10^3 \text{ psi}$

$G = \dfrac{5 \times 10^3}{7.8 \times 10^{-4}} = 6.41 \times 10^6$

$E = 2G(1+r) = 2.7(6.4 \times 10^6) = 17.32 \times 10^6 \text{ psi}$

$K = \dfrac{E}{3(1-2r)} = \dfrac{17.32 \times 10^6}{3(.3)} = 19.25 \times 10^{-6} \text{ psi}$

7.4

$E = 28.5 \times 10^6 \text{ psi}$

$\sigma_t = E\epsilon$

$\tau = G\gamma$

$\sigma_c = K\epsilon_v$

$\dfrac{\sigma_T}{\tau} = 2.82 = \dfrac{E\epsilon}{G\gamma} = \dfrac{E}{G}(1.055)$

$\dfrac{E}{G} = 2.675 = 2(1+r)$

$\qquad r = .3375$

$\sigma_c = K\epsilon_v$

$50,000 = K(.001545)$

$\qquad K = 3.34 \times 10^7 \text{ psi}$

$G = \dfrac{E}{2.675} = 1.19 \times 10^7 \text{ psi}$

$E = (3.24 \times 10^7)(3[1-2(.3375)])$

$\quad = 3.18 \times 10^7 \text{ psi}$

MATERIALS ENGINEERING SCIENCE

AN INTRODUCTION

MATERIALS ENGINEERING SCIENCE

AN INTRODUCTION

Richard W. Hanks

Brigham Young University

Harcourt, Brace & World, Inc.

New York / Chicago / San Francisco / Atlanta

See pages listed below for sources of display photographs:

Front cover, top: Fig. 4-21, page 129
Front cover, bottom: Fig. 4-17, page 125
Back cover, top: Fig. 6-18, page 212
Back cover, bottom: Fig. 4-22, page 130
Title page: Fig. 7-13, page 244
Part I: Fig. 4-36, page 143
Part II: Fig. 8-6, page 274

Figures drawn by Bertrick Associate Artists, Inc.

ISBN: 0-15-555183-3

Library of Congress Catalog Card Number: 70-113707

Printed in the United States of America

PREFACE

Many materials that the engineer encounters have similar physical properties originating from the same basic principles. If the undergraduate engineering student is to cope successfully with an ever-increasing list of new and special materials, he must have a sound knowledge of these fundamental principles. This book was written to make the theory of materials science accessible to the student upon completion of his freshman engineering prerequisites.

The subject of the properties of materials is usually covered in either too detailed or too cursory a manner to be useful to the beginner. A beginning engineering student is not (indeed should not be) expertly trained in any particular area of materials science. He may not even be aware that whole classes of materials exist—much less know their detailed properties. Consequently, he must be exposed to a wide variety of modern materials, their properties, and their interrelation and dependence upon molecular, metallic, or crystalline structure.

The chapters of the book proceed from fundamentals to applications and from the simple to the more difficult. It is assumed that the student has been introduced to molecular and atomic structure and to elementary thermodynamics in freshman chemistry. Sophomore physics is not a prerequisite, but concurrent registration in a sophomore physics course is desirable. No mathematical preparation beyond a freshman course in differential calculus is assumed.

The book is based on my one-semester, sophomore-level inter-disciplinary service course offered to students in all the engineering departments at Brigham Young University; hence it is intended for use in other one-semester or two-quarter courses, including the beginning course for students majoring in Materials Engineering Science.

Each chapter in the book contains a list of references to provide further useful reading for the student. A list of problems that can be solved in connection with the study of the textual material also follows each chapter. These problems vary in difficulty; some can be solved simply, by following the logic of the worked examples appearing occasionally in the text, while others require thought and extrapolation. Although the latter problems are more difficult, they are also more valuable in that they assist the student in assimilating and mastering the principles involved.

At the end of the book are several appendixes containing numerical data on properties of many specific materials. These appendixes are included to make the book useful as a reference as well as a text.

I make no pretense at completeness or exhaustive exposition; indeed I recognize that whole treatises have been written on each of the subjects the book introduces. If the undergraduate engineering student finds his interest whetted sufficiently to seek deeper and more scholarly understanding in any of these areas, or if he is only brought to an awareness that some reason for the apparently unending variety of materials exists, this book will have been successful. Engineering materials is a fascinating and challenging subject—one which I hope the undergraduate engineering student will appreciate more fully after studying this book.

I am indebted to many people, both directly and indirectly, for their help in the preparation of the manuscript. I especially wish to thank my colleague, Professor Dee H. Barker, for many stimulating discussions of the material. I am also grateful to the many reviewers who greatly helped to crystallize my thinking. Finally, I am very much indebted to my wife, Shirlee, who has cheerfully typed and retyped most of the manuscript and offered much encouragement to my sometimes lagging spirits. I, of course, accept the full responsibility for all failings and shortcomings of the book.

Richard W. Hanks

CONTENTS

PART

MACROSCOPIC PROPERTIES OF MATERIALS:

RESPONSE TO ENVIRONMENTAL CONDITIONS

APPENDIXES

MATERIALS
ENGINEERING
SCIENCE

AN INTRODUCTION

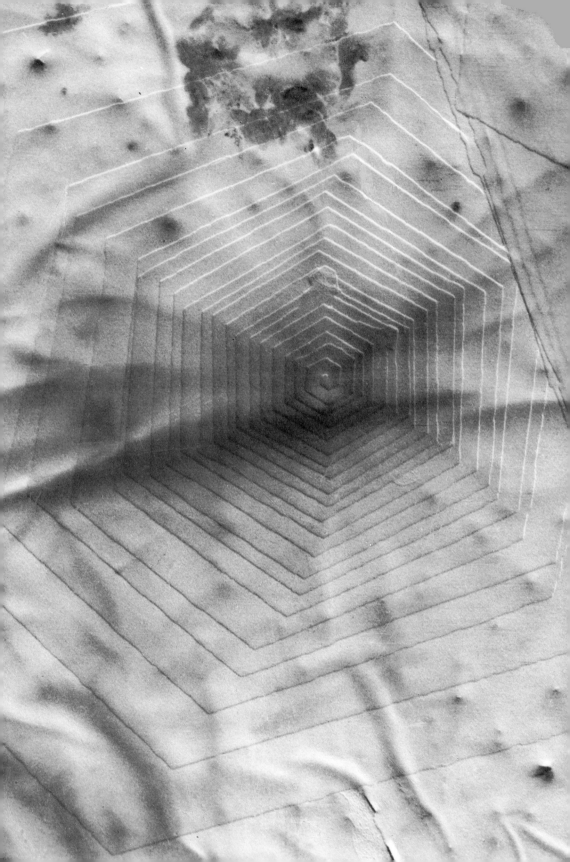

PART I

MICROSCOPIC PROPERTIES OF MATTER: MOLECULAR STRUCTURE AND PHYSICAL-CHEMICAL PRINCIPLES

1

ENGINEERING MATERIAL NOMENCLATURE

The engineer must deal with a multitude of materials, many of them new and previously unknown. Regardless of his field of specialization he must have a fundamental understanding of the principles that govern the behavior of engineering materials. The objective of a beginning course in engineering materials science is to acquaint the engineering student with the importance of microscopic and atomic characteristics in determining the macroscopic properties of engineering materials. This will involve a study of some of the basic physical and chemical aspects of matter.

Because the engineer employs numerous technical terms, it is appropriate to begin a study of engineering materials with definitions and brief descriptions of several common concepts and terms that will recur frequently throughout the book. In this chapter, the student will be introduced briefly to the many physical phenomena more completely described in subsequent chapters.

Although a much finer subdivision of properties of matter is possible, only four general classes of properties will be introduced here. They will be grouped around material responses to such environmental variables as mechanical forces, temperature fields, electromagnetic fields, and chemical environments, each of which is the subject of a separate chapter containing detailed discussions of these individual properties and their underlying causes later in the book.

Mechanical Properties

The mechanical properties of matter associated with its response to mechanical forces are generally the most familiar. A quantity more useful than force is the *engineering stress* σ which is the ratio of the magnitude of a force to the magnitude of the original undeformed area of the body upon which it is acting. This term is almost universally used in practical calculations, and the most common engineering units of stress are pounds force per square inch ($lb_f/in.^2$), abbreviated to psi.

EXAMPLE 1.1

A cylindrical rod 0.505 in. in diameter is subjected to an axial 10,800 lb_f load. A square bar 1 in. on a side is subjected to an axial load of 50,000 lb_f. Calculate the stress in each case.

Solution:

For the rod: $\sigma = \dfrac{(4)(10,800)}{(3.14)(0.505)^2} = 54,000 \text{ psi}$

For the bar: $\sigma = \dfrac{(50,000)}{(1.0)^2} = 50,000 \text{ psi}$

A stressed material undergoes deformation or *strain* ϵ, defined quantitatively as either the incremental deformation divided by the initial dimension or as a per cent of the original dimension. Since strain is a dimensionless quantity, a strain of 0.001 could equally well be in./in. or ft/ft. Two fundamentally different types of strain are observed.

Elastic strain is recoverable upon release of stress. In other words, when the causal[1] stress is removed, the resultant strain vanishes and the original dimensions of the body are recovered. Elastic strain is experimentally observed to be proportional to the applied stress. The constant of proportionality is called the *modulus of elasticity E*, where

$$\sigma = E\epsilon \qquad\qquad (1.1)$$

The modulus of elasticity, or *Young's modulus*, is a property characteristic of the stressed material. A *Hookean solid* is a material which obeys Eq. (1.1). In Chapter 3 it will be shown that this behavior is a direct result of the atomic bonding forces and that the magnitude of E is related to the type of bonds existing in the material.

---

[1]Some authors prefer to think of the strain as causal and the stress as resultant.

As the stress is increased, a value is eventually reached where a permanent or *plastic strain* occurs due to the internal realignment of the constituent particles of the body caused by the presence of various types of imperfections in the crystal structure. No simple mathematical representation of the relation between stress and plastic strain can be given. Consequently, the complete stress-strain response of a material must be determined experimentally and is usually presented graphically as an *engineering stress-strain diagram*, illustrated schematically in Fig. 1-1.

The $\overline{OP}$ portion of the curve in Fig. 1-1 is frequently linear, and corresponds to the elastic strain region described by Eq. (1.1). Point P, called the *proportional limit*, is the end of the linear curve (or Hooke's law curve). Point Y is the *yield point* and the stress σ_Y is the *yield strength*. Here the onset of significant plastic deformation occurs. In other words, σ_Y is the stress required to cause massive movement of

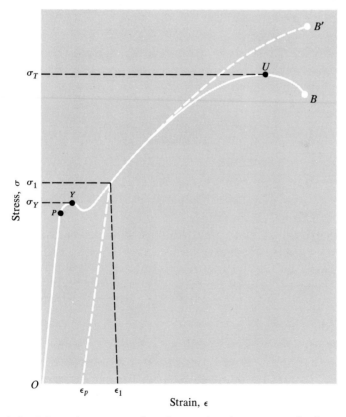

FIG. 1-1 Schematic representation of an engineering stress-strain diagram for a ductile material. σ_Y is the yield stress and σ_T is the tensile stress.

the crystal imperfections. Occasionally the $\overline{PY}$ portion of the curve is very short or negligible, and points P and Y essentially coincide.

The stress σ_T, the *tensile* or *ultimate strength* of the material, is obtained by dividing the greatest load placed on the material during the tensile test by the *original* cross-sectional area of the sample. Point B corresponds to the breaking or fracture of the material.

The experimental technique used to obtain the data for Fig. 1-1 is responsible for the shape of the $\overline{UB}$ curve. A standard[2] sample of the material, as illustrated for plate stock in Fig. 1-2, is placed in a tensile testing machine and a load is applied. The difference between the initial gage length and the length under load, divided by the initial gage length, is the *engineering strain*. The loading force divided by the original cross-sectional area is the *engineering stress*. Most solid materials are practically incompressible and, as the length of the sample increases under stress, constant volume is maintained by a decrease in the cross-sectional area. Therefore, the *true stress* (the load divided by the actual cross-sectional area under load) exceeds the engineering stress in the sample. Local imperfections in many materials cause the decrease in area to become more pronounced in one place. This results in marked *necking-down* of the sample, an increase in the stress, and the ultimate fracture of the test piece.

When necking of the sample becomes pronounced (above σ_T), smaller loads will maintain the true stress level. As the load force is reduced, the engineering stress, based on the original cross-sectional area, decreases in magnitude and this results in the downward curvature from U to B in Fig. 1-1. Therefore, the *engineering breaking stress* is less than the *engineering tensile strength* of the material. This apparent anomaly is removed if the true stress and true strain (dashed curve to B' in Fig. 1-1) are plotted. The true strain may be calculated by integrating the differential strain

$$d\epsilon' = \frac{dL}{L} \tag{1.2}$$

between the initial length L_0 and the final length L_f. Thus, the true total strain is

$$\epsilon' = \int_{L_0}^{L_f} \frac{dL}{L} = \ln \frac{L_f}{L_0} \tag{1.3}$$

If the material is essentially incompressible,

$$V_0 = A_0 L_0 = A_f L_f = V_f \tag{1.4}$$

---

[2]See T. Baumeister and L. S. Marks, *Mechanical Engineers' Handbook, 6th ed.,* McGraw-Hill Book Co., Inc., New York, 1958, p. **5-19**, for details of several standard shapes.

and Eq. (1.3) becomes

$$\epsilon' = \ln \frac{A_0}{A_f} \tag{1.5}$$

where A_0 and A_f are the initial and final cross-sectional areas of the sample, respectively. True strain is also called *natural strain*.

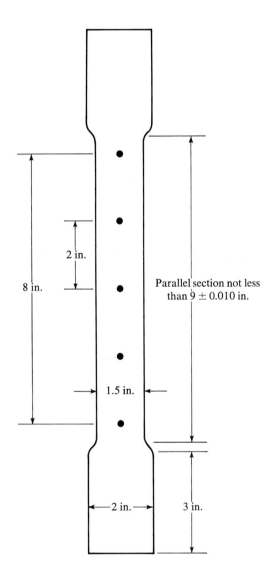

8 in.

2 in.

1.5 in.

2 in.

3 in.

Parallel section not less than 9 ± 0.010 in.

FIG. 1-2 Schematic representation of a typical tensile test specimen for plate stock (ASTM E8-54T). A common, but not universal, gage length is 2 in.

EXAMPLE 1.2

The rod in Example 1.1 is loaded to 24,000 lb$_f$ and has a final diameter of 0.35 in. Calculate the true stress and strain and the engineering stress and strain.

Solution:

$$\text{True stress} = \frac{(4)(24,000)}{(3.14)(0.35)^2} = 249,600 \text{ psi}$$

$$\text{True strain} = (2)\ln\frac{0.505}{0.35} = 0.733 \text{ in./in.}$$

$$\text{Engineering stress} = \frac{(4)(24,000)}{(3.14)(0.505)^2} = 120,000 \text{ psi}$$

$$\text{Engineering strain} = \frac{L_f - L_0}{L_0} = \frac{L_f}{L_0} - 1 = \frac{A_0}{A_f} - 1$$

$$= \frac{d_0^2}{d_f^2} - 1 = \cancel{1.433} - 1 = \cancel{0.443} \text{ in./in.}$$

[handwritten: 2.08 - 1 = 1.08]

Because the reduction in the cross-sectional area of the sample is gradual, not abrupt, the sample is not in a state of pure tension. Radial and tangential stresses exist in the necked region, and the corrections [1][3] for these give a curve intermediate to the two curves in Fig. 1-1.

Since elastic strain is recoverable but plastic strain is not, unloading the test piece at some total strain $\epsilon_1 > \epsilon_Y$ results in the material returning to a state of zero applied stress along the dashed line parallel to $\overline{OP}$ in Fig. 1-1. The residual strain ϵ_p is the plastic or permanent strain. Upon reloading, the material would follow the dashed curve to ϵ_1 elastically, at which point additional plastic strain would occur. The increase in yield value from σ_Y to σ_1 is called *strain-hardening*.

Some ductile materials, certain aluminum alloys for example, possess no sharply defined yield stress. In such cases the yield strength is defined *arbitrarily* as that stress called the *offset yield stress* which results in a certain permanent strain, usually 0.2 per cent (illustrated by the slanted dashed line in Fig. 1-3). Figure 1-1 is typical of a ductile material such as a mild carbon steel.

Ductility is the plastic strain at rupture and may be expressed either as strain or as reduction in area at rupture.

$$\text{Reduction in area} = \frac{A_0 - A_f}{A_0} \qquad (1.6)$$

---

[3]Numerals in brackets refer to literature references at the end of each chapter.

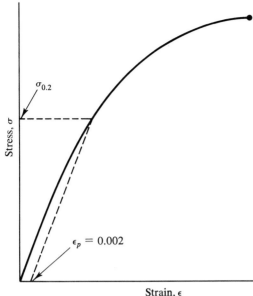

FIG. 1-3   Schematic stress-strain diagram for a completely ductile material, illustrating the 0.2% offset yield stress.

EXAMPLE 1.3

The rod in Examples 1.1 and 1.2 broke under a load of 25,000 lb$_f$ and had a diameter of 0.31 in. at rupture. Calculate the ductility of this material.

Solution:

$$\text{Ductility} = \text{Reduction in area} = \frac{100(A_0 - A_f)}{A_0} = \frac{100(d_0^2 - d_f^2)}{d_0^2}$$

$$= 100\left(1 - \left(\frac{0.31}{0.505}\right)^2\right) = 62.3\%$$

As a function of bond types, ductility varies markedly. Silver and copper are extremely ductile, while ordinary steel is only moderately ductile, and cast iron is brittle or completely nonductile under ordinary conditions.

Hardness is a measure of the resistance of a material to surface penetration. Several of the existing relative scales of hardness are compared in Fig. 1-4. In the Moh hardness test, the scratch resistance of the material is matched with that of one of ten arbitrarily chosen minerals, arranged in order of increasing hardness. The softest ma-

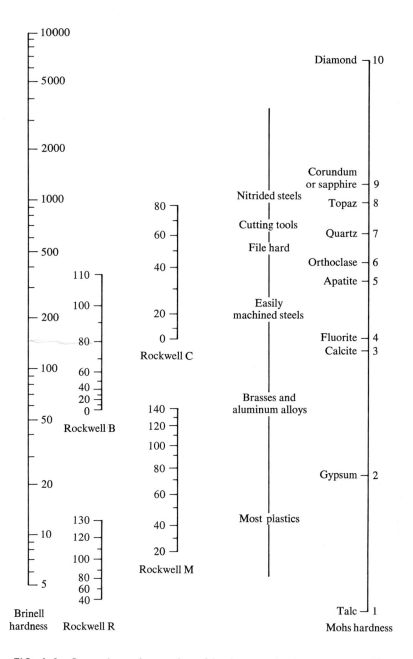

FIG. 1-4 Comparison of a number of hardness scales in current use. (From *Engineering Properties and Applications of Plastics*, by G. F. Kinney, copyright 1957, John Wiley and Sons, Inc. Used by permission.)

terial, talc, is given the arbitrary hardness number 1, and the hardest, diamond, is assigned the number 10. All materials fall somewhere between 1 and 10. The Moh scale is nonlinear, since some minerals in it are very close to one another in relative hardness; it is used only by field mineralogists for qualitative analyses.

In most widely used hardness tests, the resistance of the material to surface indentation under rigidly standardized conditions is measured by pressing a hardened intender of standard shape into the surface of the material under a specified load. The area of indentation or the depth of penetration is measured and assigned a numerical value. For metals and certain plastics the most common methods bear the names Brinell, Vicker, and Rockwell. As illustrated in Fig. 1-5, some correla-

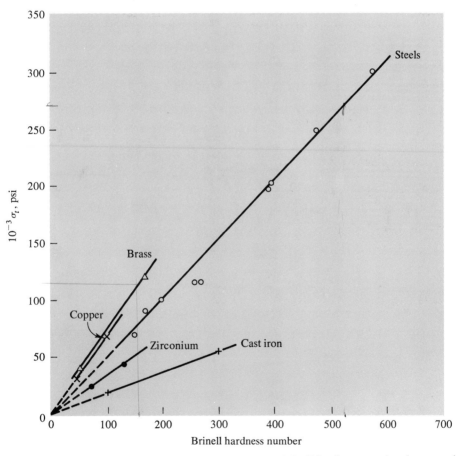

FIG. 1-5 Approximate relation of tensile strength to Brinell hardness number for several metals. (From Table 1, Section 5, p. 5, *Mechanical Engineers Handbook, 6th ed.*, by T. Baumeister and L. S. Marks, copyright 1958, McGraw-Hill Book Co., Inc. Used by permission.)

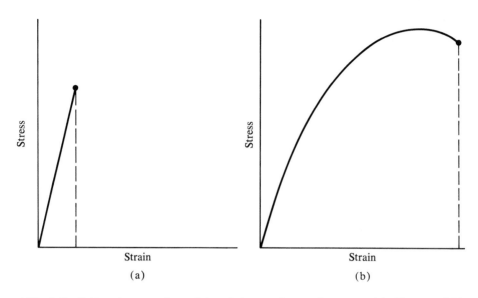

FIG. 1-6 Schematic comparison of the relative toughness of two materials. The material in part (b) is tougher than that in part (a) because the area under the stress-strain curve up to the breaking point is much greater in (b) than in (a). Consequently, more energy is absorbed by (b) than (a).

tion exists between hardness and tensile strength. (This correlation is only approximate, however, and extrapolations beyond the data given in the graph should not be made.)

Toughness is a measure of the energy required to rupture a material in a tensile test, and is proportional to the area under the stress-strain curve. The greater this area, the more strain energy is absorbed and the

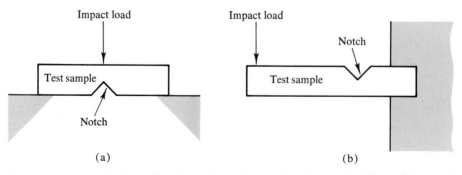

FIG. 1-7 Schematic illustration of the shape of test pieces for use in (a) the Charpy test and (b) the Izod test. The impact load is applied at the indicated point by swinging a specified weight on an arm from a specified height. The notches are carefully machined to specified dimensions.

tougher the material is. The relative toughness of two materials is illustrated schematically in Fig. 1-6.

A quantitative measure of toughness is the toughness modulus $\hat{T}$, which represents the maximum amount of energy a unit volume of the material can absorb without fracture, and hence the area under the

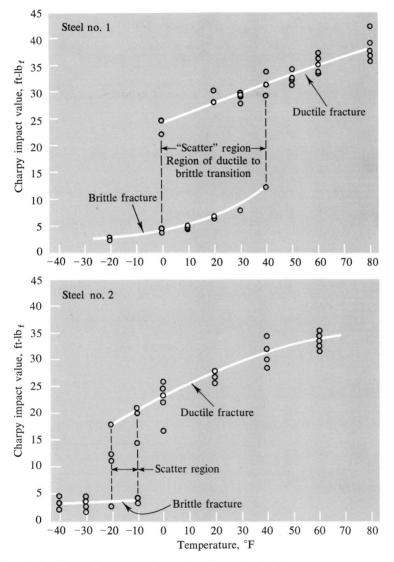

FIG. 1-8 Charpy impact toughness data illustrating brittle to ductile transition for two steels using keyhole shaped notch in specimens. (From an article by N. A. Kahn and E. A. Imbembo, [2]. Used by permission.)

$\sigma = \epsilon$ curve. It is expressed as stress times strain, and has units of ft lb_f/ft^3 or energy per unit volume. If the complete stress-strain curve is not available, $\hat{T}$ may be estimated as

$$\hat{T} = \tfrac{1}{2}(\sigma_Y + \sigma_T)\epsilon_B \qquad (1.7)$$

Some materials such as concrete have nearly parabolic stress-strain curves, for which $\hat{T}$ is approximately

$$\hat{T} = \tfrac{2}{3}\sigma_T\epsilon_B \qquad (1.8)$$

Toughness is determined by standardized high-speed tensile tests in which an impact load is applied to a notched specimen by swinging a weight from a specified height, as illustrated in Fig. 1-7. The energy required to break the specimen is the *impact toughness*. The two different sample arrangements indicated in Fig. 1-7 yield totally different data; the arrangement must be specified before the data are meaningful.

A single test at room temperature is inadequate in determining the ductile or brittle nature of a material under actual service conditions. Several tests over a range of temperatures are required because some, although not all, materials show a sharp transition from brittle to ductile behavior at certain temperatures [2], as illustrated in Fig. 1-8 for a certain type of steel. If this brittle transition temperature lies within the range of operating conditions, considerable difficulties may be encountered. An unfortunate example occurred in a number of the Liberty ships used by the United States in World War II. As the ships entered cold, arctic waters, their steel hulls passed through the ductile to brittle transition and were easily fractured upon contact with solid objects or other severe stress concentrations.

Thermal Properties

In any discussion of thermal properties a careful distinction must be made between the concepts of heat and temperature. *Heat* is correctly thought of as energy in transit from one region of space to another under the influence of a temperature gradient. The internal energy content of materials is often erroneously equated with heat. *Temperature* is a measure of the *level* of thermal energy within a body which is caused by the kinetic motion of its atoms or molecules. Bodies having more thermal energy are at higher temperatures.

At the *melting* or *softening temperature*, the constituent particles of a material assume a different arrangement. Pure crystalline solids exhibit sharply defined melting temperatures, while materials such as metallic alloys, glasses, plastics, rubbers, and clays do not melt at definite temperatures, but soften over a range of temperatures. The energy changes associated with these molecular rearrangements are called *latent heat changes*.

To be serviceable at high temperatures, a material must not only have a high melting or softening point, but must retain its mechanical properties at elevated temperatures as well. Although pure iron melts near 2800°F, for example, it loses its useful mechanical strength above about 1000°F. The addition of alloying materials decreases the melting point, but increases the high temperature strength. Certain nickel, chromium, or cobalt alloys can be used at temperatures [3] as high as 1500°F. Nickel base alloys such as Inconel, Incoloy, Nimonic, and a series of Hastelloys are used in the 1500–1800°F range [3], and cobalt alloys are available for service in the 1600–2200°F range [3].

Some pure metals such as molybdenum, titanium, tungsten, and zirconium have melting points in excess of 3000°F. However, these metals have rather poor high temperature oxidation resistance and are difficult to fabricate.

Hybrid materials called cermets (ceramic-metal combinations) have been developed to withstand high temperatures and rapid, large temperature fluctuations.

Organic polymeric materials such as plastics, rubbers, and woods have poor heat resistance, and many of them show a tendency to decompose without melting. Most thermoplastics (polymeric materials which soften markedly with increased temperatures) must be used below 150°F. Some thermosetting plastics (polymeric materials which do not soften, but usually decompose with increased temperatures) can be used up to 300°F. This property is useful in designing heat shields for high heat input over a small time increment. Certain thermosetting silicones and fluorocarbons can be used at temperatures as high as 500°F.

Heat capacity is the amount of heat energy which must be added to a unit mass in order to raise its temperature level one degree. The only difference between heat capacity and the Btu or the calorie is that a particular temperature and material are specified in defining the latter two energy units. Hence, by definition, the heat capacity of water at 39°F or 15°C is 1 Btu/lb°F or 1 cal/g°C.

Specific heat is the ratio of the heat capacity of a material to the heat capacity of water at some reference temperature. Two types of heat capacities are commonly encountered in technical work: the constant-volume and constant-pressure heat capacities C_v and C_p, respectively. Most materials expand with the addition of heat energy at constant pressure. Some of the heat energy must be used to expand the volume against the constant pressure and, therefore, C_p exceeds C_v.

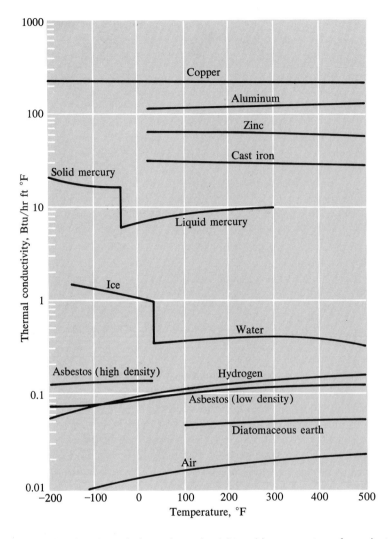

FIG. 1-9 Variation of thermal conductivity with temperature for selected materials. (From *Heat and Thermodynamics, 3rd ed.*, by M. W. Zemansky, copyright 1951, McGraw-Hill Book Co., Inc., and *Heat Transmission, 3rd ed.*, by W. H. McAdams, copyright 1954, McGraw-Hill Book Co., Inc. Used by permission.)

Heat capacity is dependent on temperature level. The rule of Dulong and Petit [4] states that the constant-pressure, atomic heat capacities of solid crystalline elements approximate 6.4 cal/g-atom°C, the g-atom being the mass in grams numerically equal to the atomic weight of the element. At room temperature and slightly above, this rule holds reasonably well for elements with atomic weights greater than about 40. However, if the temperature is lowered, the atomic heat capacities of all elements will drop rapidly and approach zero at the absolute zero of temperature.

Nearly all solid materials expand when heated and contract when cooled. Isotropic materials (those whose properties are not different in different directions) expand equally in all directions. The resulting strain per degree of temperature rise is called the *coefficient of thermal expansion*, defined by

$$\epsilon = \alpha_T \, \Delta T \tag{1.9}$$

Temperature gradients (temperature differences per unit distances) induce thermal stresses in an object which, if large enough, can cause mechanical failure of the material. The coefficient of thermal expansion varies with temperature in nearly the same manner as does heat capacity.

Thermal conductivity relates *heat flux* (the flow of heat energy per unit area, per unit time) to the temperature gradient causing the heat flux, and is defined by

$$q_x = -k \frac{dT}{dx} \tag{1.10}$$

where q_x is the heat flux in the x-direction, $-dT/dx$ is the temperature gradient in the same direction, and k is the thermal conductivity. The units of thermal conductivity are Btu/hr ft^2 (°F/ft) or cal/sec cm^2 (°C/cm). Thermal conductivity is temperature dependent. Its magnitude varies over a wide range and is determined by the material in question (see Fig. 1-9 [5]).

EXAMPLE 1.4

An insulating material is placed inside the walls of a house. The walls are 6 in. thick and the material has a thermal conductivity $k = 0.03$ Btu/hr ft °F. Calculate the heat flux through the wall if the inside temperature is 70°F and the outside temperature is 0°F.

Solution:

For this system Eq. (1.10) can be written

$$q_x = -k \frac{dT}{dx} = -\frac{k \, \Delta T}{\Delta x}$$

Assuming the dimensions and temperature quoted apply to the insulation

$$q_x = -\left(\frac{0.03 \text{ Btu}}{\text{hr ft °F}}\right) \frac{(0 - 70)°F}{6 \text{ in.}} \frac{(12 \text{ in.})}{\text{ft}} = \frac{4.2 \text{ Btu}}{\text{hr ft}^2}$$

The *thermal diffusivity* of a material is defined as the ratio $k/\rho C_p$, where k is the thermal conductivity, C_p is the constant-pressure heat capacity, and ρ is the mass density of the material. The units of thermal diffusivity are ft^2/hr or cm^2/sec. The term C_p does not vary appreciably with temperature, while k does. Consequently, thermal diffusivity is temperature dependent in a manner similar to thermal conductivity.

Rapid heating generates large temperature gradients in a material resulting in differential thermal expansion and the development of high mechanical stresses. If the stress exceeds the breaking strength of the material, surface failure or *spalling* usually occurs. Spalling results in portions of the surface layers of the material flaking off and causes general surface deterioration.

A *thermal shock resistance* parameter may be defined [6] as

$$TSR = \frac{k\sigma_T}{E\alpha_T} \qquad (1.11)$$

where k is thermal conductivity, σ_T is tensile strength, E is Young's modulus, and α_T is the coefficient of linear thermal expansion. High values of TSR, for one material relative to another, indicate relative resistance to spalling. Brittle materials such as glass and ceramics are particularly sensitive to thermal shock because they experience brittle failure instead of plastic yield. Materials such as Pyrex glass and Pyroceram are highly shock resistant because of a low coefficient of thermal expansion.

Heating or cooling rates are important in that they determine temperature gradients. For example, a coke oven must be cooled or heated slowly to prevent excessive temperature gradients which would cause the failure of its fire-brick lining.

The interiors of large massive objects do not respond as quickly to large or rapid temperature fluctuations as do their surfaces. Thus, large surface thermal gradients can cause large thermal stresses and brittle failure of the surface layers. This sometimes occurs in the heat treatment of large metal parts.

EXAMPLE 1.5

Compare the thermal shock resistance parameter for alumina, porcelain, and berylia ceramics, and for fiberglass, ASTM-A7 steel, and 6061-T6 aluminum. Then evaluate their relative thermal shock resistances.

Solution:

The properties of these materials and their TSR values are tabulated:

MATERIAL	k (Btu in./hr ft^2 °F)	σ_T (psi)	E (psi)	α_T (°F^{-1})	TSR
Al$_2$O$_3$ ceramic	114	28,500	39.5(10^6)	4.1 (10^{-6})	2.01 (10^4)
Porcelain ceramic	9	7,750	12.5(10^6)	3.06(10^{-6})	0.182(10^4)
BeO ceramic	1,742	30,000	51 (10^6)	3.2 (10^{-6})	32.0 (10^4)
ASTM-A7 steel	460	60,000	28 (10^6)	6 (10^{-6})	16.4 (10^4)
6061-T6 aluminum	1,190	38,000	10 (10^6)	12 (10^{-6})	37.6 (10^4)
Fiberglass	3	20,000	1.5(10^6)	8 (10^{-6})	0.5 (10^4)

Clearly BeO ceramic and 6061-T6 aluminum can stand much greater thermal shocks than can porcelain ceramics or fiberglass.

Electrical Properties

Some materials are good electrical conductors, some called semi-conductors are relatively poor electrical conductors, and some called insulators are very poor electrical conductors. These properties are governed by the energies of electrons in the atoms and bonds of the material. Insulators are essentially nonconductive. *Electrical conductivity* is a measure of the ability of a material to conduct electrical energy and is defined by

$$j_{e_x} = -k_e \frac{dV}{dx} \qquad (1.12)$$

where j_{e_x} is the *electrical flux* (current per unit area), k_e is the electrical conductivity, and $-dV/dx$ is the voltage gradient. The common units of k_e are (ohm-cm)$^{-1}$. Equation (1.12) is a form of Ohm's law.

In metals the conduction mechanism for thermal and electrical energy is essentially the same. These two conductivities are closely related by the Wiedemann–Franz–Lorenz law

$$L_0 = \frac{k}{k_e T} \qquad (1.13)$$

where k is thermal conductivity, k_e is electrical conductivity, T is absolute temperature, and L_0 is the Lorenz constant (2.45 × 10^{-8} watt-ohm/°K^2).

The *electrical resistivity* ($\rho_e = 1/k_e$) is used to calculate electrical resistance

$$R = \rho_e \frac{L}{A} \qquad (1.14)$$

where L is the conductor length, A is its cross-sectional area, and R is the resistance.

EXAMPLE 1.6

The resistivity of an aluminum alloy is 2.8 micro-ohm-cm. Calculate the heat generated in a wire 0.01 cm² in cross-section and 1000 ft in length by a current of 0.25 amperes.

Solution:

The power dissipated in a wire by a current I is given by

$$P = I^2 R = \frac{I^2 \rho_e L}{A} = \frac{(0.25 \text{ amp})^2 (2.8)(10^{-6} \text{ ohm-cm})(10^3 \text{ ft})(30.48 \text{ cm/ft})}{(10^{-2} \text{ cm}^2)}$$

$$= 0.533 \text{ watts}$$

A number of other electromagnetic properties are also important, but their definition requires a more detailed understanding of the molecular phenomena involved so we will defer them to a later chapter.

Chemical Properties

All real materials are subject to chemical deterioration under the proper environmental conditions. This phenomenon is due primarily to the type of intermolecular forces present within the specific material. The form of attack encountered varies according to the material and its bond types. For example, due to the electrical nature of their bonds thermoplastic polymers are susceptible to dissolution by some solvents, but not others. Many metals are readily oxidized in their natural condition (the rusting of iron, for example), but not when alloyed with other metals (stainless steel, an alloy of iron with chromium, is highly resistant to oxidation). Still another form of attack important to understand is galvanic or electrolytic corrosion. Here the difference in the electronic energies of two dissimilar metals results in a difference in their gross electrical potentials. A galvanic cell or *battery* may then be formed, resulting in a flow of electric current which could rapidly deteriorate one of the metals. All of these phenomena are determined by the electronic nature of the atoms and the bonds which are formed.

In this brief introduction to the various types of material properties of interest to engineers, we have suggested that the underlying atomic and molecular structure is important in determining the gross or macroscopic properties of these materials. It is for this reason that we shall be concerned at some length with molecular phenomena in succeeding chapters.

References Cited

[1] Jastrzebski, Z. D., *Nature and Properties of Engineering Materials*, John Wiley and Sons, Inc., New York, 1962, p. 204.

[2] Kahn, N. A., and E. A. Imbembo, *A.S.T.M. Bull.*, **146**, p. 66 (1947).

[3] Levy, A., *Materials and Methods*, **41**, No. 4, Manual 115, p. 117 (April 1955).

[4] Daniels, F., and R. A. Alberty, *Physical Chemistry*, John Wiley and Sons, Inc., New York, 1955, p. 107.

[5] Zemansky, M. W., *Heat and Thermodynamics, 3rd ed.*, McGraw-Hill Book Co., Inc., New York, 1951, p. 85; W. H. McAdams, *Heat Transmission, 3rd ed.*, McGraw-Hill Book Co., Inc., New York, 1954, p. 445*ff*.

[6] Manson, S. S., *Heat Transfer Symposium*, University of Michigan Press, Ann Arbor: Engineering Research Institute, 1953, pp. 9–75. *See also* W. G. Lidman and A. R. Bobrowsky, *NACA RM E*, 9BO7, 1949.

PROBLEMS

1.1 A steel bar 0.75 in. in diameter is loaded to 20,000 lb$_f$. If Young's modulus for this steel is 31.2(10^6) psi, calculate the strain assuming only elastic deformation occurs.

1.2 The steel in Problem 1.1 begins to undergo plastic deformation when the load reaches 35,000 lb$_f$. Calculate (a) the yield strength and (b) the strain at yield for this steel.

1.3 A load of 45,000 lb$_f$ on the bar in Problem 1.1 causes failure. The final diameter of the bar is 0.475 in. Calculate (a) the true stress; (b) the true strain; (c) the ductility of the steel; (d) the engineering strain.

1.4 A copper alloy has the properties $E = 16(10^6)$ psi, $\sigma_Y = 48,000$ psi, and $\sigma_T = 51,000$ psi.

   (a) How much stress is required to produce an elastic deformation of 0.01 in. in a 5 ft bar?
   (b) What diameter round bar will support a 10,000 lb$_f$ load without permanent deformation?
   (c) What minimum load is required to deform an equilateral triangular cross-sectional bar with 1 in. sides plastically?

1.5 What size square bar of the alloy in Problem 1.4 is required to support (without plastic deformation) a load which just caused plastic deformation of the rod in Problem 1.2?

1.6 A certain steel has a Brinell hardness number (BHN) of 525. Estimate its ultimate strength.

1.7 A zirconium alloy has a tensile strength of 72,000 psi. Estimate its Rockwell C hardness.

1.8 A brass alloy has a hardness of 80 R_B. Estimate its tensile strength.

1.9 Estimate the toughness modulus $\hat{T}$ for the steel in Problem 1.3, given that $\sigma_T = 90,000$ psi.

1.10 The following data were obtained with a standard 0.505 in. test specimen of a certain alloy:

Load, lb$_f$	Gage length, in.
0	2.000
4,000	2.010
7,900	2.020
10,300	2.030
11,600	2.040
12,600	2.060
13,200	2.080
13,000	2.120
11,000	2.160
10,800	Broke (diameter = 0.21 in.)

Determine (a) engineering stress-strain diagram; (b) true stress-strain diagram; (c) Young's modulus; (d) yield strength; (e) tensile strength; (f) true and engineering breaking strengths; (g) toughness modulus $\hat{T}$ from Eq. (1.7); (h) $\hat{T}$ from graphical integration under the engineering stress-strain curve.

1.11 The heat capacity of iron is given by the empirical equation $C_p = 3.04 + 7.58(10^{-3})T + 0.60(10^5)T^{-2}$ (cal/mole °C), where T is in °K. Calculate the heat capacity of iron at 350°F. What error would have been made by using the Dulong–Petit rule?

1.12 A housewife finds the children's room cold in the winter. She feels the walls and they are cold to the touch. She reasons that application of some cork to the walls would help to keep the room warmer. The outside wall is 8 ft high and 15 ft long, with an ordinary glass window 4 ft wide by 2 ft high. The average room temperature is 65°F and the average outside temperature is 0°F. The wall is 6 in. thick and filled with rockwool insulation ($k = 0.030$ Btu/hr ft °F). The window is $\frac{1}{8}$ in. thick ($k = 0.45$ Btu/hr ft °F). Calculate:

(a) The heat flux through the wall.
(b) The heat flux through the window.

   (c)   The total heat flow rate through both wall and window separately.

   (d)   Was the housewife's reasoning sound?

1.13   A material is placed in a test apparatus to determine its thermal conductivity. A 100°F temperature difference across a 0.5 in. thick slab results in a 180 Btu/hr ft^2 flux. What is the thermal conductivity of the material (a) in Btu/hr ft °F? (b) in cal/sec cm °C?

1.14   The thermal conductivity of pure copper at 572°F is 212 Btu/hr ft °F. Calculate the electrical resistance in ohms of a copper wire 1000 ft long and 0.005 in. in diameter at this temperature. (Note: 1 Btu = 0.293 watt hr.)

1.15   In the accompanying table, the thermal conductivity of steel is given as a function of temperature. Determine the temperature coefficient of resistivity α_e from the equation $\rho_e = \rho_{0e} + \alpha_e(t - t_0)$, where ρ_{0e} is the electrical resistivity at temperature t_0. Select $t_0 = 200°C$ for this case.

k(Btu/hr ft °F)	t(°C)
26	200
25	300
23	400
22	500
21	600

1.16   The resistivity of brass is 3.2(10^{-6}) ohm cm. $R = \rho_e \dfrac{L}{A}$

   (a)   What is the resistance of 10 ft of a brass wire 0.04 in. in diameter? $P = I^2 R$

   (b)   How much heat is lost from this wire if a current of 5 amperes is passed through it?

   (c)   How much decrease in heat loss would occur if the wire were replaced by a copper wire of the same dimensions ($\rho_e = 1.7(10^{-6})$ ohm cm)?

1.17   How many feet of 10 mil diameter copper wire ($\rho_e = 1.7 \times 10^{-6}$ ohm cm) are required to give a resistance of 5.0 ohms? If this wire were replaced with an aluminum alloy wire ($\rho_e = 2.8 \times 10^{-6}$ ohm cm), how much weight savings would occur?

ATOMIC THEORY
AND THE FORMATION
OF BONDS

Electronic Nature of Atoms

Atomic properties and interatomic attractions provide the basis for an understanding of the macroscopic properties of matter. Early scientists believed atoms to be the smallest subdivision of matter, but we now know that atoms can contain even smaller particles. Current models of the atom envision a massive, positively charged nucleus, composed of positive protons and slightly more massive uncharged neutrons, and surrounded by negative electrons[1] equal in number to the number of protons. Nuclear radii are of the order of 10^{-12} or 10^{-13} cm, while atomic radii are of the order of 10^{-8} cm. Atomic dimensions are expressed in angstrom units, where $1 \text{ Å} = 10^{-8}$ cm.

The *atomic number* is the number of protons in the nucleus. The total number of protons and neutrons in an atomic nucleus is the *atomic mass number*. *Isotopes* are atoms having the same atomic numbers but different atomic mass numbers. Atomic masses are expressed in terms of a relative scale made up by *arbitrarily* [1] assigning to some element an integral number of atomic mass units equal to its atomic

---

[1]The mass of the electron is about $\frac{1}{1836}$ that of the proton, although its size is roughly comparable to both the proton and the neutron.

mass number. This relative mass scale, together with the absolute mass measured for the reference element, determines the magnitude of the atomic mass unit (amu).

The *physical scale* of atomic masses, based on the O^{16} isotope of oxygen having the integral weight 16 amu, is often encountered in physics literature. The absolute magnitude of one amu in this scale is [1]

$$1 \text{ amu} = (1.65983 \pm 0.00010) \times 10^{-24} \text{ g} \qquad \textbf{(2.1)}$$

and the relative weight of the hydrogen atom is 1.008141 amu. In the *chemical scale*, often encountered in the older chemical literature, a weight of 16 amu is assigned to the naturally occurring mixture of oxygen isotopes (99.1758% O^{16}, 0.0373% O^{17}, and 0.2039% O^{18} by weight). The ratio of the amu values of the chemical to the physical scale is 1.000279. In most engineering this difference is negligible, but whenever precise calculations are performed the basis of the amu scale used must be carefully specified.

A compromise scale, based on the assignment of 12 amu to the C^{12} isotope of carbon, is presently in use by both chemists and physicists and is used here. The O^{16} isotope in this scale has the relative weight 15.9994 amu.

Avogadro's number is the number of atoms in a gram-atom, and is assigned the value 6.023×10^{23} atoms/g-atom. The weight of an atom may be calculated by dividing its gram-atomic weight by Avogadro's number.

With the exception of density and specific heat, engineering properties of materials are relatively insensitive to atomic masses. *By far the most important atomic characteristic of matter is the number of extranuclear electrons in the atom. In particular, the outermost or valence electrons determine the majority of the important macroscopic properties of matter.*

Atomic Models

Of the numerous models of atomic structure which have been proposed we will consider two here: the Bohr–Sommerfeld orbital model and the Schrödinger–Heisenberg–Pauli wave mechanical model. The former introduces several significant concepts in the context of the popular, albeit incorrect, visualization of the atom as a micro-solar system. The latter will only be considered in a general, conceptual outline because a complete study of its mathematical details would far exceed the scope of this book. However, the quantum notions associated with this model are indispensable in understanding much of the bonding theory and subsequent material.

The Bohr–Sommerfeld Atom

Because of its relative simplicity, the hydrogen atom has been the subject of considerable theoretical study by physicists. The first quantitatively satisfactory model of the hydrogen atom is due to Bohr [2], who visualized it as a single nuclear proton orbited by a single electron in the same fashion that the planets of the solar system orbit the sun. To avoid the dilemma raised by the classical electrodynamic law which requires all accelerated charges to emit radiation, Bohr introduced the notion of the *quantum energy state*. He suggested that the electron could occupy certain stable circular orbits with certain allowed energies and not emit any radiation; such radiation would only be emitted by the electron when it changed from one allowed orbit to another. (The energies associated with these allowed orbits are now called the quantum energy states of the electron.) Bohr utilized this quantum hypothesis to suggest that the energy emitted (or absorbed) by the electron in changing energy states be given by

$$\Delta E = h\nu \tag{2.2}$$

where h is Planck's constant and ν is the frequency of the electromagnetic quantum of energy emitted or absorbed.

Bohr derived expressions for the radii and energies of the circular orbits as

$$a_n = \frac{n^2 h^2}{4\pi^2 m e^2 Z} \tag{2.3}$$

$$E_n = \frac{-2\pi^2 Z^2 e^4 m}{h^2 n^2} \tag{2.4}$$

where h is Planck's constant, m is the electron mass, e is the electron charge, Z is the atomic number, and n is a nonzero integer called the *quantum number* which assumes the values 1, 2, 3, ... Thus, the energy emitted by the electron as a result of one of its possible quantum transitions was

$$\Delta E_n = \frac{-2\pi^2 Z^2 e^4 m}{h^2} \left[\frac{1}{n_1^2} - \frac{1}{n_2^2} \right] \tag{2.5}$$

Equation (2.5) agreed reasonably well with the experimental observations for hydrogen, lending weight to Bohr's proposed model of the atom.

EXAMPLE 2.1

Calculate the ionization potential of hydrogen (the energy required to remove the electron to infinity) using Eq. (2.5).

Solution:

$$\Delta E_\infty = \frac{(2)(3.14)^2(4.80 \times 10^{-10} \text{ esu})^4(9.107 \times 10^{-28} \text{ g})}{(6.6625 \times 10^{-27})^2 \text{ erg}^2 \text{ sec}^2}\left[\frac{1}{1^2} - \frac{1}{\infty^2}\right]$$

$$= 2.178 \times 10^{-11} \text{ erg} = 13.6 \text{ eV}$$

This result is in good agreement with the experimental value 13.575 eV, where one eV = 1.602×10^{-12} erg is the energy acquired by an electron as it passes through a potential field of one volt.

Bohr's theory of the hydrogen atom in principle could be applied to any hydrogen-like ion, for example, He^+, Li^{++}, Be^{+++}.

Small, but significant discrepancies between Bohr's theory and experimental measurements led Sommerfeld [3] to consider the presence of elliptical as well as circular orbits in the atom. He found that the total energy of the atom depends on a *total quantum number* $n = k' + n'$, where k' is an integer that assumes all values from 1 to n and describes the angular dependence of the energy. The

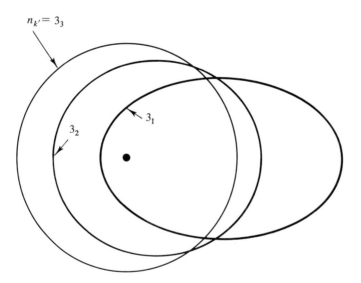

FIG. 2-1   Schematic representation of the Bohr–Sommerfeld orbitals of the hydrogen atom for a principal quantum number $n = 3$.

Sommerfeld orbits for a principal or total quantum number $n = 3$ are illustrated schematically in Fig. 2-1.

The Wave Mechanical Model

Bohr and his successors visualized extranuclear electrons as solid bits of matter following geometrically well-defined paths with specifiable velocities and positions. In 1924, de Broglie [4] postulated that matter in particular electrons possessed wave properties similar to light, and calculated the *wavelength* of an electron as

$$\lambda_m = \frac{h}{p} \tag{2.6}$$

where h is Planck's constant and p is the linear momentum of the electron.

EXAMPLE 2.2

An atom of hydrogen in a gas moves at its mean thermal speed $\bar{v} = (8kT/\pi m)^{1/2}$, where $k = 1.38(10^{-16})$erg/°C is Boltzmann's constant and $m = 1.671(10^{-24})$g is its mass. Calculate the de Broglie wavelength of the hydrogen atom at 25°C.

Solution:

$$\lambda_m = \frac{h}{m\bar{v}} = \frac{h}{m\sqrt{8kT/\pi m}} = \frac{h}{\sqrt{8kTm/\pi}}$$

$$= \frac{(6.62)(10^{-27})\text{erg sec}}{\left[\dfrac{8(1.38)(10^{-16})\text{erg/°C}(298°K)(1.671)(10^{-24})\text{g}}{(3.142)}\right]^{1/2}}$$

$$= 1.579(10^{-8} \text{ cm}) = 1.579 \text{ Å}$$

Because of the small, but observable, discrepancies between the most refined forms of Bohr's theory and experimental observations, many felt a new theory was needed. Recognizing the implications of de Broglie's work, Schrödinger [5] and Heisenberg [6] developed a theory of hydrogen based on its wave properties, giving rise to a completely different physical picture of the atom than Bohr's simple, solar system model. Because of the wave-like nature of the Schrödinger–Heisenberg model, only the *probability* of finding the electron in space could be computed. Thus, instead of well-defined orbits in the Bohr–Sommerfeld sense, the wave mechanical model resulted in probability density distributions somewhat analogous to the stationary wave forms of a vibrating string. The description of electron energy distributions according to this theory now requires the specification of three quantum numbers designated as n, l, and m, which assume the values:

$$n = 1, 2, 3, 4, \ldots$$

$$l = 0, 1, 2, 3, \ldots, n - 1 \tag{2.7}$$

$$m = -l, -(l - 1), \ldots, 0, \ldots, (l - 1), l$$

An allowed quantum energy state of the electron corresponds to each quantum number. The numbers n, called the *principal quantum numbers*, are similar to the quantum numbers of the Bohr–Sommerfeld theory. The numbers l are the *azimuthal quantum numbers* related to the Sommerfeld k' numbers. The numbers m are the *magnetic quantum numbers*, so called because the energy states to which they correspond are observed only when the atom is placed in a magnetic field.

One could compute the ellipticity of the Bohr–Sommerfeld type orbits as

$$e = \frac{n}{(l(l + 1))^{1/2}} \tag{2.8}$$

One fundamental difference between the Bohr–Sommerfeld orbital theory and the Schrödinger–Heisenberg wave mechanical theory of the hydrogen atom can now be seen. In the lowest energy or *ground state* ($l = 0$), the angular momentum of the hydrogen electron, given according to the wave mechanical theory by $(l(l + 1))^{1/2}h/2\pi$, vanishes. This presents a different impression than does the Bohr–Sommerfeld model.

Bohr's electronic orbits had been circular paths in which the electron possessed *nonzero* angular momentum. An alternate orbit with zero angular momentum would be the elliptical [7] orbit of infinite eccentricity (set $l = 0$ in Eq. (2.8)) which is shown in Fig. 2-2. This orbit is the limiting case of classical, central force motion. The conditions of zero angular momentum and spherical symmetry of charge distribution lead to a visualization of the electron moving in and out from the nucleus equally in all directions of space. Thus, the $l = 0$ orbit is not circular in the Bohr–Sommerfeld sense. Figure 2-1 must be modified [7] slightly (as in Fig. 2-3) if the angular momentum is to be given by $(l(l + 1))^{1/2}h/2\pi$. Increasing l to its maximum value decreases the ellipticity of the orbit to a minimum value of $e = n/((n - 1)n)^{1/2}$.

Although the solar system analogy of the Bohr–Sommerfeld model

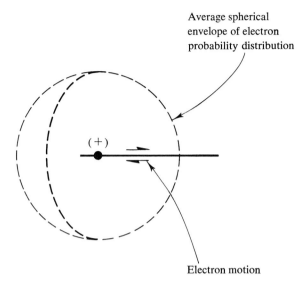

FIG. 2-2 Schematic illustration of limiting central force motion. Electron path (solid line) is ellipse of infinite eccentricity. Motion occurs in direction of arrows, and orbit traces out spherical envelope in such a manner that average angular momentum vanishes.

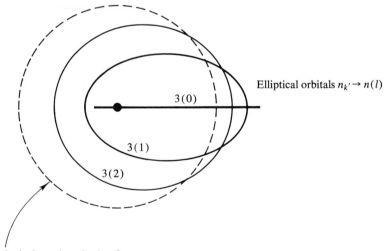

FIG. 2-3 Modification of schematic Bohr–Sommerfeld elliptical orbits. The 3_3 orbit is replaced by the $3(0)$ spherical envelope; the $3(0)$ orbit is infinitely eccentric.

is not accurate, it persists because it is so easy to visualize. The quantization of electron spin vectors, one final result of quantum theory, is much clearer in terms of this model. As the earth rotates about its own axis while orbiting the sun, so the electron can be visualized spinning about its own axis as it orbits the nucleus. A spinning electric charge has an electromagnetic *spin angular momentum* characterized by a vector pointing along the axis of spin.

In 1925, Pauli [8] proposed that no more than two electrons in an atom can have the same energy simultaneously and that the spin angular momentum vectors of two such electrons must be oppositely directed. This condition, now known as the *Pauli exclusion principle*, is satisfied by assigning to each electron a spin quantum number having the value

$$S = \pm \frac{1}{2} \tag{2.9}$$

An alternate statement of the Pauli exclusion principle is that no two electrons in an atom can have the same set of four quantum numbers (n, l, m, S) simultaneously.

The numerical values of these four quantum numbers have quantitative significance only when used in detailed wave mechanical equations. Their only significance here is to aid in qualitatively understanding the underlying reasons for the electronic behavior of atoms.

Electronic Notation

Electronic notation is used to describe the orbital distribution of extranuclear electrons. The principal quantum number n is a coefficient to the azimuthal quantum number l; l is then raised to an exponent representing the number of electrons in the n, l energy states that have differing m and S values.

One part of the old spectroscopic notation has survived and is still used in electronic notation. The symbols s, p, d, and f are assigned[2] to the azimuthal quantum numbers $l = 0$, 1, 2, and 3, respectively. For example, a $3s$ electron is in the $n = 3$ principal energy state and the $l = 0$ substate. In this case, $m = 0$ and $S \pm \frac{1}{2}$, so that two antiparallel,

---

[2]These symbols are derived from the first letters of the names of four series of spectral emission lines characteristic of various elements. These series were named *s*harp, *p*rincipal, *d*iffuse, and *f*undamental because of their physical appearance in spectral photographs.

spin-oriented electrons can occupy this state. Such a case would be designated as $3s^2$.

Periodic Classification of the Elements

Prior to the development of the quantum theory, chemists had observed empirically that certain groups of elements showed marked similarities in chemical and physical properties. Mendeleev had proposed a systematic table in which all the elements appeared in order of increasing atomic weights. On the basis of this data, he accurately predicted the properties of elements not then discovered. The key results of the quantum theory permit one to understand this periodic arrangement of the elements.

The maximum number of electrons with the same principal quantum number is determined by the Pauli exclusion principle. For example, when $n = 1$, $l = m = 0$, and $S = \pm\frac{1}{2}$, N (the total number of electrons in the basic *shell*) is 2. When $n = 2$, $l = 0, 1$, $m = -1, 0, 1$, and $S = \pm\frac{1}{2}$ for each m value, N is 8. Similarly, for $n = 3$, $N = 18$, etc. The maximum number of electrons corresponding to a principal quantum number n is easily seen to be

$$N = 2n^2 \tag{2.10}$$

As n increases, some orbitals that are more highly elliptic in the Bohr–Sommerfeld sense correspond on *higher* principal quantum levels to lower *total* electron energies than do the nearly circular Bohr–Sommerfeld orbits in lower levels. An example of this inversion is seen in the 4f-5s orbitals. From Eq.(2.8) we compute the Bohr–Sommerfeld ellipticity of the 4f orbital to be 1.153, while that of the 5s orbital is infinity. The energy of the 4f state is observed to exceed that of the 5s state so that the 5s orbitals are filled in some atoms while the 4f orbitals are unoccupied. The close differences between the f and d energy levels in the $n = 4, 5,$ or 6 principal energy states results in many interesting physical and chemical characteristics of elements. The experimentally observed order [9] of increasing electronic energy states is 1s, 2s, 2p, 3s, 3p, 4s, 3d, 4p, 5s, 4d, 5p, 6s, 4f, 5d, 6p, 7s, 5f, 6d.

Table 2-1 contains a systematic tabulation of experimental electron energy state configurations of the elements in terms of electronic notation. The periodic table of the elements in Table 2-2 is most easily understood in terms of the Pauli exclusion principle and the experimental fact that each electron occupies the *lowest available total energy state*.

TABLE 2-1

Experimental Electron Distribution of Atoms

ELEMENT		$(n=1)$	$(n=2)$		$(n=3)$			$(n=4)$				$(n=5)$				$(n=6)$			$(n=7)$
Symbol	Number	1s	2s	2p	3s	3p	3d	4s	4p	4d	4f	5s	5p	5d	5f	6s	6p	6d	7s
H	1	1																	
He	2	2																	
Li	3	2	1																
Be	4	2	2																
B	5	2	2	1															
C	6	2	2	2															
N	7	2	2	3															
O	8	2	2	4															
F	9	2	2	5															
Ne	10	2	2	6															
Na	11	2	2	6	1														
Mg	12	2	2	6	2														
Al	13	2	2	6	2	1													
Si	14	2	2	6	2	2													
P	15	2	2	6	2	3													
S	16	2	2	6	2	4													
Cl	17	2	2	6	2	5													
Ar	18	2	2	6	2	6													
K	19	2	2	6	2	6		1											
Ca	20	2	2	6	2	6		2											
Sc	21	2	2	6	2	6	1	2											
Ti	22	2	2	6	2	6	2	2											
V	23	2	2	6	2	6	3	2											
Cr	24	2	2	6	2	6	5	1											
Mn	25	2	2	6	2	6	5	2											
Fe	26	2	2	6	2	6	6	2											
Co	27	2	2	6	2	6	7	2											
Ni	28	2	2	6	2	6	8	2											
Cu	29	2	2	6	2	6	10	1											
Zn	30	2	2	6	2	6	10	2											
Ga	31	2	2	6	2	6	10	2	1										
Ge	32	2	2	6	2	6	10	2	2										
As	33	2	2	6	2	6	10	2	3										
Se	34	2	2	6	2	6	10	2	4										
Br	35	2	2	6	2	6	10	2	5										
Kr	36	2	2	6	2	6	10	2	6										
Rb	37	2	2	6	2	6	10	2	6			1							
Sr	38	2	2	6	2	6	10	2	6			2							
Y	39	2	2	6	2	6	10	2	6	1		2							
Zr	40	2	2	6	2	6	10	2	6	2		2							
Nb	41	2	2	6	2	6	10	2	6	4		1							
Mo	42	2	2	6	2	6	10	2	6	5		1							
Te	43	2	2	6	2	6	10	2	6	6		1							
Ru	44	2	2	6	2	6	10	2	6	7		1							
Rh	45	2	2	6	2	6	10	2	6	8		1							
Pd	46	2	2	6	2	6	10	2	6	10									

(*continued*)

ELEMENT		(n=1)	(n=2)		(n=3)			(n=4)				(n=5)				(n=6)			(n=7)
Symbol	Number	1s	2s	2p	3s	3p	3d	4s	4p	4d	4f	5s	5p	5d	5f	6s	6p	6d	7s
Ag	47	2	2	6	2	6	10	2	6	10		1							
Cd	48	2	2	6	2	6	10	2	6	10		2							
In	49	2	2	6	2	6	10	2	6	10		2	1						
Sn	50	2	2	6	2	6	10	2	6	10		2	2						
Sb	51	2	2	6	2	6	10	2	6	10		2	3						
Te	52	2	2	6	2	6	10	2	6	10		2	4						
I	53	2	2	6	2	6	10	2	6	10		2	5						
Xe	54	2	2	6	2	6	10	2	6	10		2	6						
Cs	55	2	2	6	2	6	10	2	6	10		2	6			1			
Ba	56	2	2	6	2	6	10	2	6	10		2	6			2			
La	57	2	2	6	2	6	10	2	6	10		2	6	1		2			
Ce	58	2	2	6	2	6	10	2	6	10	2	2	6			2			
Pr	59	2	2	6	2	6	10	2	6	10	3	2	6			2			
Nd	60	2	2	6	2	6	10	2	6	10	4	2	6			2			
Pm	61	2	2	6	2	6	10	2	6	10	5	2	6			2			
Sm	62	2	2	6	2	6	10	2	6	10	6	2	6			2			
Eu	63	2	2	6	2	6	10	2	6	10	7	2	6			2			
Gd	64	2	2	6	2	6	10	2	6	10	7	2	6	1		2			
Tb	65	2	2	6	2	6	10	2	6	10	8	2	6	1		2			
Dy	66	2	2	6	2	6	10	2	6	10	10	2	6			2			
Ho	67	2	2	6	2	6	10	2	6	10	11	2	6			2			
Er	68	2	2	6	2	6	10	2	6	10	12	2	6			2			
Tm	69	2	2	6	2	6	10	2	6	10	13	2	6			2			
Yb	70	2	2	6	2	6	10	2	6	10	14	2	6			2			
Lu	71	2	2	6	2	6	10	2	6	10	14	2	6	1		2			
Hf	72	2	2	6	2	6	10	2	6	10	14	2	6	2		2			
Ta	73	2	2	6	2	6	10	2	6	10	14	2	6	3		2			
W	74	2	2	6	2	6	10	2	6	10	14	2	6	4		2			
Re	75	2	2	6	2	6	10	2	6	10	14	2	6	5		2			
Os	76	2	2	6	2	6	10	2	6	10	14	2	6	6		2			
Ir	77	2	2	6	2	6	10	2	6	10	14	2	6	7		2			
Pt	78	2	2	6	2	6	10	2	6	10	14	2	6	8		2			
Au	79	2	2	6	2	6	10	2	6	10	14	2	6	10		1			
Hg	80	2	2	6	2	6	10	2	6	10	14	2	6	10		2			
Tl	81	2	2	6	2	6	10	2	6	10	14	2	6	10		2	1		
Pb	82	2	2	6	2	6	10	2	6	10	14	2	6	10		2	2		
Bi	83	2	2	6	2	6	10	2	6	10	14	2	6	10		2	3		
Po	84	2	2	6	2	6	10	2	6	10	14	2	6	10		2	4		
At	85	2	2	6	2	6	10	2	6	10	14	2	6	10		2	5		
Rn	86	2	2	6	2	6	10	2	6	10	14	2	6	10		2	6		
Fr	87	2	2	6	2	6	10	2	6	10	14	2	6	10		2	6		1
Ra	88	2	2	6	2	6	10	2	6	10	14	2	6	10		2	6		2
Ac	89	2	2	6	2	6	10	2	6	10	14	2	6	10		2	6	1	2
Th	90	2	2	6	2	6	10	2	6	10	14	2	6	10		2	6	2	2
Pa	91	2	2	6	2	6	10	2	6	10	14	2	6	10	2	2	6	1	2
U	92	2	2	6	2	6	10	2	6	10	14	2	6	10	3	2	6	1	2
Np	93	2	2	6	2	6	10	2	6	10	14	2	6	10	5	2	6		2
Pu	94	2	2	6	2	6	10	2	6	10	14	2	6	10	6	2	6		2
Am	95	2	2	6	2	6	10	2	6	10	14	2	6	10	7	2	6		2
Cm	96	2	2	6	2	6	10	2	6	10	14	2	6	10	7	2	6	1	2

TABLE 2-2
Periodic Table of the Elements*

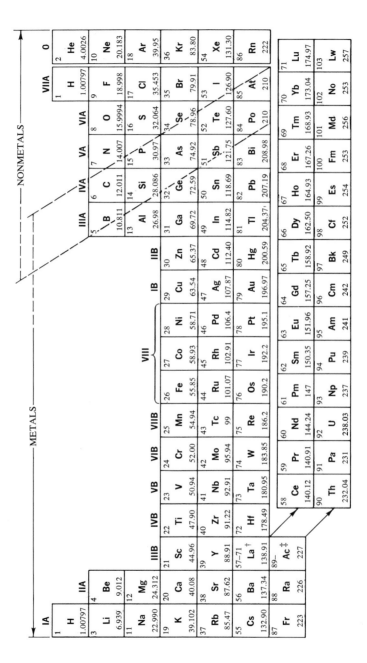

*The atomic number and the atomic weight (carbon = 12.000) are shown for each element.
†Lanthanide series
‡Actinide series

Interatomic Attractions and Bond Formation

The atoms of most materials are bound together in one way or another. If this were not true, all matter would be in the gaseous state due to thermal vibrations. Since many materials not only exist as liquids and solids, but possess great mechanical strength in their solid states, we know that atoms sometimes form very strong bonds.

Hydrogen and helium liquefy only when the temperature is reduced to practically absolute zero, suggesting the existence of very weak bonds different in nature from those which can either lend strength to steel or liquefy gases. The different types of bonds are best understood in terms of the electronic distributions of the atoms involved. We shall consider the two groups of bonds, primary and secondary, separately.

Primary or Chemical Bonds

The strongest bonds between atoms are the primary or chemical bonds, called ionic, covalent, and metallic.

IONIC BONDS   Interatomic attraction and repulsion forces are electrostatic, and the tendency of atoms to form positive or negative ions depends upon their relative positions in the periodic table. For example, sodium is easily ionized to produce a positive ion, while chlorine readily accepts an electron to form a negative ion. The spatial distribution of the net excess electrical charge in these ions is spherically symmetric about the nucleus, so that any spatial direction is equally probable for attraction by an oppositely charged ion.

Electrically charged bodies attract each other with the force given by Coulomb's law of electrostatic attraction

$$f_a = \frac{Z_1 Z_2 e^2}{r_{12}^2} \qquad (2.11)$$

where Z_1, Z_2 are the numbers of charges per body, e is the elementary electronic charge, and r_{12} is the center-to-center separation distance between the two bodies.

The Coulombic attraction force causes the oppositely charged ions to draw closer together. The relative potential energy[3] of the ions

---

[3]The potential energy is the amount of work which would have to be done on the ion pair to separate them from their position r_{12} to an infinitely great distance away. In general, work $= \int$ (force) d (distance).

$V_a(r_{12})$ as they approach each other, given by

$$V_a(r_{12}) = \int_{\infty}^{r_{12}} f_a \, dr = e^2 Z_1 Z_2 \int_{\infty}^{r_{12}} \frac{1}{r^2} \, dr = \frac{-e^2 Z_1 Z_2}{r_{12}} \quad \textbf{(2.12)}$$

decreases corresponding to a lowering of the total energy of the ion pair.

However, as the ions get very close together, the interaction of their respective electron clouds gives rise to a strong repulsive force of the form

$$f_R = \frac{-ne^2 B_{12}}{r_{12}^{n+1}} \quad \textbf{(2.13)}$$

where B_{12} is an empirical coefficient, called a *repulsion coefficient*, and n is the Born exponent [10]. Experimental values of the Born exponent n are tabulated in Table 2-3. This repulsion force gives rise to a *repulsive potential energy* $V_R(r_{12})$ given by

$$V_R(r_{12}) = \int_{\infty}^{r_{12}} f_R \, dr = -ne^2 B_{12} \int_{\infty}^{r_{12}} \frac{1}{r^{n+1}} \, dr = \frac{e^2 B_{12}}{r_{12}^n} \quad \textbf{(2.14)}$$

When $f_a + f_R = 0$, the ion pair achieves a stable equilibrium configuration. The separation $r_{12} = r_0$ when this occurs is called the *bond length of the pair*. The total energy $V = V_a + V_R$ of the pair at this condition is at a minimum, as illustrated in Fig. 2-4. This potential energy E_{min} is called the *bond strength* or *bond energy* and is the energy which would have to be added to the bond to rupture it.

In the formation of the ionic bond, the spherical symmetry of charge distribution introduces a factor not considered in the above

TABLE 2-3

Experimental Values of the Born Exponent n*

ION TYPE	n
He	5
Ne	7
Ar, Cu^+	9
Kr, Ag^+	10
Xe, Au^+	12

*From *The Nature of the Chemical Bond, 3rd ed.*, by L. Pauling, copyright 1960, Cornell University Press. Used by permission of Cornell University Press.

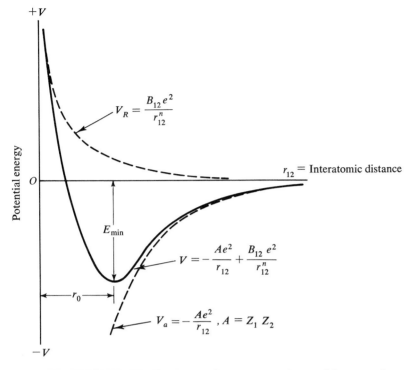

FIG. 2-4   Schematic illustration of minimum in curve of potential energy of an ion pair as a function of ion separation distance. The minimum value $V(r_0) = E_{min}$ is the bond strength, and the distance $r_{12} = r_0$ is the bond length.

analysis of the ion pair potential energy. Because of this spherical symmetry, each ion can attract more than one ion of the opposite charge simultaneously. Therefore, in forming an ionically bonded solid, each charged spherical ion surrounds itself with as many oppositely charged spheres as is consistent with the geometrical problem[4] of packing different sized spheres. For example, in NaCl each Na^+ ion finds itself surrounded by six equidistantly spaced Cl^- ions and conversely. Thus, the crystal is composed of a three-dimensional array of alternating Na^+ and Cl^- ions. This three-dimensional character of the crystal causes perturbations in the forces acting between any given ion pair due to the presence of neighboring ions of both signs. This perturbing influence can be computed [11] from purely geometrical considerations and is expressed as

$$V(R) = \frac{-Ae^2Z^2}{R} + \frac{Be^2}{R^n} \qquad (2.15)$$

[4]Discussed at greater length in Chapter 3.

TABLE **2-4**

Values of the Madelung Constants (A) for Several Crystal Structures

STRUCTURE	A
NaCl, M^+X^-	1.74756
CsCl, M^+X^-	1.76267
Sphalerite, M^+X^-	1.63806
Wurtzite, M^+X^-	1.64132
Fluorite, $M^{++}X_2^-$	5.03878
Cuprite, $M_2^+X^{--}$	4.11552
Rutile, $M^{++}X_2^-$	4.816
Anatase, $M^{++}X_2^-$	4.800
CdI_2, $M^{++}X_2^-$	4.71
β-Quartz, $M^{++}X_2^-$	4.4394
Corundum, $M_2^{++}X_3^-$	25.0312
Perovskite, $M^+M^{++}X_3^-$	12.3774

where R represents the interionic separation distance, B is a *crystal repulsion coefficient*, and A is a dimensionless constant, called the *Madelung constant* [11] and calculated by straightforward geometrical methods. Numerical values of the Madelung constant are tabulated in Table 2-4. Values of n are determined experimentally from crystal compressibility measurements, R_0 is determined by x-ray analysis, and values of B are calculated by using this data and Eq. (2.17).

EXAMPLE 2.3

Derive expressions for R_0, $B(R_0)$, and $E_{min}(R_0)$ for ionic crystals.

Solution:

At equilibrium, $dV/dR = 0$. From Eq. (2.15) we have

$$0 = \frac{dV}{dR}\bigg|_{R=R_0} = \frac{e^2}{R_0^2}\left[AZ^2 - \frac{nB}{R_0^{n-1}}\right]$$

$$\therefore R_0 = \left[\frac{nB}{AZ^2}\right]^{1/(n-1)} \tag{2.16}$$

From this result we can also solve for $B(R_0)$ as

$$B = \frac{AZ^2}{n}R_0^{n-1} \tag{2.17}$$

which, when introduced into Eq. (2.15), gives

$$V(R_0) = E_{min} = \frac{-Ae^2Z^2}{R_0}\frac{(n-1)}{n} \tag{2.18}$$

In addition to simple ionic bonds formed by individual charged ions, numerous complex ionic compounds are formed where several atoms are covalently bound into a distinct ionic group which maintains its identity in the solid, liquid, and dissolved states. Examples of such ionic groups are CO_3^{--}, SO_4^{--}, and PO_4^{---} ions.

COVALENT BONDS   While ionic bonds are spatially spherically symmetric or isotropic and involve a number of ions, covalent bonds are spatially anisotropic and involve only two atoms. The solutions of the Schrödinger wave equation for the hydrogen atom electron distributions provide some insight into the anisotropy of the covalent bond. These solutions, or wave functions, take on the form of a product of three functions, each of which depends only on a single variable

$$\psi(r, \phi, \theta) = R(r)\Phi(\phi)\Theta(\theta) \tag{2.19}$$

where r is the radial variable, ϕ is the planar polar angle, and θ is the conical angle of a spherical polar coordinate system centered in the hydrogen nucleus. Although hydrogen only has a single $1s$ electron, the wave functions permit the calculation of orbital probabilities for other energy states. Consider, for example, the $2p$ state wave functions for hydrogen, the angular portions of which (corresponding to $m = -1$, 0, 1) are shown schematically in Fig. 2-5.

The quantum number l determines the number of planar and conical nodes[5] in the wave function which pass through the origin. Only orbitals with $l = 0$ (s-orbitals) are spherically symmetric. The schematic curves in Fig. 2-5 represent the angular probability factors of the $2p$ state. The radial probability factor R^2 is illustrated schematically in Fig. 2-6. The radius vector $\overline{OP}$ gives the magnitude of $\Phi^2\Theta^2$ by which the radial probability factor R^2 must be multiplied to obtain ψ^2 (the probability per unit volume of finding the electron at any distance r along a line in the direction of $\overline{OP}$). For the direction defined by the vector $\overline{OP'}$ in Fig. 2-6, the total probability curve will have the same *shape* as it will for the direction $\overline{OP}$, but will be decreased in magnitude by the ratio of the two angular probability factors.

These figures show that the electron distribution probabilities of the $2p$ orbitals are highly directional. When such orbital distributions in more complex atoms overlap to form a bond through the mutual sharing of orbitals, the resultant bond is itself highly directional. However, when all p orbitals of hydrogen are considered together, *the total ψ^2 distribution is spherically symmetric.* (Note the difference between the idea of three-dimensional spherical symmetry here and the

---

[5]A node is a point of zero amplitude in a wave. In this case the nodes are points of zero probability density.

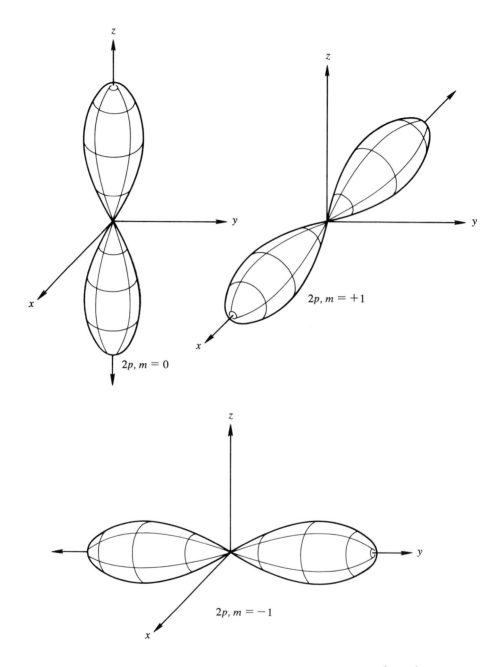

FIG. 2-5 Schematic representation of angular probability functions $\Phi^2(\phi)\Theta^2(\theta)$ for the three $2p$ states of hydrogen. (From *The Structure and Properties of Materials, Vol. I, Structure,* by W. G. Moffatt, G. W. Pearsall, and J. Wulff, copyright 1964, John Wiley and Sons, Inc. Used by permission.)

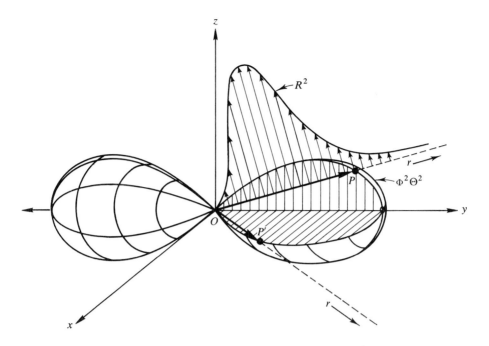

FIG. 2-6   Schematic representation of radial probability function $R^2(r)$, superimposed on angular probability function $\Phi^2(\phi)\Theta^2(\theta)$, for the $2p$ state of atomic hydrogen. (Product of $R^2\,\Phi^2\Theta^2 = \psi^2$.) (From *The Structure and Properties of Materials, Vol. I, Structure,* by W. G. Moffatt, G. W. Pearsall, and J. Wulff, copyright 1964, John Wiley and Sons, Inc. Used by permission.)

shapes of the Bohr–Sommerfeld orbits which were planar orbits of varying ellipticity.)

The normal or ground state of the hydrogen atom is the $1s$ state. The total energy of a hydrogen atom electron can be decreased in accord with the exclusion principle by the sharing of a $1s$ orbital of another atom which has an oppositely spin-oriented $1s$ electron. As the atoms approach each other, their electron clouds overlap, permitting dual orbital sharing and reducing the total energy of both atoms. Upon very close approach, the electrostatic repulsion of the nuclei overpowers the attraction potential created by the orbital sharing and forces the atoms apart. As in the ion-pair case, an equilibrium distance or bond length is attained corresponding to a minimum total energy condition. This combination of two bound hydrogen atoms is a hydrogen molecule and the bond thus formed is called a *covalent bond*.

In the hydrogen molecule bond, the probability of the electron pair being found between the two nuclei is large because this position represents the lowest energy state for the molecule. In this condition the bond is covalent and highly directional along the line joining the centers of the two nuclei.

A small, but finite probability exists for one of the nuclei to be bereft of electrons while the other has both, the result being that the real bond is partially ionic.[6] Although all real bonds are partially ionic, it is useful to think of the limiting cases of purely ionic and purely covalent bonds.

Nonmetallic atoms such as nitrogen, oxygen, carbon, fluorine, chlorine, and hydrogen form nearly pure covalent bonds, while elements such as silicon, germanium, arsenic, and selenium form bonds which are partially covalent and partially metallic. If each atom of a pair possesses at least one half-filled orbital, the total energy of the pair can be lowered through the sharing of orbitals and a covalent bond is usually formed. The more two orbitals can overlap, the greater the energy decrease of the total combination and the stronger the bond formed. The degree of overlap is limited either by electrostatic repulsion, orbital symmetry, or by the exclusion principle since filled orbitals cannot overlap.

Covalent bonds formed by overlapping orbitals lie in the direction in which the orbitals are concentrated in order to achieve maximum overlap. Thus, if ψ^2 is large in a given direction, bonds will be strong and concentrated in that direction; for small ψ^2 in a given direction, weak bonds will be formed. If ψ^2 is spherically symmetrical, bonds will be nondirectional.

Free atom wave functions may not give reliable indications as to the type, spatial distribution, or strength of the bonds which the atom will form. For example, the free carbon atom [12] has three $2p$ and one $2s$ valence electrons. These might be supposed to form three strong mutually perpendicular covalent bonds from the $2p$ orbitals and one weaker bond from the spherically symmetric $2s$ orbital. However, carbon forms four equally strong covalent bonds which point to the corners of a regular tetrahedron. Apparently, a $2s$ electron is promoted to a $2p$ orbital, the necessary expenditure of energy being compensated by the energy decrease which accompanies bonding. Since the resultant four bonds are equal in strength and highly directional, the four electrons actually occupy not four atomic $2p$ orbitals, but *four hybrid molecular orbitals* referred to by Pauling [12] as tetrahedral orbitals. Mathematically, these hybrid molecular orbitals must correspond to an equivalent set of solutions to the time-independent wave equation for $n = 2$. Each molecular orbital has a larger maximum value of ψ^2 than the atomic orbitals and can overlap other orbitals to a greater extent, resulting in even lower energy levels for the bonding electrons.

---

[6]This phenomenon is discussed later in this chapter in connection with electro-negativity.

In addition to the above mechanisms, covalent bonds can arise when one atom contributes *both* of its electrons to the bond, forming a *coordinate covalent bond*. This type of covalent bonding can be observed in the ammonium ion. Three hydrogen atoms are covalently bonded to one nitrogen atom in the electrically neutral ammonia molecule. However, the energy of the remaining two of the five valence electrons of the nitrogen atom could be lowered if bonded to another hydrogen nucleus. Thus, a proton (hydrogen ion) can be covalently bound to the ammonia molecule to produce a positive, monovalent ammonium ion by the formation of a coordinate covalent bond with the nitrogen atom.

Compounds containing covalently bonded atoms often exhibit the phenomenon of stereoisomerism, where two materials of identical atomic compositions have certain atoms arranged in different spatial relations to one another by virtue of their covalent bond angles. Stereoisomers often have radically different properties. For example, of the two stereoisomers *cis-* and *trans*-polyisoprene, the former is the springy, elastic material called rubber, while the latter is a hard brittle resin called gutta percha, frequently used as a potting material. This phenomenon will be discussed in more detail in Chapter 12.

METALLIC BONDS   The metallic[7] bond is of considerable importance to engineers because it occurs in the crystalline phase of metallic elements. The majority of the chemical elements are metallic and, unlike nonmetals, are opaque, lustrous, good conductors of heat and electricity, and plastically deformable. The explanation of all of these properties lies in the nature of the metallic bond itself.

Whereas ionic bonds exist between a number of oppositely charged ions and covalent bonds exist between only two atoms, the metallic bond can exist between a *relatively large aggregate* of atoms of the *same* type. Like the ionic bond, the metallic bond is essentially nondirectional.

The valence electrons of free metal atoms are observed to have mean orbital radii greater than the observed interatomic distance in solid metal. In a solid, therefore, the valence electrons of a given atom will be closer to another nucleus than to the parent nucleus, resulting in the lowering of their total energy from that of the free atomic state. This produces bonding between the atoms of the aggregate. The kinetic energy of the electron in the metallically bound state is reduced over the free state because the ψ^2 function is extended in space [13].

_____

[7]The student is referred to Chapter 11 of L. Pauling, *The Nature of the Chemical Bond*, *3rd ed.*, Cornell University Press, Ithaca, New York, 1960, for a detailed discussion beyond the scope of this book.

Each electron loses its association with any given nucleus by getting closer to another nucleus. Thus, it is free to wander about in the solid as a free *electron gas*, permeating the regular geometrical arrangement of the positive ion cores (nuclei plus nonvalence electrons) and preserving the overall electrical neutrality of the material.

The various metal properties mentioned above can be explained qualitatively in terms of this model of the metallic bond. For example, high thermal and electrical conductivities result from rapid energy transport by the free electron gas under the influence of relatively small external temperature gradients or electric fields. Opacity and lustrousness are caused by the scattering and reflection of light quanta by the electron gas. The plastic deformability of the solid is due to the looseness of the electron bonding since this permits permanent rearrangements in the basic lattice structure of the positive cores to occur without fracturing the material. (This model of the metallic bond is only qualitative; a detailed discussion of the bonds requires a complex quantum mechanical analysis.)

The tendency toward covalent bonding increases as the atomic number increases for a given principal quantum number. A gradual change is thus observed in the properties of materials ranging from metallic to nonmetallic. For example, the fourth period transition elements with incomplete $3d$-shells show a significant fraction of covalent-type bonding, accounting in part for the very high melting points of these metals. A division between covalent and metallic bonding is particularly evident in the fourth group of the periodic table: Carbon in the form of diamond exhibits almost purely covalent bonding; silicon and germanium are increasingly metallic in nature; tin exists in two different *allotropic* forms—one mostly covalent and the other mostly metallic; and lead is metallic.

Since the valence electrons of a metal are relatively free to move about, the metallic bond is spherically symmetrical with respect to any given core, and highly coordinated[8] structures similar to those of ionic compounds can therefore be observed in metallic elements. However, the metallic bond occurs between identical atoms in pure metals or between chemically similar atoms in alloys, while the ionic bond acts between chemically dissimilar ions with opposite electric charges.

Secondary Bonds

A secondary class of bonds exists between atoms or molecules. They are weak in comparison with primary bonds (Secondary bond

---

[8]Large numbers of nearest neighbors. This subject is treated in detail in Chapter 3.

strengths are of the order of magnitude of 1–5 kcal/mole; primary bond strengths are of the order of 20–50 kcal/mole.), but are still very significant in determining certain physical characteristics of materials. Secondary bonding forces, frequently referred to as *van der Waals forces*, are electrostatic in nature like the primary bonding forces. Secondary bonding forces result from molecular polarizability and permanent dipole moments, rather than from electronic orbital interactions.

POLARIZABILITY   As two atoms or molecules approach closely, the outer electron orbitals of each are perturbed by the nucleus of the other. If this perturbation is of sufficient magnitude, the atoms form a primary bond. However, in many cases these perturbations result in only a slight separation of the centers of positive and negative charges causing a weak, temporarily induced *dipole moment* (defined in the next section). This, in turn, results in very weak electrostatic, interatomic attractions. If the thermal energy of the atoms is sufficiently reduced, these weak, attractive forces result in a condensation of the gas to a liquid or solid state. In the noble gases, the melting and boiling temperatures are very near the absolute zero because of the weakness of these induced dipoles. Many elements and compounds in addition to noble gases have weak van der Waals forces and resultant low boiling points. In general, as the number of electrons increases the magnitude of the above-mentioned perturbation increases, so that heavy molecules attract one another more strongly than light molecules. That substances with high molecular weights usually have high boiling points is illustrated in Table 2-5.

TABLE 2-5

Thermal Data for Some Simple Molecules

MOLECULE	MOLECULAR WEIGHT	MELTING POINT (°C)	BOILING POINT (°C)
H_2	2.016	−259	−252
CH_4	16.043	−183	−161
NH_3	17.031	−78	−33
N_2	28.014	−209	−195
O_2	31.9988	−218	−183
Cl_2	70.906	−102	−34
CF_4	88.003	−185	−128
CCl_4	153.823	−23	+76

DIPOLE MOMENTS   Some molecules possess permanent dipole moments due to asymmetries in the electron attracting power of different atoms. The dipole moment is defined as the product of the unit of electric charge e and the distance between the average centers of positive and negative charges $\bar{r}$. When e is expressed in esu and $\bar{r}$ in cm, the electric dipole moment

$$m = e\bar{r} \tag{2.20}$$

has units of esu-cm. For example, a positive and a negative charge, each of magnitude e and separated by a distance of 1 Å, would have a dipole moment $m = (4.80 \times 10^{-10}\,\text{esu})(10^{-8}\,\text{cm}) = 4.80 \times 10^{-18}$ esu-cm. Because of the small numerical magnitudes, the Debye (1 Debye $= 10^{-18}$ esu-cm), a more convenient unit, is defined. Thus, the above dipole moment would be 4.80 Debyes [14].

Molecules with permanent dipoles, called *polar molecules*, align themselves in an electric field, making measurement of the magnitudes of their dipole moments possible. From the experimental dipole moments for several compounds listed in Table 2-6 [15], it is apparent that molecules with a center of symmetry have no resultant dipole moment. Therefore, CO_2 and CS_2 must be linear molecules, but because SO_2 has a moment it cannot be linear. In this way the planar characteristics of molecules can be detected. NH_3, for example, has a dipole moment of 1.46 Debyes and is not planar, while BCl_3, having no dipole moment at all, is assumed to be a symmetrical planar molecule.

It is sometimes convenient to assign average dipole moments to

TABLE **2-6**

Experimental Magnitudes of Dipole Moments of Various Compounds in Debyes*

COMPOUND	m	COMPOUND	m
HCl	1.03	CH_3NH_2	1.30
HBr	0.78	$C_2H_5NH_2$	1.30
HI	0.38	CH_3Cl	2.00
H_2O_2	2.10	CH_2Cl_2	1.60
C_2H_5OH	1.70	$CHCl_3$	1.10
NH_3	1.46	C_6H_5Cl	1.73
H_2O	1.84	C_6H_5OH	1.70
H_2S	1.10	$C_6H_5NH_2$	1.56
SO_2	1.60	H_2	0.00
$C_6H_5NO_2$	4.23	N_2	0.00
HCN	2.93	CO_2	0.00
CH_3CN	3.20	CS_2	0.00
C_2H_5CN	3.40	CCl_4	0.00
CH_3OH	1.68	C_6H_6	0.00

*From *The Solid State for Engineers*, by M. W. Sinnott, copyright 1961, John Wiley and Sons, Inc. Used by permission.

TABLE **2-7**

Dipole Moments of Atom Groups in Debyes*

C–H	C–O	C–Cl	NO_2
0.4	0.7	1.5	−3.9
O–H	C=O	N–H	CN
1.6	2.3	1.7	−1.7
$N–H_2$	N–O	NO	$C–H_3$
1.5	0.1	3.2	0.4

*From *The Solid State for Engineers,* by M. W. Sinnott, copyright 1961, John Wiley and Sons, Inc. Used by permission.

various molecular groups that occur frequently. When these values are added vectorially for a compound, an estimate of the dipole moment of the compound is obtained. Dipole moments for several such groups are listed in Table 2-7 [15]. There, although the CH bond has a dipole moment, vectorial addition of four such bonds makes CH_4 with a net dipole moment of zero.

EXAMPLE 2.4

Using the group contribution data given in Table 2-7, calculate the dipole moments of water and methanol.

Solution:

These molecules have the structures

Taking a reference axis as the bisector of the angles, we find

$$m(H_2O) = 2m(OH) \cos \frac{\theta}{2} = (2)(1.6) \cos \frac{104.5}{2} = (3.2) \cos (52.3°) = 1.96 \text{ Debyes}$$

$$m(CH_3OH) = m(CH_3) + m(OH) = 0.4 + 1.6 = 1.64 \text{ Debyes}$$

Comparison with Table 2-6 shows that this method does not always give the exact experimental value, as in the H_2O example, although the methanol value is reasonably close.

TABLE 2-8

Extra Ionic Energy of Bonds and
Electronegativity Differences of Atoms*

BOND	Δ'_{AB} (KCAL/MOLE)	$0.18 \sqrt{\Delta'_{AB}}$ EQ. (2.21)	$X_A - X_B$ FIG. 2-7
C–H	5.8	0.4	0.4
Si–H	4.0	.4	.3
N–H	30.1	1.0	.9
P–H	3.3	0.3	.0
As–H	0.8	.2	.1
O–H	41.8	1.2	1.4
S–H	8.3	0.5	0.4
Se–H	−1.6	—	.3
Te–H	−1.9	—	.0
H–F	72.9	1.5	1.9
H–Cl	25.4	0.9	0.9
H–Br	18.2	.8	.7
H–I	10.1	.6	.4
C–Si	10.0	.6	.7
C–N	13.2	.7	.5
C–O	31.5	1.0	1.0
C–S	−2.4	—	0.0
C–F	50.2	1.3	1.5
C–Cl	9.1	0.5	0.5
C–Br	4.0	.4	.3
C–I	2.6	.3	.0
Si–O	50.7	1.3	1.7
Si–S	7.8	0.6	0.7
Si–F	90.0	1.7	2.2
Si–Cl	36.2	1.1	1.2
Si–Br	25.0	0.9	1.0
Si–I	11.8	.6	0.7
Ge–Cl	50.8	1.3	1.2
N–F	27.0	0.9	1.0
N–Cl	0.5	.1	0.0
P–Cl	24.5	.9	.9
P–Br	16.7	.7	.7
P–I	8.3	.5	.4
As–F	77.0	1.6	2.0
As–Cl	25.8	0.9	1.0
As–Br	18.0	.8	0.8
As–I	7.5	.5	.5
O–F	9.3	.5	.5
P–Cl	4.6	.4	.5
S–Cl	5.3	.4	.5
S–Br	2.2	.3	.3
Se–Cl	9.1	.63	.6
Cl–F	14.5	.7	1.0
Br–Cl	0.6	.1	0.2
I–Cl	4.5	.4	.5
I–Br	1.7	.3	.3

*From *The Nature of the Chemical Bond, 3rd ed.*, by L. Pauling, copyright 1960, Cornell University Press. Used by permission of Cornell University Press.

Permanent molecular dipole moments cause secondary intermolecular van der Waals attractions which are much stronger than those attractions due to induced temporary dipoles. This effect is primarily responsible for the dissolution of ionic salts by polar liquid solvents. Solvent molecules (water, for example) are aligned about the ionic salt (NaCl, for example) in such a way that the negative ends of the dipoles attract and surround the positive salt ion and shield it from the negative ions. The positive ends of the solvent dipoles behave in a like manner toward the negative ions so that the salt ions are segregated. The salt dissolves into the solvent with each ion encapsulated by appropriately-oriented solvent dipoles.

The separation of the centers of the positive and negative charges, resulting in the dipole moment of a molecule, arises from the dissimilarities in the electron attracting powers of different atoms. This electron attracting power can be expressed quantitatively in terms of the *electronegativity* of the atom [16]. The electronegativity of an atom differs from its ionization potential (the energy which must be added to an electron to remove it completely from the atom) and from the electron affinity (the energy released by an atom when an electron is added to it to form a negative ion), although all of these properties are interrelated [16].

Pauling [16] defines the electronegativity scale, illustrated in Fig. 2-7, by the equations

$$X_A - X_B = \sqrt{\frac{\Delta'_{AB}}{30}} \qquad (2.21)$$

$$X(\text{fluorine}) = 4.0 \qquad (2.22)$$

where X_A and X_B are the electronegativities of elements A and B, respectively, and Δ'_{AB} is the A–B bond energy in kcal/mole units. Tables 2-8 and 2-9 contain, respectively, a comparison of experimental ΔX values calculated [16] from Eq. (2.21) and the ΔX values from Fig. 2-7, and the tabulation of numerical values [16] of X for most of the chemical elements relative to the fixed value of $X(\text{fluorine})$ given in Eq. (2.22).

The partial ionic behavior of the covalent bonds mentioned above is easily understood in terms of the electronegativity differences of the elements involved in the bonds. The greater the electronegativity difference of the atoms, the more ionic the bond. This is illustrated [16] in Fig. 2-8 in which experimental data are compared with the per cent ionic character of the bond, as calculated [16] from a simple formula proposed by Pauling

$$\% \text{ ionic character} = 100 \left[1 - \exp\left\{-\tfrac{1}{4}(X_A - X_B)^2\right\}\right] \qquad (2.23)$$

where, as before, X_A and X_B are the electronegativity values of atoms A and B, respectively. From this figure one can see that the predicted

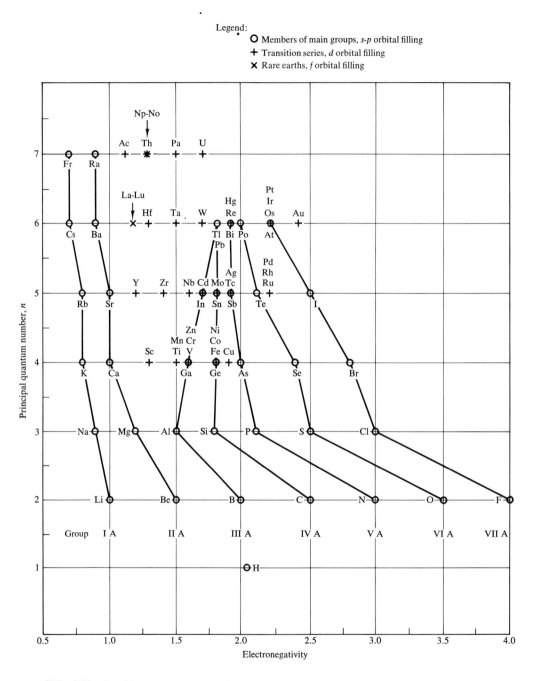

FIG. 2-7   Graphical representation of the electronegativities of the elements. Ordinate is the principal quantum number *n*, the period number. Solid lines connect elements of primary groups.

TABLE **2-9**

Pauling's Electronegativity Values of the Chemical Elements*

X		X		X		X		X		X		X	
H	2.1	Si	1.8	Mn	1.5	Br	2.8	Rh	2.2	Ba	0.9	Hg	1.9
Li	1.0	P	2.1	Fe	1.8	Rb	0.8	Pd	2.2	La–Lu	1.1–1.2	Tl	1.8
Be	1.5	S	2.5	Co	1.8	Sr	1.0	Ag	1.9	Hf	1.3	Pb	1.8
B	2.0	Cl	3.0	Ni	1.8	Y	1.2	Cd	1.7	Ta	1.5	Bi	1.9
C	2.5	K	0.8	Cu	1.9	Zr	1.4	In	1.7	W	1.7	Po	2.0
N	3.0	Ca	1.0	Zn	1.6	Nb	1.6	Sn	1.8	Re	1.9	At	2.2
O	3.5	Sc	1.3	Ga	1.6	Mo	1.8	Sb	1.9	Os	2.2	Fr	0.7
F	4.0	Ti	1.5	Ge	1.8	Tc	1.9	Te	2.1	Ir	2.2	Ra	0.9
Na	0.9	V	1.6	As	2.0	Ru	2.2	I	2.5	Pt	2.2	Ac	1.1
Mg	1.2	Cr	1.6	Se	2.4			Cs	0.7	Au	2.4	Th	1.3
Al	1.5											Pa	1.5
												U	1.7
												Np–No	1.3

*From *The Nature of the Chemical Bond, 3rd ed.,* by L. Pauling, copyright 1960, Cornell University Press. Used by permission of Cornell University Press.

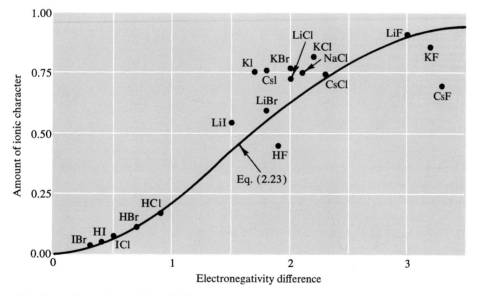

FIG. 2-8 Comparison of Eq. (2.23) with experimental values of the partial ionic characters of bonds. (Figure reprinted from L. Pauling, *The Nature of the Chemical Bond.* Copyright 1939 and 1940 by Cornell University. Third edition © 1960 by Cornell University. Used by permission of Cornell University Press.)

results are in rough qualitative agreement with the experimental observations.

EXAMPLE 2.5

Calculate the partial ionic character of the C–H bond and compare it with the experimental value.

Solution:

From Table 2-9, $X_C = 2.5$ and $X_H = 2.1$. Using Eq. (2.23) we find, for the theoretical value,

$$100[1 - \exp(- (0.25)(2.5 - 2.1)^2)]$$
$$= 100(1 - \exp(-0.04)) = 100(1 - 0.961)$$
$$= 3.9\%$$

To calculate the experimental value, we note that the C–H bond length r_0 is 1.095 Å in the compound CH_4, and that the dipole moment of the C–H bond group is 0.4 Debyes. From Eq. (2.20) we calculate

$$m = er_0 = (4.803)(10^{-10} \text{ esu})(1.095)(10^{-8} \text{ cm}) = 5.26 \text{ Debyes}$$

Hence the

$$\% \text{ ionic character} = \frac{(100)(0.4)}{9.15} = 7.6\%$$

This is an approximation which is in fair agreement with the theoretical value.

HYDROGEN BONDS   An important secondary bond which markedly affects the physical properties of the many substances in which it is formed is the hydrogen bridge or bond, of the type found in the water molecule's liquid phase. The formation of hydrogen bonds between water molecules is directly responsible for many of the striking physical properties of this uniquely important substance.

The hydrogen bond is a weak bond consisting of one hydrogen atom between two molecules. The hydrogen atom was originally thought to form two stable covalent bonds with the atoms of both molecules. However, it is now recognized that a hydrogen atom, having a single stable orbital, can form only one purely covalent bond. Therefore, hydrogen bond formation must be due to ionic attraction.

In a hydrogen bond, the number of the hydrogen atom's nearest neighbors is two. Only the most electronegative atoms form hydrogen bonds, and the strength of these bonds is observed to decrease with the decreasing electronegativity of the participating atoms. For example, fluorine forms strong hydrogen bonds, oxygen weaker ones, and nitrogen even weaker ones. Chlorine, which has the same electronegativity as nitrogen although larger in size, hardly forms any hydrogen bonds. The ammonium ion and its derivatives form stronger hydrogen bonds than do ammonia or the normal amines. Similarly, phenols form stronger hydrogen bonds than the aliphatic alcohols (compounded of linear chains of singly bound carbon atoms) because of an increase in the electronegativity of the oxygen atom due to the *bond resonance* [17] *phenomenon* (where electrons oscillate or *resonate* between two or more possible bond configurations, resulting in a net energy decrease which stabilizes the entire structure).

In some chemically bonded groups such as the OH⁻ ion or the water molecule, the electronegative oxygen atom strongly attracts the $1s$ electron, leaving the hydrogen nucleus with a nearly unshielded proton ionically bound to the negative oxygen ion. Because of the spherical symmetry of the electrostatic forces of ionic attraction, this hydrogen nucleus can also attract other negative charges. In water, the oxygen atom forms the negative end of a strong, permanent dipole and is attracted to the positively charged hydrogen ions of another water molecule, thus forming the weak hydrogen bonds responsible for the high boiling point of water and many of its other unique physical properties. Hydrogen bonding also plays a significant role in polymeric materials, discussed later in Chapter 12.

Bond Properties

BOND LENGTHS   In different molecules and crystals covalent bond lengths between two atoms, A and B, are very nearly the same, making the assignment of definite values to the A–B covalent bond length possible. This bond length constancy is illustrated well by the C–C bond length, determined from several different compounds to have the value [18] of 1.536 ± 0.006 Å. Various experimenters have quoted the experimental bond lengths used to obtain this value as being precise to ±0.01 Å.

If a single C–C bond is adjacent to a double bond, an interaction resulting in the shortening of the single bond by about 0.03 Å is observed. This bond shortening is even greater (0.08 Å) when the single bond is adjacent to a triple bond. Significant shortening also occurs when a single bond is placed between two double bonds, between two aromatic nuclei, or in a small ring structure. (The C–C bond length in cyclopropane is only 1.524 Å. This is probably due to the bending of bonds in forming the small ring structure.)

Covalent bond lengths are frequently additive; the A–B bond length is often equal to the arithmetic mean of the A–A and B–B bond lengths. For example, the arithmetic mean (1.765 Å) of the C–C bond (1.542 Å, diamond) and the Cl–Cl bond (1.988 Å) is in close agreement with the experimental value of 1.766 ± 0.003 Å found for the C–Cl bond.

Due to these observations, the various elements have been assigned *covalent radii*, the sum of two covalent radii being approximately equal to the corresponding experimental covalent bond length between the two atoms. Covalent radii must be used in molecules where the atoms form covalent bonds in numbers given by

$$NB = 8 - NG \qquad (2.24)$$

where NB is the number of covalent bonds formed and NG is the main group number in the periodic table. Covalent radii are applicable to covalent bonds with considerable partial ionic characters, but other values must be used for extreme ionic bonds.

A selected set [18] of experimental single bond lengths and corresponding covalent radii for several elements is given in Table 2-10. The entries X-ray and ED indicate that x-ray diffraction crystallography and the electron diffraction study of gas molecules, respectively, were used to measure bond lengths; Sp indicates spectroscopy was used.

EXAMPLE 2.6

Calculate the C–F bond length in polytetrafluoroethylene.

Solution:

From Table 2-10, the C–F bond length is $0.772 + 0.640 = 1.412$ Å.

Of the different atomic aggregates, ionic crystals are the most amenable to theoretical treatment due to their simple Coulomb attraction and repulsion terms and the spherical symmetry of their ionic electron distributions. Consequently, approximate quantum-mechanical calculations can be made from which values of the bond length and other quantities of interest may be determined. Based on such calculations, a set of ionic radii have been formulated from the five experimental [11] values of interionic distances in NaF, KCl, RbBr, CsI, and Li_2O. Quantitative values of ionic radii are dependent upon the physical property under consideration. Because they are different for different properties, care must be exercised in their use [11]. Pauling [11] and others [19, 20] have devised tables of ionic crystal radii to compute interionic distances in ionic solids. Table 2-11 contains values [11] of two radii—univalent and crystal radii; the second entries are the univalent radii.

Univalent radii are determined by dividing the A^+X^- distances in alkali halides between the two ions in the inverse ratio of their *effective nuclear charges* (the actual nuclear charges corrected [11] for the screen-

TABLE **2-10**

Single Bond Distances and Covalent Radii for Several Elements

BOND	SUBSTANCE	METHOD	$\frac{1}{2}$ BOND LENGTH	COVALENT RADIUS
C–C	Diamond	X-Ray	0.772 Å	0.772 Å
Si–Si	Si (c)	X-Ray	1.17	1.17
Ge–Ge	Ge (c)	X-Ray	1.22	1.22
Sn–Sn	Sn (c)	X-Ray	1.40	1.40
P–P	P (g)	ED	1.10	1.10
As–As	As$_4$ (g)	ED	1.22	1.21
Sb–Sb	Sb (c)	X-Ray	1.43	1.41
S–S	S$_8$ (g)	ED	1.04	1.04
Se–Se	Se$_8$ (c, α), Se$_8$ (c, β)	X-Ray	1.17	1.17
	Se (c, gray)	X-Ray	1.16	—
Te–Te	Te (c)	X-Ray	1.38	1.37
F–F	F$_2$ (g)	ED	0.718	0.64
Cl–Cl	Cl$_2$ (g)	Sp	0.994	0.99
Br–Br	Br$_2$ (g)	Sp	1.140	1.14
I–I	I$_2$ (g)	Sp	1.333	1.33
H–	—	—	0.300	0.30*

*This value for hydrogen is the average of a large number of radii taken from many compounds involving covalently bound hydrogen. The covalent radius of the H–H molecule itself is 0.37 Å, and some variability is observed. In most compounds, however, the 0.30 Å radius is more closely observed. (Data taken from a number of literature sources; see [18], p. 225 for a tabulation.)

ing effect of the nonvalence electrons). Thus, the tabulated univalent radii agree with observed anion-cation distances for monovalent ions, but not with those for multivalent ions. Pauling [11] gives as the correction equation

$$R_m = R_u Z^{-2/(n-1)} \qquad (2.25)$$

where R_u is the univalent radius taken from Table 2-11, Z is the valence of the ion, n is the Born exponent, and R_m is the multivalent crystal radius. It is apparent from Eq. (2.25) that interionic distances are smaller for multivalent ions than the corresponding univalent radii. This is due to the closer distance of approach caused by larger attractive forces associated with more highly charged ions.

The assignment of effective atomic radii to atoms in metallic solids is complicated by the various ways in which atoms of a given type are sometimes packed together.[9] Also, the apparent valence of an atom involved in the metallic bond is often quite different from the valence of the same atom when it is involved in either the ionic or the covalent bond. Table 2-12 contains a listing [21] of apparent metallic valences

---

[9]This problem of atomic coordination will be discussed at greater length in Chapter 3.

TABLE 2-11
Crystal Radii and Univalent Radii of Ions*

Charge	−4	−3	−2	−1	0	+1	+2	+3	+4	+5	+6	+7
				H^- 2.08 (2.08)	He (0.93)	Li^+ 0.60 (0.60)	Be^{++} 0.31 (0.44)	B^{3+} 0.20 (0.35)	C^{4+} 0.15 (0.29)	N^{5+} 0.11 (0.25)	O^{6+} 0.09 (0.22)	F^{7+} 0.07 (0.19)
	C^{4-} 2.60 (4.14)	N^{3-} 1.71 (2.47)	O^{--} 1.40 (1.76)	F^- 1.36 (1.36)	Ne (1.12)	Na^+ 0.95 (0.95)	Mg^{++} 0.65 (0.82)	Al^{3+} 0.50 (0.72)	Si^{4+} 0.41 (0.65)	P^{5+} 0.34 (0.59)	S^{6+} 0.29 (0.53)	Cl^{7+} 0.26 (0.49)
	Si^{4-} 2.71 (3.84)	P^{3-} 2.12 (2.79)	S^{--} 1.84 (2.19)	Cl^- 1.81 (1.81)	Ar (1.54)	K^+ 1.33 (1.33)	Ca^{++} 0.99 (1.18)	Sc^{3+} 0.81 (1.06)	Ti^{4+} 0.68 (0.96)	V^{5+} 0.59 (0.88)	Cr^{6+} 0.52 (0.81)	Mn^{7+} 0.46 (0.75)
						Cu^+ 0.96 (0.96)	Zn^{++} 0.74 (0.88)	Ga^{3+} 0.62 (0.81)	Ge^{4+} 0.53 (0.76)	As^{5+} 0.47 (0.71)	Se^{6+} 0.42 (0.66)	Br^{7+} 0.39 (0.62)
	Ge^{4-} 2.72 (3.71)	As^{3-} 2.22 (2.85)	Se^{--} 1.98 (2.32)	Br^- 1.95 (1.95)	Kr (1.69)	Rb^+ 1.48 (1.48)	Sr^{++} 1.13 (1.32)	Y^{3+} 0.93 (1.20)	Zr^{4+} 0.80 (1.09)	Cb^{5+} 0.70 (1.00)	Mo^{6+} 0.62 (0.93)	
						Ag^+ 1.26 (1.26)	Cd^{++} 0.97 (1.14)	In^{3+} 0.81 (1.04)	Sn^{4+} 0.71 (0.96)	Sb^{5+} 0.62 (0.89)	Te^{6+} 0.56 (0.82)	I^{7+} 0.50 (0.77)
	Sn^{4-} 2.94 (3.70)	Sb^{3-} 2.45 (2.95)	Te^{--} 2.21 (2.50)	I^- 2.16 (2.16)	Xe (1.90)	Cs^+ 1.69 (1.69)	Ba^{++} 1.35 (1.53)	La^{3+} 1.15 (1.39)	Ce^{4+} 1.01 (1.27)			
						Au^+ 1.37 (1.37)	Hg^{++} 1.10 (1.25)	Tl^{3+} 0.95 (1.15)	Pb^{4+} 0.84 (1.06)	Bi^{5+} 0.74 (0.98)		

*From *The Nature of the Chemical Bond*, 3rd ed., by L. Pauling, copyright 1960, Cornell University Press. Used by permission of Cornell University Press.

TABLE **2-12**

Metallic Valences and Radii of the Elements*

Period 2

	Li	Be	B
v	1	2	3
$R(L12)$	1.549	1.123	0.98
R	1.225	0.889	.80

Period 3

	Na	Mg	Al	Si	P	S
v	1	2	3	2.56	(3)	(2)
$R(L12)$	1.896	1.598	1.429	1.375	1.28	1.27
R	1.572	1.364	1.248	1.173	1.10	1.04

Period 4

	K	Ca	Sc	Ti	V	Cr	Mn	Fe	Co	Ni	Cu	Zn	Ga	Ge	As	Se
v	1	2	3	4	5	6	6	6	6	6	5.56	4.56	3.56	2.56	1.56	(2)
$R(L12)$	2.349	1.970	1.620	1.467	1.338	1.276	1.268	1.260	1.252	1.244	1.276	1.339	1.404	1.444	1.476	1.40
R	2.025	1.736	1.439	1.324	1.224	1.186	1.178	1.170	1.162	1.154	1.176	1.213	1.246	1.242	1.210	1.17

Period 5

	Rb	Sr	Y	Zr	Nb	Mo	Tc	Ru	Rh	Pd	Ag	Cd	In	Sn	Sb	Te
v	1	2	3	4	5	6	6	6	6	6	5.56	4.56	3.56	2.56	1.56	(2)
$R(L12)$	2.48	2.148	1.797	1.597	1.456	1.386	1.361	1.336	1.342	1.373	1.442	1.508	1.579	1.623	1.657	1.60
R	2.16	1.914	1.616	1.454	1.342	1.296	1.271	1.246	1.252	1.283	1.342	1.382	1.421	1.421	1.391	1.37

Period 6

	Cs	Ba	La†	Hf	Ta	W	Re	Os	Ir	Pt	Au	Hg	Tl	Pb	Bi
v	1	2	3	4	5	6	6	6	6	6	5.56	4.56	3.56	2.56	1.56
$R(L12)$	2.67	2.215	1.871	1.585	1.457	1.394	1.373	1.350	1.355	1.385	1.439	1.512	1.595	1.704	1.776
R	2.35	1.981	1.690	1.442	1.343	1.304	1.283	1.260	1.265	1.295	1.339	1.386	1.437	1.502	1.510

	Th	U
v	4	6
$R(L12)$	1.795	1.516
R	1.652	1.426

Lanthanide series

	†Ce	Pr	Nd	Pm	Sm	Eu	Gd	Tb	Dy	Ho	Er	Tm	Yb	Lu
v	3.2	3	3	3	3	3	3	3.5	3	3	3	3	2	3
$R(L12)$	1.818	1.824	1.818	1.834	1.804	2.084	2.804	1.773	1.781	1.762	1.761	1.759	1.933	1.738
R	1.646	1.643	1.637	1.633	1.623	1.850	1.623	1.613	1.600	1.581	1.580	1.578	1.699	1.557

*From The Nature of the Chemical Bond, 3rd ed., by L. Pauling, copyright 1960, Cornell University Press. Used by permission of Cornell University Press.
†Lanthanide series

v, metallic radii R for the free metal atoms, and solid metallic radii $R(L12)$ for the condition of packing with twelve nearest neighbors.

BOND ANGLES Of two given orbitals in an atom, the one that can overlap to a greater extent with an orbital of another atom will form the stronger covalent bond with that atom. Consequently, the p–p bond is expected to be stronger and more directional than the s–s bond. According to Fig. 2-5, it seems that a pair of p–p bonds should be orthogonal. This is approximately true for many compounds, as illustrated [22] by the data in Table 2-13.

The formation of the hybrid molecular orbitals of carbon is interesting. Four linear combinations of the regular atomic $2s$ and $2p$ orbitals can be constructed that will form bonds of maximum strength when oriented at the tetrahedral angle of 109° 28' with respect to one another. The experimental bond angles between carbon and other elements listed by Pauling [22] for 37 different compounds all lie within 2° of this tetrahedral value.

BOND ENERGIES The bond energies of ionic solids can be computed from Eq. (2.18), developed in Example 2.3, and the results obtained are in fair agreement with experimental values. In addition to the interactions considered in developing Eq. (2.18), the weaker van der Waals forces also influence ionic bond energies. These forces, of a secondary influence, are usually ignored, but more refined treatments by Born and Mayer [10] and Mayer and Helmholtz [23] take them into account and show that the Born expression is reliable up to about 2%. Table 2-14 contains a comparison [24] of computed and observed ionic crystal energies and illustrates the reasonably good degree of agreement between theory and experiment.

TABLE 2-13

Observed Bond Angles in Hydrides

SUBSTANCE	BOND ANGLE	EXPERIMENTAL VALUE
H_2O	HOH	104.45 $\pm$ 0.10°
H_3N	HNH	107.3. $\pm$ 0.2
H_2S	HSH	92.2 $\pm$ 0.1ˈ
H_3P	HPH	93.3 $\pm$ 0.2
H_2Se	HSeH	91.0 $\pm$ 1
H_3As	HAsH	91.8 $\pm$ 0.3
H_2Te	HTeH	89.5 $\pm$ 1
H_3Sb	HSbH	91.3 $\pm$ 0.3

TABLE **2-14**

**Comparison of Calculated and Experimental
Lattice Energies for Ionic Crystals***

COMPOUND	V (CALCULATED, KCAL/MOLE)	V (EXPERIMENTAL, KCAL/MOLE)
LiCl	193.3	198.1
NaCl	180.4	182.8
KCl	164.4	164.4
RbCl	158.9	160.5
CsCl	148.9	155.1
LiBr	183.1	189.3
NaBr	171.7	173.3
KBr	157.8	156.2
RbBr	152.5	153.3
CsBr	143.5	148.6
LiI	170.7	181.1
NaI	160.8	166.4
KI	149.0	151.5
RbI	144.2	149.0
CsI	136.1	145.3

*From *The Modern Theory of Solids,* by F. Seitz, copyright 1940, McGraw-Hill Book Co., Inc. Used by permission.

The calculation of covalent bond energies involves the use of molecular orbital wave functions. For example, the wave function representing the single bond in a symmetric molecule AA may be a linear combination of the wave functions ψ_{AA}, $\psi_{A^+A^-}$, and $\psi_{A^-A^+}$, representing, respectively, the purely covalent AA bond, the ionic A^+A^- bond, and the ionic A^-A^+ bond.[10] Similar expressions are written for the wave function of the BB bond. The form of these functions is

$$\psi(AA) = a\psi_{AA} + b(\psi_{A^+A^-} + \psi_{A^-A^+}) \qquad \text{(2.26)}$$

The ratio b/a, which determines the partial ionic character of the bond, is small, and is probably about the same for all bonds between like atoms.

For similar atoms, one might expect the wave function $\psi(AB)$ to be similar in form to Eq. (2.26), perhaps an average of $\psi(AA)$ and $\psi(BB)$. Such a bond would be called a *normal covalent bond.*

[10]The combination involving these possible structures is based on the concept of a *resonance* of the molecule between the various structures. This resonance phenomenon results in a lower total energy than any of the pure states possess. The interested student is referred to Pauling's book for a detailed discussion and explanation of the resonance phenomenon.

If the atoms **A** and **B** are dissimilar (that is, if one has greater electronegativity than the other), a more general wave function

$$\psi(AB) = a\psi_{AB} + c\psi_{A^+B^-} + d\psi_{A^-B^+} \qquad (2.27)$$

must be chosen to represent the bond, where a, c, and d are constants. The ratios c/a and d/a are chosen to minimize the total energy of the molecule and maximize the energy of the bond. The values of these ratios in general differ from the ratio b/a in Eq. (2.26), with one being smaller and one being larger. The energy of the bond between unlike atoms is greater than or equal to the energy of a normal covalent bond between unlike atoms. This additional energy, called the *ionic resonance energy* of the bond, is due to the partial ionic character of the real bonds.

The energy of a normal covalent bond is the arithmetic mean of the normal bond energies of two like molecules AA and BB. Because of this *postulate of the additivity of normal covalent bonds*, the energy difference Δ_{AB}, defined as

$$\Delta_{AB} = D(AB) - \tfrac{1}{2}(D(AA) + D(BB)) \qquad (2.28)$$

should never be negative. In Eq. (2.28), $D(AB)$ is the experimental AB

TABLE 2-15

Comparison of Bond Energies, Normal Covalent Bond Energies, and Δ_{AB} for Hydrogen Halogenide and Halogen Halogenide Molecules (kcal /mole)*

COMPOUND	BOND ENERGY	NORMAL COVALENT BOND ENERGY	Δ_{AB}
HH	104.2	—	—
FF	36.6	—	—
ClCl	58.0	—	—
BrBr	46.1	—	—
II	36.1	—	—
HF	134.6	70.4	64.2
HCl	103.2	81.1	22.1
HBr	87.5	75.2	12.3
HI	71.4	70.2	1.2
ClF	60.6	47.3	13.3
BrCl	52.3	52.1	0.2
ICl	50.3	47.1	3.2
IBr	42.5	41.1	1.4

*From *The Nature of the Chemical Bond, 3rd ed.,* by L. Pauling, copyright 1960, Cornell University Press. Used by permission of Cornell University Press.

TABLE **2-16**

Experimental Energy Values for Single Bonds (kcal /mole)*

BOND	ENERGY	BOND	ENERGY	BOND	ENERGY
H–H	104.2	P–H	76.4	Si–Cl	85.7
C–C	83.1	As–H	58.6	Si–Br	69.1
Si–Si	42.2	O–H	110.6	Si–I	50.9
Ge–Ge	37.6	S–H	81.1	Ge–Cl	97.5
Sn–Sn	34.2	Se–H	66.1	N–F	64.5
N–N	38.4	Te–H	57.5	N–Cl	47.7
P–P	51.3	H–F	134.6	P–Cl	79.1
As–As	32.1	H–Cl	103.2	P–Br	65.4
Sb–Sb	30.2	H–Br	87.5	P–I	51.4
Bi–Bi	25	H–I	71.4	As–F	111.3
O–O	33.2	C–Si	69.3	As–Cl	68.9
S–S	50.9	C–N	69.7	As–Br	56.5
Se–Se	44.0	C–O	84.0	As–I	41.6
Te–Te	33	C–S	62.0	O–F	44.2
F–F	36.6	C–F	105.4	O–Cl	48.5
Cl–Cl	58.0	C–Cl	78.5	S–Cl	59.7
Br–Br	46.1	C–Br	65.9	S–Br	50.7
I–I	36.1	C–I	57.4	Cl–F	60.6
C–H	98.8	Si–O	88.2	Br–Cl	52.3
Si–H	70.4	Si–S	54.2	I–Cl	50.3
N–H	93.4	Si–F	129.3	I–Br	42.5

*From *The Nature of the Chemical Bond, 3rd ed.,* by L. Pauling, copyright 1960, Cornell University Press. Used by permission of Cornell University Press.

bond energy; $D(AA)$ and $D(BB)$ are the energies of the AA and BB bonds, respectively. Table 2-15 compares values of the experimental bond energies, the normal covalent bond energies [25], and Δ_{AB} for each of the hydrogen halogenides and the halogen halogenides.

For the alkali metal hydrides, all the values of Δ_{AB} calculated from Eq. (2.28) are negative. Pauling [25] corrected this difficulty by introducing the *concept of the geometric mean,* which states that the proper average of the individual bond energies necessary to provide the normal covalent bond energy is the geometric, not the arithmetic mean. For small or moderate differences between $D(AA)$ and $D(BB)$, the two averages are nearly the same. However, in the cases of large differences, as in the alkali metal hydrides, the smaller geometric mean is required to produce positive values for Δ_{AB}. The empirical values used in calculations for the bond energies of diatomic molecules are given directly by the energies of dissociation of the atoms, and are tabulated [25] in Table 2-16.

EXAMPLE 2.7

Calculate the normal covalent bond energy for the HF molecule, using the arithmetic and geometric mean concepts. Calculate Δ_{AB} for each case.

Solution:

Arithmetic mean case   (see Table 2-15)

$$D(HF)(normal) = \tfrac{1}{2}(104.2 + 36.6) = 70.4 \text{ kcal/mole}$$

$$\Delta_{AB} = 134.6 - 70.4 \quad = 64.2 \text{ kcal/mole}$$

Geometric mean case

$$D(HF)(normal) = ((104.2)(36.6))^{1/2} = 61.6 \text{ kcal/mole}$$

$$\Delta_{AB} = 134.6 - 61.6 \quad = 73.0 \text{ kcal/mole}$$

Many of the entries in Table 2-16 represent averages for several bonds in polyatomic molecules, instead of individual bond dissociation energies. In the water molecule (which has two OH bonds in its structure), for example, 119.9 kcal/mole is required to remove the first hydrogen while only 101.2 kcal/mole is needed to disrupt the second bond. The average of these two values, 110.6 kcal/mole, is the value which is tabulated. This and other techniques have been used to determine the values listed in Table 2-16.

References Cited

[1]   Richtmeyer, F. K., E. H. Kennard, and T. Lauritsen, *Introduction to Modern Physics*, 5th ed., McGraw-Hill Book Co., Inc., New York, 1955, pp. 461–63.

[2]   Bohr, N., *Phil. Mag.*, **26**, p. 1 (1913).

[3]   Sommerfeld, A., *Atomic Structure and Spectral Lines*, Methuen & Co., Ltd., London, 1929.

[4]   de Broglie, L., *Phil. Mag.*, **47**, p. 446 (1924); *Ann. Physik*, **3**, p. 22 (1925).

[5]   Schrödinger, E., *Ann. Physik*, **79**, pp. 361, 489, 734 (1926); **80**, p. 437, (1936); **81**, p. 190 (1936).

[6]   Heisenberg, W., *Zeit. Physik*, **43**, p. 172 (1927).

[7]   Pauling, L., *The Nature of the Chemical Bond*, 3rd ed., Cornell University Press, Ithaca, New York, 1960, pp. 35–36.

[8]   Pauli, W., *Zeit. Physik*, **31**, p. 765 (1925).

[9]   Pauling, *op. cit.*, p. 49.

[10] Born, M., and J. E. Mayer, *Zeit. Physik*, **75**, p. 1 (1932).

[11] Pauling, *op. cit.*, Chapter 13.

[12] Pauling, *op. cit.*, p. 111*ff*.

[13] Moffatt, W. G., G. W. Pearsall, and J. Wulff, *The Structure and Properties of Materials, Volume 1, Structure*, John Wiley and Sons, Inc., New York, 1964, p. 15.

[14] Daniels, F., and R. A. Alberty, *Physical Chemistry*, John Wiley and Sons, Inc., New York, 1955, p. 78.

[15] Sinnott, M. W., *The Solid State for Engineers*, John Wiley and Sons, Inc., New York, 1961, pp. 189–90.

[16] Pauling, *op. cit.*, pp. 88–99.

[17] Pauling, *op. cit.*, Chapters 1 and 8.

[18] Pauling, *op. cit.*, Chapter 7.

[19] Wasastjerna, J. A., *Soc. Sci. Fenn. Comm. Phys. Math.*, **38**, p. 1 (1923).

[20] Goldschmidt, V. M., "Geochemische Verteilungsgesetze der Elemente," *Skrifter Norske Videnskaps-Akad. Oslo, I. Mat.-Naturv. Kl.*, 1926.

[21] Pauling, *op. cit.*, Chapter 11.

[22] Pauling, *op. cit.*, Chapter 4.

[23] Helmholtz, L., and J. E. Mayer, *Zeit. Physik*, **75**, p. 19 (1932).

[24] Seitz, F., *The Modern Theory of Solids*, McGraw-Hill Book Co., Inc., New York, 1940, p. 80.

[25] Pauling, *op. cit.*, Chapter 3.

PROBLEMS

2.1   The covalent bond energy of the NH bond is 93.4 kcal/mole. Calculate the wave length of the photon which must be absorbed to break a single NH bond.

2.2   A photon of wavelength 3000 Å is absorbed by a substance. Calculate the energy absorbed in (a) electron volts; (b) cal; (c) joules; (d) Btu.

2.3   A hydrogen atom electron jumps from the third ($n = 3$) to the second ($n = 2$) Bohr orbit. What is the wavelength of the radiation omitted?

2.4   Calculate the radius of the first three Bohr orbits for the Li^{++} ion.

2.5   What is the wavelength of the photon which must be absorbed by an electron in the ground state of the He^+ ion to raise it to the second Bohr orbit?

2.6   Calculate the de Broglie wavelength of an electron moving at a speed $v = 2.5(10^8)$cm/sec.

2.7   Calculate the Bohr–Sommerfeld ellipticities of the $3p$, $3d$, $4s$, $4p$, $4d$, $4f$, $5s$, and $5d$ states.

2.8   How many electrons are required to fill the $n = 5$ principal quantum shell?

2.9   A certain divalent ion pair has values of $n = 9$ and $B_{12} = 5.77(10^{-62})$cm^8. Calculate the equilibrium distance of separation in Å and the energy of the ion-pair bond in erg.

2.10   What is the magnitude of the attractive and repulsive forces acting on the ion pair in Problem 2.9?

2.11   Write the electron distribution for the elements Na (atomic number $Z = 11$), Zn($Z = 30$), Kr($Z = 36$), and Rb($Z = 37$), using only the atomic number and the exclusion principal as guides. Compare your results with Table 2-1.

2.12   Derive an expression for the bond energy of a monovalent ion pair which involves only the bond length r_0, the electronic charge e, and the repulsion exponent n.

2.13   Calculate the force necessary to decrease the ion-pair separation distance by 10% from the equilibrium value r_0. How much work must be done to effect this compression? (Use the constants given in Problem 2.9.)

2.14   Using the data in Tables 2-7 and 2-13, calculate the dipole moment of the NH_3 molecule.

2.15   Calculate the dipole moment of CH_3Cl using the data of Table 2-7.

2.16   Calculate the per cent of ionic character in the Si–F, C–Cl, P–O, S–O, and C–N bonds.

2.17   The experimental dipole moments of HF, HCl, HBr, and HI are 1.98, 1.03, 0.78, and 0.38 Debye units, respectively. Using the data of Table 2-10, calculate the experimental values of the per cent of partial ionic character in each of these hydrogen halogenide bonds. Then compare your answers with the values calculated from Eq. (2.23).

2.18   Calculate the C–H, Si–H, Ge–H, and Sn–H bond lengths. (The experimental values are 1.095, 1.48, 1.53, and 1.70 Å, respectively.)

2.19   For Na^+Cl^- crystals, the Born exponent is $n = 8$ and the anion-cation distance is $R_0 = 2.814$ Å. The crystal energy is defined as $V_0 = -N_0V(R_0)$, where N_0 is Avogadro's number and $V(R_0)$ is the potential energy calculated from Eq. (2.18). Calculate V_0 for Na^+Cl^- crystals.

2.20   For Na^+F^- crystals, the Born exponent is $n = 7$ and $R_0 = 2.307$ Å. Calculate the crystal energy $V_0 = -N_0 V(R_0)$, and compare it with the experimental value 216.8 kcal/mole. (See Problem 2.19 for definition of $V(R_0)$.)

2.21   Pauling postulated that electronegativity differences were related to bond energies by $D(AB) = [D(AA)D(BB)]^{1/2} + 30(X_A - X_B)^2$, where D's are expressed in kcal/mole. Use this relation to calculate $D(AB)$ for the pairs C–H, C–F, C–Cl, Si–H, and Si–F. Compare your results with the experimental values listed in Table 2-16.

2.22   Calculate the energy required to break a single Cl–Cl bond (that is, the bond between just two Cl atoms). What wavelength photon is required for this purpose?

2.23   The energy of the C=C double bond is 146 kcal/mole. Calculate the energy liberated if one g mole of ethylene ($CH_2=CH_2$) is polymerized to form $(C_2H_4)_n$, with no double bonds remaining.

THE THEORY
OF IDEAL
CRYSTALLINE
SOLIDS

Most materials can be classified as gaseous, liquid, or solid. However, glasses and some high-molecular weight polymers possess a number of the characteristics of both liquid and solid states, and are classed separately as members of the vitreous or glassy state. Many of the specific characteristics of these states of aggregation will be discussed at various places in the following chapters. Here only brief descriptions of their general characters should suffice in introducing the concepts of phases. The bulk of this chapter will then be devoted to a study of the properties of the geometrical arrays of points in space, in preparation for a detailed study of the solid state. Such a study will give a theoretical estimate of the limiting properties of the "perfect" crystalline solid.

States of Aggregation

The Gaseous State

The atoms or molecules of a gaseous substance move about essentially independently of one another, as if they were tiny, hard spheres possessing a certain amount of temperature-dependent energy in the form of kinetic energy of motion and internal potential energy of vibration and rotation. At very high temperatures, gas molecules may also possess energy in excited electronic states.

The gas molecules move about with a random distribution of velocities. By a statistical averaging process [1] a mean thermal speed $\bar{v}$ may be calculated from this distribution

$$\bar{v} = \left(\frac{8kT}{\pi m}\right)^{1/2} \tag{3.1}$$

where k is the Boltzmann constant (1.380×10^{-16} erg/°K), T is the absolute temperature, and m is the mass of the molecule. The pressure P exerted by molecules impinging on a plane wall exposed to the gas is

$$P = \tfrac{1}{3}nm\overline{v^2} \tag{3.2}$$

where n is the number of molecules per unit volume and $\overline{v^2}$ is the mean-square speed of the molecules. (Note here that the average of v^2 will be different from that of $\bar{v}^2$.) Equation (3.2) is easily converted to

$$PV = \frac{2}{3}\frac{nV}{N_0}\frac{N_0 m\overline{v^2}}{2} \tag{3.3}$$

If we let $\nu = nV/N_0$ be the number of moles of gas in the container and $U = N_0 m\overline{v^2}/2$ be the average kinetic energy of translation of the molecules per mole, then

$$PV = \tfrac{2}{3}\nu U \tag{3.4}$$

When this result is compared with the experimentally determined perfect-gas law

$$PV = \nu RT \tag{3.5}$$

it is evident that

$$T = \frac{2U}{3R} \tag{3.6}$$

which may be interpreted as the kinetic theory definition of absolute temperature. Since temperature is related to molecular motion, it follows that raising the temperature of a gas increases the mean kinetic energy of the gas molecules.

EXAMPLE 3.1

Calculate the mean thermal speed of hydrogen gas molecules at a temperature of 25°C.

Solution:

The mass of a hydrogen molecule is

$$m_{H_2} = \frac{(2)(1.008)}{(6.023)(10^{23})} = 3.34(10^{-24}\text{ g})$$

From Eq. (3.1) we have

$$\bar{v} = \left\{ \frac{(8)(1.380)(10^{-16} \text{ erg } °K^{-1})(298.2°K)}{(3.1416)(3.34)(10^{-24} \text{ g})} \right\}^{1/2}$$

$$= [3.14(10^{10})]^{1/2} \text{ cm/sec}$$

$$= 1.771(10^5) \text{ cm/sec}$$

A gas is characterized by complete lack of shape or size, and its volume is sensitive to changes in temperature and pressure as indicated by Eq. (3.5). At a constant temperature, the pressure may be increased continuously until it reaches a point where the van der Waals forces are of sufficient magnitude to cause condensation. It is possible, however, for the temperature to be high enough that the intermolecular forces of attraction cannot overcome the kinetic momentum transfer *regardless of the pressure*, and, consequently, the gas cannot be liquefied. Thus, a *critical temperature* exists for each gas where these two effects are balanced. The liquefaction pressure at this temperature is called the *critical pressure*, and the density under these conditions is the *critical density*. Values of critical constants for several common gases are listed in Table 3-1.

TABLE **3-1**

Critical Constants for Selected Gases*

GAS	$T_c(°C)$	$P_c(\text{atm})$	$\rho_c(\text{g/cm}^3)$
Acetylene	36.0	62.0	0.231
Air	−140.7	37.2	0.35(0.31)
Ammonia	132.4	111.5	0.235
n-Butane	153	36.0	—
Carbon dioxide	31.1	73.0	0.460
Carbon disulfide	273.0	76.0	0.441
Chlorine	144.0	76.1	0.573
Ethane	32.1	48.8	0.21
Fluorine	−155.0	25.0	—
Helium	−267.9	2.26	0.0693
Hydrogen	−239.9	12.8	0.0310
Krypton	−63.8	54.3	1.10
Methane	−82.5	45.8	0.162
Neon	−228.7	25.9	0.484
Nitrogen	−147.1	33.5	0.3110
Oxygen	−118.8	49.7	0.430
Sulfur dioxide	157.2	77.7	0.52
Water	374.15	218.4	0.323
Xenon	16.6	58.2	1.155

*From *Chemical Engineer's Handbook, 4th ed.,* by R. H. Perry, *et al.,* copyright 1963, McGraw-Hill Book Co., Inc. (section 3, p. 100).

The Liquid State

The liquid state of aggregation of matter *normally* results from gas condensation. (In some systems gas will condense directly into a solid, and conversely. This phenomenon, called *sublimation*, is exhibited when carbon dioxide forms "dry" ice at *atmospheric pressure*.) Unlike a gas, a liquid assumes the shape of its container, but does not always fill the container volume entirely. The total energy of the liquid state must be lower than that of the gaseous state of the same material because energy must be *removed* from the gas to cause liquefaction. As the energy of a gas is reduced, its temperature decreases until the first drops of liquid appear at a temperature called the *dew point*. At the dew point for a pure substance, the temperature decrease is arrested and energy is removed isothermally until all of the gas has been condensed; after this, further removal of energy results in a decrease of the temperature of the liquid. Energy removed isothermally during the condensation of a pure substance is called the *latent heat of condensation*. The same quantity of energy, the *latent heat of vaporization*, must be added to the liquid at this particular temperature and pressure in order to cause vaporization. At atmospheric pressure the vaporization temperature is called the normal boiling point. The latent heat of vaporization is the energy necessary to overcome the intermolecular forces of attraction and to perform the work of expanding the vapor against an external pressure of gas. Normal boiling points and latent heats of vaporization for several liquids are listed in Table 3-2.

The molecular arrangement of the liquid state is more or less regularly ordered for the distances of a few molecular diameters. However, long-range disorder prevails because of vacancies [2] or *holes* in the liquid structure, caused by incomplete bonding of the molecules or atoms due to thermal molecular vibration and agitation. Decreasing the temperature of the liquid will cause the number of these vacancies to decrease gradually as more bonds are formed. (The concept of vacancies will be discussed in more depth in Chapter 4.)

The holes in the liquid structure allow it to flow and deform permanently under the action of very small mechanical shear forces. If there are many holes, the liquid structure shears easily and its *fluidity* is large. Conversely, decreasing the number of holes will decrease the fluidity of the liquid. As the temperature is decreased, therefore, the fluidity of the liquid phase decreases. (Although a detailed discussion [2, 3] of the theory of the liquid state lies outside the scope of this book, this phenomenon of fluidity will be considered at some length in Chapter 8.)

TABLE **3-2**

Thermal Data for Selected Liquids*

LIQUID	NORMAL BOILING POINT (°C)	LATENT HEAT OF VAPORIZATION (CAL/G MOLE)
CCl_4	77	7,280
CH_4	−161.4	2,040
Cl_2	−34.1	4,878
Cu	2595	72,810
HCl	−85.0	3,860
HF	33.3	7,460
H_2O	100.0	9,729
H_2O_2	158	10,270
Pb	1744	42,060
LiCl	1382	35,960
Hg	361	13,980
KCl	1407	38,840
NaCl	1465	40,810
NaF	1704	53,260
S(rhombic)	444.6	2,200
$TiCl_4$	136	8,350
UF_6	55.1	9,990
Acetic acid	118.1	5,810
Acetone	56.5	7,223
CS_2	46.5	6,403
Chloroform	61.3	7,044
Ethyl alcohol	78.3	9,410
Ethylene glycol	197	11,860
Methyl alcohol	64.7	8,421
n-propyl alcohol	97.2	9,878
n-octane	125.6	8,361
Benzene	80.1	7,354
Toluene	110.6	7,998

*From various tables in Section 3 of *Chemical Engineer's Handbook, 4th ed.*, by R. H. Perry, *et al.*, copyright 1963, McGraw-Hill Book Co., Inc.

The Solid State

As temperature decreases, the number and strength of the inter-molecular secondary bonds in the liquid increases, corresponding to the decrease in the number of holes in the liquid structure. Eventually the temperature is sufficiently reduced for the intermolecular forces to overcome most of the thermal molecular vibration, and a rapid increase in the ordering of molecular arrangement occurs. This condition, known as *solidification* or *crystallization*, occurs for pure substances at the melting point of the solid. Energy, called *latent heat*, is removed isothermally from the melt at this temperature as the liquid is completely converted to solid. When solidification is complete, further removal of energy results in a decrease in the temperature of the solid. The amount of energy that must be added to the solid at the melting temperature in order to cause melting or fusion, the *latent heat of fusion*, is used largely to rearrange the fixed structure of the solid into the more or less disordered structure of the liquid. This is accomplished by breaking sufficient interatomic bonds to introduce the requisite number of holes into the structure thus giving rise to the fluidity of the liquid phase. Typical melting points and latent heats of fusion for several solids are listed in Table 3-3.

Solids have both definite volumes and shapes and are able to maintain these shapes to varying extents under the influence of external mechanical forces. Solids occur in two more or less distinct forms: crystalline and amorphous. A *crystalline solid* possesses an ordered, three-dimensionally repetitive atomic arrangement; an *amorphous solid* is characterized by a random, but rigid, molecular arrangement. Cellulose, rubber, bakelite, and glass are solids in the sense that they possess definite volume and form, but do not conform to the definition of true crystalline solids just mentioned. These materials are sometimes classed as supercooled liquids, and occasionally as molecular solids.

The molecules of a solid are permanently located relative to their neighbors, and vibrate constantly about their average or equilibrium positions due to their thermal energy. This internal energy may be variously distributed as kinetic vibrations of the atoms about the equilibrium positions (called *lattice vibrations*), internal atomic vibrations, translational energy of free electrons, partial molecular rotation, or electron energy-level excitation energy. Not all of these modes necessarily exist simultaneously in each solid. Nonmetals, for example, have no free electrons and cannot possess energy of translation of free electrons. In metals, single atoms comprise the basic lattice and, therefore, some of the rotational and vibrational modes do not exist.

The total internal energy of the solid increases as temperature increases. Consequently, the frequency and amplitude of the atomic lattice vibrations also increase, causing the molecules or atoms to move farther away from their equilibrium positions. This increased vibration and movement of the atoms from their equilibrium lattice positions (thus storing energy) is responsible for the thermal expansion and heat capacity of solids. Thermal lattice vibrations provide one mechanism for the conduction of heat through solids.

In metallic solids, the degree of heat conductivity is due primarily to the ease with which the "gas" of valence electrons can move under the influence of an external temperature gradient, rather than to the lattice vibrations of the cores. Thus, metals usually have high thermal conductivities.

TABLE 3-3

Thermal Data for Selected Solids*

SOLID	NORMAL MELTING POINT (°C)	LATENT HEAT OF FUSION (CAL/G MOLE)
Al	660.0	2,550
Al_2O_3	2045	(26,000)†
Be	1280	2,500
Bi	271.3	2,505
Cd	320.9	1,460
CaO	2707	(12,240)
C(graphite)	3600	11,000
CsCl	642	3,600
Cr	1550	3,930
Co	1490	3,660
Cu	1083.0	3,110
CuO	1447	2,820
Cu_2S	1127	5,500
Ge	959	(8,300)
Au	1063.0	3,030
H_2O	0.0	1,436
Fe	1530	3,560
FeO	1380	(7,700)
FeS	1195	5,000
Pb	327.4	1,224
PbO	890	2,820
PbS	1114	4,150
LiCl	614	3,200
Mg	650	2,160
Mo	2622	(6,660)
Ni	1455	4,200
Pt	1773.5	4,700
KCl	770	6,410
Re	(3000)	—
Si	1427	9,470
SiO_2(quartz)	1470	3,400
SiO_2(cristobalite)	1700	2,100
Ag	960.5	2,700
NaCl	800	7,220
W	3390	(8,400)
Zn	419.5	1,595
ZnO	1975	4,470
ZrO_2	2715	20,800

*From *Chemical Engineer's Handbook, 4th ed.,* by R. H. Perry, *et al.,* copyright 1963, McGraw-Hill Book Co., Inc. (section 3, pp. 107–109).
†Values enclosed in parentheses are uncertain.

The Vitreous or Glassy State

Glasses are currently thought to consist of random, three-dimensional networks where the formation of definite chains or sheets of atoms is not possible. A typical silicate glass structure is illustrated [4] schematically in Fig. 3-1. The relatively large, irregular openings in the structure permit the inclusion of various modifying network ions (such as the Na^+ ion in Fig. 3-1) which serve to change the basic properties of the glass. The strongest bonds in glasses are the Si–O bonds; due to the absence of any symmetry in the overall structure, these vary in strength and length throughout. (The *average* Si–O bond length in glasses (1.61 Å) is very similar to the 1.60 Å value found in crystalline silica [5].) Because of this variation in bond lengths and strengths, glasses do not melt at a single temperature, but soften over a range of temperatures and gradually become more liquid-like.

Glasses are frequently compared with supercooled liquids, but may be distinguished from them by following the volume changes which occur upon cooling (illustrated schematically in Fig. 3-2). At high temperatures a glass behaves as a true liquid, with its atoms free to move about and respond to shear stresses. As the liquid is cooled, the atoms rearrange and pack themselves more efficiently, causing a decrease in volume. A temperature is eventually attained at which the rate of change of the expansion coefficient of the glass alters markedly, "freezing" the structure and preventing further atomic rearrangement.

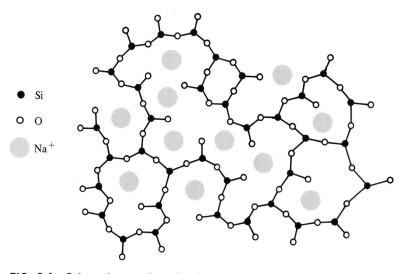

● Si

○ O

Na$^+$

FIG. 3-1   Schematic, two-dimensional representation of the structure of sodium silicate glass.

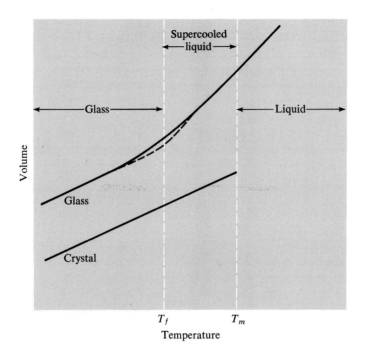

FIG. 3-2   Schematic representation of the difference in the volume changes that accompany the cooling of a glass and the solidification of a liquid into crystal at a temperature T_m. Between the temperatures T_f and T_m, the glass behaves as a supercooled liquid. Below T_f (the fictive temperature of the glass), it behaves as a true solid.

The temperature at which this abrupt change in the thermal expansion occurs is called the *fictive temperature*.

Thermodynamically, a glass is stable only if its free energy[1] is less than or equal to the free energy of the corresponding crystal structure. Therefore, glasses can only be formed by compounds which tend to form network-type structures. (The single element which exists normally as a glass is selenium, although one modification of sulfur, called plastic sulfur, is similar to a glass [5].) These elements solidify in chain and ring structures so that the glassy state can be pictured simply as a disordered array of atomic groupings. Certain other compounds form structures composed of discrete groups, and these, in turn, readily form glasses. Examples of such compounds are the inorganic oxides SiO_2, BeF_2, B_2O_3, GeO_2, P_2O_5, As_2O_5, Sb_2O_5, P_2O_3, and Sb_2O_3.

---

[1]See the introduction to Chapter 5 for a detailed definition of this quantity.

The effect of disorder of the structural pattern leading to glasslike behavior is illustrated schematically in Fig. 3-3 for B_2O_3. Each small boron atom fits in among three larger oxygen atoms. Since boron has a valence of three and oxygen a valence of two, the electrical balance is maintained if each oxygen atom is located between two boron atoms thus forming a continuous structure of strongly bonded atoms. The relative long-range order or disorder of this structure determines whether the solid is glassy or crystalline.

The network-forming ions of a glass are too tightly bound to contribute significantly to electrical conductivity; instead such conductivity is due primarily to the diffusion of the modifying metal ions entrapped in the holes. The arrangement of these vacancies and, consequently, the electrical conductivity are highly dependent upon the thermal history of the glass structure. Similarly, the refractive index of a glass is dependent upon its thermal history because thermal stress causes local *anisotropy* in the vicinity of the induced strain. (In an isotropic medium, properties are independent of spatial orientations.)

The open structure of inorganic oxide glasses is due to strong cation-cation repulsion, most pronounced when the anion polyhedra are small and the cation charge is high. An anion polyhedron is an imaginary geometrical body whose shape corresponds to the geometrical arrangement of the cation's nearest neighbors with the cation at the center of the body. For example the SiO_4^{4-} anion polyhedra of silica are tetrahedra, with each corner corresponding to an O^{2-} ion center and the Si^{4+} ion located at the center of the tetrahedron. (Anions are negative ions, so called because, in the terminology of the physical chemist, they are formed at the negative cathode in an electrochemical reaction and migrate toward the positive anode in solution. Cations, conversely, are positive ions which form at the positive anode and migrate in solution toward the negative cathode. Although this convention is exactly the reverse of the electrochemist's approach used in Chapter 11 to discuss corrosion, the terminology will be retained here because it is so firmly established.)

An inorganic compound tends to be noncrystalline if the following conditions are met [5]:

1. Each anion is bounded to only two cations.
2. No more than four anions surround a cation.
3. The anion polyhedra share corners, but not edges or faces (as illustrated schematically in Fig. 3-4 for silica glass).
4. The compound has a large number of constituents distributed irregularly throughout its network.

The typical compositions [6] of several common glasses, expressed as weight per cent of the constituent oxides, are listed in Table 3-4.

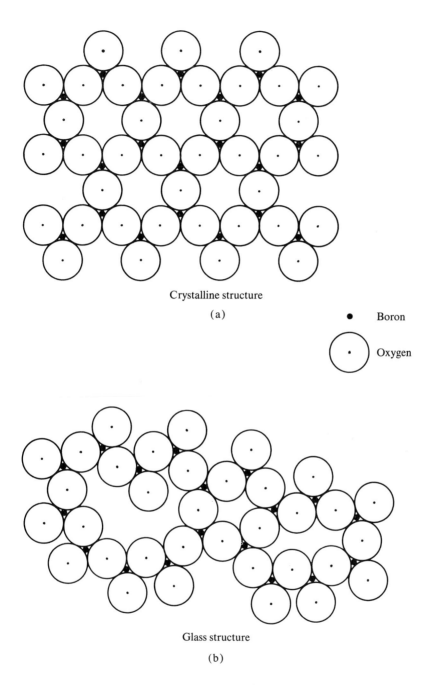

Crystalline structure

(a)

● Boron

Oxygen

Glass structure

(b)

FIG. 3-3 Schematic representation of structure of B_2O_3 in forming (a) a crystalline solid and (b) a glass. The glassy state is simply a disorganized form of the crystalline state.

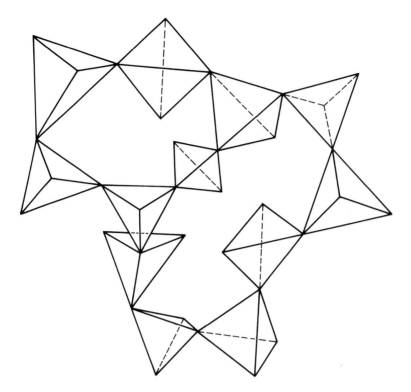

FIG. 3-4   Schematic representation of the three-dimensional structure of silica glass. Each tetrahedron represents an SiO_4^{4-} anion polyhedron in which the oxygen ions occupy the corners of the tetrahedron which surrounds the silicon ion at its center. Only anion polyhedral corners are shared in the glass structure.

TABLE **3-4**

Composition of Common Glasses (weight %)*

GLASS	NETWORK FORMERS			NETWORK MODIFIERS				
	SiO_2	B_2O_3	Al_2O_3	Na_2O	K_2O	MgO	CaO	PbO
Fused silica	99.8	—	—	—	—	0.1	0.1	—
Vycor	96.0	3	1	—	—	—	—	—
Pyrex	80.0	14	2	3.5	0.5	—	—	—
Soda-silica	72.0	—	1	20.0	—	3.0	4.0	—
Lead-silica	63.0	—	1	8.0	6.0	—	1.0	21

*From *Introduction to Solids*, by L. V. Azaroff, copyright 1960, McGraw-Hill Book Co., Inc. Used by permission.

SPACE LATTICES

UNIT CELL

Ideal Crystal Structures in the Solid State

Because most construction materials exist in either single crystals or crystal aggregates, crystalline solids and the basic principles of crystallography are of considerable importance to the engineer.

Crystallography (the study of crystal structures) can be considered from two different viewpoints: one concerned with the external shape and symmetry of *macrocrystals* and one concerned with the orderly arrangement of atoms on a microscopic scale.

The external appearance of a crystal depends on its internal structure as well as on the physical-chemical environment in which it was formed [7]. But only a skilled crystallographer can obtain significant information about the internal structure of a crystal from its external appearance. For our purposes here, it is more desirable to study the internal structure of the crystal, from which many external characteristics may also be deduced. This study can conveniently be divided into two parts: geometrical crystallography and the study of real crystals. We shall treat only the first here, deferring the study of real crystalline materials to Chapter 4.

Geometrical Crystallography

Geometric crystallography is essentially an exercise in solid geometry and, as such, requires certain special nomenclature which must be rather carefully defined.

SPACE LATTICES   A space lattice is a set of straight lines constructed to divide all space into small volumes of equal size. Alternatively, it can be a set of points extending indefinitely in space in a regular arrangement so that each point has identical surroundings [8]. (Note that the second is equivalent to the first definition, since sets of points can be described by a set of linear reference axes.) Regardless of the definition used, the basic notion is the same—all of space is divided up into small, identifiable regions of regular size and distribution by means of some system of reference measures.

Although it might be assumed that a great many space lattices are possible, Bravais proved [9] mathematically that only fourteen different ways exist in which the points of a space lattice can be arranged so that each point has identical surroundings [8].

THE UNIT CELL   The space lattice, as defined, extends indefinitely in all directions, excluding no portion of space. However, it is incon-

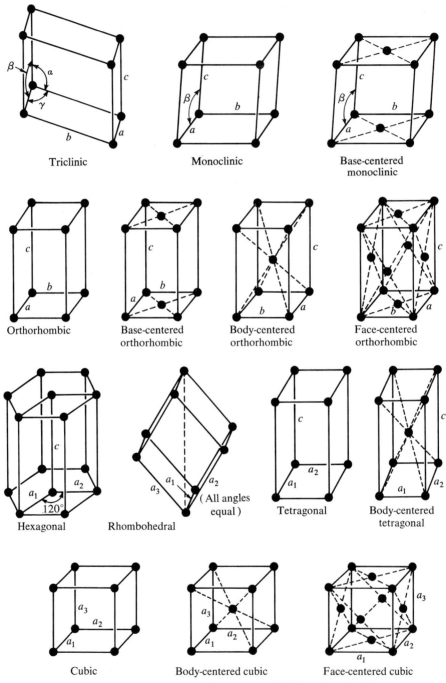

FIG. 3-5   Schematic representation of the fourteen standard Bravais unit cells.

venient to deal with all of space, so one deals with the *unit cell*, or that portion of a space lattice which will generate the lattice when translated along its axes. Thus, the fourteen different Bravais lattices can be represented in terms of the fourteen Bravais unit cells illustrated in Fig. 3-5.

The edge length or *lattice parameter* of the unit cell (also called the *lattice constant* or *primitive translation*) is the linear translation of the cell in a given direction which is required to repeat its structure. Lattice parameters are found experimentally to be of the order of a few angstrom units. Therefore, very large numbers of unit cells are required to make up even microscopically small crystals of the solid.

Bravais' proof of the uniqueness of the fourteen possible unit cells was positive. Although different unit cells are occasionally proposed by some authors, they are reducible to one of the basic fourteen (as illustrated in Fig. 3-6 for the body-centered and face-centered tetragonal unit cells).

The concepts of crystal lattices and actual crystal structures should not be confused. Although only fourteen Bravais unit cells are possible

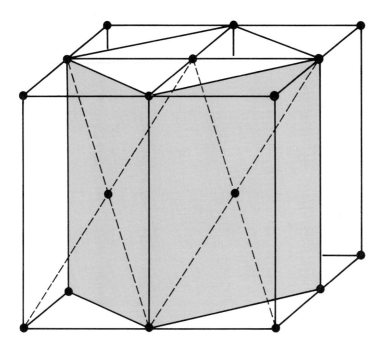

FIG. 3-6 Construction of a face-centered tetragonal unit cell from four body-centered tetragonal unit cells, illustrating the equivalence of the two forms. Shaded faces show face-centered cell.

and each one possesses certain characteristic geometrical parameters, in actual crystals these parameters assume different values for each substance and can form hundreds of thousands of varieties of individual crystal structures—as many as there are crystalline solid materials.

CRYSTAL AXES AND CRYSTAL SYSTEMS   Once a proper unit cell has been defined, a set of reference axes is required to describe relative positions of points within the unit cell. These axes are illustrated in Fig. 3-7. Only the seven crystal axes systems listed in Table 3-5 are required to describe the fourteen Bravais lattices. The parameters, measured in Angstrom units, are written in the order (a, b, c) and the angles, measured in degrees, are written in the order (α, β, γ).

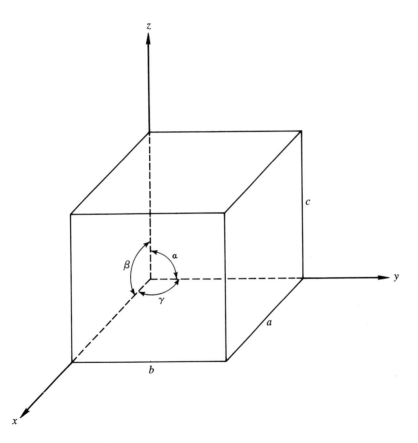

FIG. 3-7 Crystal coordinates and nomenclature. *a, b, c* are lattice parameters in *x, y, z* directions, respectively. α, β, γ are angles between *xy–xz, xy–yz,* and *xz–yz* planes, respectively.

TABLE **3-5**   KNOW   RELATIONSHIPS

The Seven Basic Crystal Systems

NAME	LATTICE PARAMETER RELATIONS	ANGULAR RELATIONS
Cubic	$a = b = c$	$\alpha = \beta = \gamma = 90°$
Tetragonal	$a = b \neq c$	$\alpha = \beta = \gamma = 90°$
Orthorhombic	$a \neq b \neq c$	$\alpha = \beta = \gamma = 90°$
Monoclinic	$a \neq b \neq c$	$\alpha = \beta = 90° \neq \gamma$
Triclinic	$a \neq b \neq c$	$\alpha \neq \beta \neq \gamma \neq 90°$
Rhombohedral	$a = b = c$	$\alpha = \beta = \gamma \neq 90°$
Hexagonal	$a = b \neq c$	$\alpha = \beta = 90°, \gamma = 120°$

Some authors describe the rhombohedral system in terms of an hexagonal set of axes, thus reducing the required number of crystal systems to six. The relationship which makes this substitution possible is illustrated in Fig. 3-8. In order to visualize the transferral, imagine the *z* axis of the hexagonal system to pass through the opposite corners *a* and *o* in Fig. 3-8(a). The projection of the unit cell onto the plane in Fig. 3-8(b) results in angles of 120° between the edges *ab, ac, ad,* as well as between *oe, of,* and *og*. In addition to the rhombohedral cell

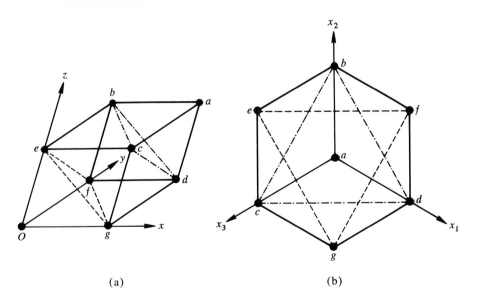

(a)   (b)

FIG. 3-8 Illustration of rhombohedral unit cell viewed as relative to (a) rhombohedral and (b) hexagonal axes. Although the cell appears to be hexagonal in (b), the planes *bcd* and *efg* are not coplanar.

shown, this is true for any cubic crystal cell oriented with the hexagonal *z* axis as a body diagonal of the cube. Such a reorientation of the cubic cell transforms its faces into pyramids, obscuring the high degree of cubic symmetry. Points *b*, *c*, and *d* are coplanar, as are points *e*, *f*, and *g*, but the planes *bcd* and *efg*, although parallel, are not coplanar with the axes x_1, x_2, x_3 of the hexagonal coordinates. This makes the identification of the faces of the unit cell and other significant planes through it difficult, and obscures the relatively high degree of symmetry indicated in Fig. 3-8(a). Therefore, it is convenient to emphasize the natural differences between the rhombohedral and hexagonal arrangements of points by using separate sets of axes to describe each.

SYMMETRY CLASSES   In terms of symmetry properties and the seven crystal axes systems, all crystals must fall into 32 different external symmetry classes [7, 10]. Thirteen of these are commonly observed, seventeen are rarely observed, and two have never been observed in either natural or artificial crystals [10].

The external symmetry properties of crystals involve points or centers, axes or lines, and planes of symmetry. A center of symmetry or *symmetry point* is so situated that all straight lines drawn through it pass through symmetric pairs of points mutually equidistant from the center, as in the center of a cube.

A line of symmetry or *symmetry axis* is situated so that specified angular rotations of the crystal about the axis reproduce the original crystal configuration. If rotations of 180°, 120°, 90°, or 60° restore the initial configuration, the axes are 2-, 3-, 4-, or 6-fold symmetric, respectively. The *z* axis of a hexagonal crystal is 6-fold symmetric, while the principal axes of a cube are 4-fold symmetric.

When a *plane of symmetry* passes through a crystal, two equal parts result, each the mirror-image of the other with respect to the plane— as in the six diagonal symmetry planes of a cube.

In addition to the three external symmetry concepts above, several "point" operations applicable to the internal symmetry of the crystal lattice are possible: *rotation, reflection, inversion, rotation-inversion, gliding,* and *screw axes.* (The latter two operations are not analogous to any of the external symmetry operations.) A *point symmetry operation* is one which, when applied to a point in a lattice, reproduces it in order to build up the structure of the cell.

The rotation operation is essentially the same as an external axis of symmetry, and may be 2-, 3-, 4- or 6-fold symmetric. A reflection operation produces the same effect as a symmetry plane. A crystal undergoes inversion if every point on one of its sides corresponds to a position on the other side which is equidistantly located from and on the same line through a center. A rotation-inversion operation combines the two operations described above. A crystal possesses a *glide plane* when it combines a reflection plane with a translation parallel to this plane, bringing the structure into self-coincidence by the movement and reflection across the specified plane. In such operations, not all translations are permissible [10]. A crystal possesses a *screw axis* when a point combines clockwise or counter-clockwise rotation with translation parallel to the axis. This axis may be either 2-, 3-, 4-, or 6-fold symmetric.

KNOW   MILLER   INDICES   WELL

CRYSTAL INDICES OF PLANES   Infinitely, many planes can pass through the points of a space lattice and subdivide the space further. Frequently planes other than the fundamental unit cell planes have greater crystallographic importance, necessitating a positive identification scheme for crystal planes. The most widely used procedure involves a set of indices which are defined in terms of the intercepts of a given plane and measured in fractions of lattice parameters with the reference axes of the crystal system.

The shaded plane in Fig. 3-9 intersects the axes at $a/2$, $b/2$, and infinity. Although these intersections will be at a different absolute

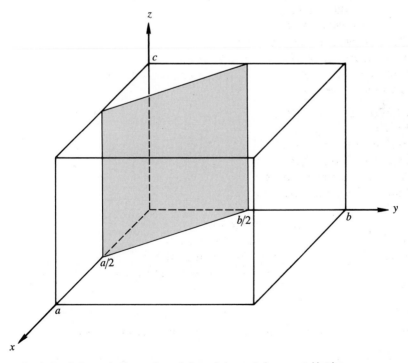

FIG. 3-9   Schematic illustration of planar intercepts in a crystal lattice.

distance from the origin if $b \neq a$, the plane has the intercepts $(\frac{1}{2}, \frac{1}{2}, \infty)$; the appropriate lattice parameters are understood to be (a, b, c), respectively. By similar reasoning, the plane parallel to the y–z coordinate plane which passes through point a on the x axis has intercepts $(1, \infty, \infty)$, and the plane which is a single lattice parameter from and parallel to the x–y plane has intercepts $(\infty, \infty, 1)$.

Detailed crystallographic studies of real crystal systems have led to a very useful generalization regarding crystal planes known as the *law of rational indices* or Hauy's law [11], which states that *the ratios among the various intercepts of any significant plane through a crystal are rational numbers*. A more useful form of this rule is that *the reciprocals of the intercepts of any significant plane are always rational fractions whose common denominator is a relatively small integer*. This latter form of the law of rational indices provides both the justification for and the derivation of the very commonly used *Miller indices*.[2] These provide a concise identification scheme for crystallographically equivalent planes, or planes differing in absolute orientation, but not in crystallographic properties.

The reciprocals $1/m$, $1/n$, and $1/p$ of the planar intercepts (m, n, p) possess a least common denominator q. These reciprocal intercepts can thus be written in the form $(h/q, k/q, l/q)$, where $h = q/m$, $k = q/n$, and $l = q/p$. The values (h, k, l) calculated are *always* integers and *may* include zero. The denominator q is only algebraically significant and is omitted. The integers (hkl), enclosed in parentheses without distinguishing punctuation, are the *Miller indices of the plane*.

The above derivation establishes the following rules for calculating the Miller indices of a plane:

1. Select some point outside the plane of interest as the origin of the reference axes *to avoid intercepts of zero*.
2. Determine the planar intercepts on the reference axes as fractions or multiples of the lattice parameters a, b, and c.
3. Calculate the reciprocal intercepts.
4. Reduce these reciprocal intercepts to a least common denominator.
5. Enclose the *numerators* of the reduced fractions in parentheses in the order[3] (hkl) to identify them as the Miller indices. (Except in rare cases where the numerical value of indices exceeds nine, the three indices are not separated by punctuation marks.)

---

[2] A different set, called the *Miller–Bravais indices*, is sometimes used for hexagonal systems.

[3] In this grouping, the order hkl corresponds to the xyz order of the axes.

EXAMPLE 3.2

Calculate the Miller indices of the plane which intersects a set of axes at $a/2$, b, and $3c$.

Solution:

Intercepts are ($\frac{1}{2}$, 1, 3); reciprocals are 2, 1, $\frac{1}{3}$; and the least common denominator is 3. Hence, the reciprocals become

$$\frac{6}{3}, \frac{3}{3}, \frac{1}{3}$$

and the Miller indices are (631).

A plane intersecting any axis on the negative side of the origin has a negative intercept and, correspondingly, a negative Miller index for that axis. Negative Miller indices are written with the minus sign *above* the index rather than before it; thus, a plane whose intercepts are 2, -1, ∞ has Miller indices ($1\bar{2}0$). The fourth rule involves the reduction of intercepts to a least common denominator. As a result of this operation, *all planes which lie on the same side of the origin and which are parallel and crystallographically equivalent will have the same Miller indices.*

EXAMPLE 3.3

Prove that parallel planes which are crystallographically equivalent have identical Miller indices.

Proof:

Consider the planes having intercepts a, $2b$, $6c$, and $2a$, $4b$, $12c$, both of which are parallel to the (631) plane of Example 3.2 and pass through identical sequences of points.

$$\text{Take reciprocals:} \quad 1, \frac{1}{2}, \frac{1}{6} \quad \text{and} \quad \frac{1}{2}, \frac{1}{4}, \frac{1}{12}$$

$$\text{Place over lowest common denominator (LCD):} \quad \frac{6}{6}, \frac{3}{6}, \frac{1}{6} \quad \text{and} \quad \frac{6}{12}, \frac{3}{12}, \frac{1}{12}$$

$$\text{Miller indices are:} \quad (631) \quad \text{and} \quad (631)$$

Q.E.D.

This will be true for *any* crystallographically equivalent plane that lies on the same side of the origin parallel to these planes.

Since the location of the origin of the reference axes is arbitrary, there is no difference between an (hkl) plane and a ($\bar{h}\bar{k}\bar{l}$) plane. Therefore, for those parallel planes which lie on the opposite side of the origin from the (hkl) plane, the Miller indices will be obtained by simultaneously reversing the signs of *all three* indices to give $(\bar{h}\bar{k}\bar{l})$. This simultaneous reversal of the signs of the Miller indices of a given plane does not represent an actual change in the indices, but an arbitrary translation of the origin of the reference axes. (Although the artificiality of distinguishing between a (100) and a $(\bar{1}00)$ plane should be evident, this distinction is often made, particularly with the (111) planes, and can result in real confusion.)

SETS OF PLANES IN THE CUBIC SYSTEM   All parallel sets of crystallographically equivalent planes through the points of the same space lattice contain the same number of points per unit area, identically arranged and similarly related to other non-coplanar points. A *set of planes* is defined as a series of parallel planes spaced so that each point in a space lattice is contained within one of the planes of the set. For example, all planes parallel to the principal (100) plane are crystallographically identical. Similarly, the entire (010) series of planes in the cubic system is identical to the (100) series with respect to number of points per unit area and all other crystallographic properties, as is the (001) series. Therefore, all three of these series of planes are crystallographically equivalent, that is, the Miller indices of one can be converted into those of another by means of a simple rotation of the reference axes through 90°. These three series are conveniently classed as a *family of planes* and designated by the notation {100}, where the braces indicate the entire family. Thus, the {100} family of planes, including the (100), (010), (001) planes, represents the cube faces for the *cubic system* in general.

The Miller indices of all planes included in a given family in the cubic system *can* be derived from any member of the family by permutation of the three indices or their signs to obtain all possible combinations. (This is not true in general, as is easily proved by computing the planar atomic density for the (110) and (011) planes of the orthorhombic system.) Both such changes are equivalent to rotations of the structure about its reference axes.

CRYSTAL INDICES OF DIRECTION   In addition to the specification of crystallographic planes, it is often desirable to specify crystallographic directions, which requires a system of *indices of direction*. In form the Miller indices of direction resemble the Miller indices of planes, but their calculations and significances are quite different.

In dealing with directions in a crystal, the concept of a vector is very useful. When a vector is translated *parallel to itself*, it still defines the *same direction* in space. Thus, in a given space lattice, any one of an infinite number of parallel vectors define the same crystallographic direction. For convenience the vector from the origin of the reference axes to the point of interest is selected so the indices of the given direction relate simply to the components of the vector with respect to the reference axes. Although the indices of direction are determined from this specific vector, *they apply equally well to any vector which is parallel to it.*

Indices of a given crystallographic direction are derived from the coordinates of a given point (in terms of the numbers of lattice parameters) which lies on a line through the origin and points in the desired direction. These coordinates, when reduced to the *smallest set of integers* which still retain the original ratios among them, are the *Miller indices of direction*. The integers *uvw* are enclosed in square brackets without punctuation to distinguish them from the planar indices (*hkl*). For example, the vector illustrated in Fig. 3-10 has coordinates *a*, *b*, and 2*c*, and the Miller indices of direction are evidently [112]. The same vector

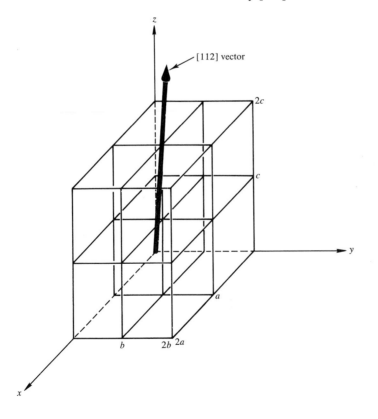

FIG. 3-10 Illustration of the [112] direction vector in a crystal lattice.

also passes through the points [$\frac{1}{2}$ $\frac{1}{2}$ 1], [224], [336], etc., and placing any of these sets of intercepts over a common denominator will result in the same set of Miller indices of direction, namely [112]. Thus, just as in the case of the indices of planes, the same set of indices applies to all sets of points which lie along this line, *or along parallel lines.*

Some authors utilize the planar index symbols [*hkl*] with square brackets to designate crystallographic directions and (*hkl*) with parentheses to indicate the Miller indices of planes. Since the concepts involving direction and planar indices are very different, it would be confusing to use the same letters for each. Therefore, in this book (*hkl*) will designate the Miller indices of planes and [*uvw*] the Miller indices of direction.

EXAMPLE 3.4

Prove that the Miller indices of direction of two parallel vectors passing through different points are identical.

Proof:

Consider the two parallel vectors in Fig. 3-11. The coordinates of point P are a, 0, c. The coordinates of point Q *relative* to point R are $2a - a = a$, $2b - 2b = 0$, $2c - c = c$. Hence, the Miller indices of both vectors are [101].   Q.E.D.

If any of the coordinates of a point are negative, the indices will be negative. Thus, the indices of the vector pointing in the opposite direction to that of [101] are [$\bar{1}0\bar{1}$].

SPECIAL RELATIONS FOR CUBIC SYMMETRY   As in the case of crystallographic planes, general families of crystallographic directions can be defined for the cubic system. The [100], [010], and [001] directions and their negatives are interconvertible by means of 90° rotations and, therefore, are equivalent to one another. This set of six directions constitutes the $\langle 100 \rangle$ family. The caret $\langle \rangle$ notation is used to designate families of directions in the *cubic* system and can be deduced from any member of the set by permutation of the indices and signs to obtain all possible combinations.

The Miller indices of direction of the vector in the *cubic lattice*[4] which are normal or perpendicular to a given plane have the same numerical value as the indices of that plane. Thus, the [111] direction is normal to the (111) plane, as illustrated in Fig. 3-12.

---

[4]Note that all of these special properties apply *only* to the cubic system, arising because of its high degree of symmetry.

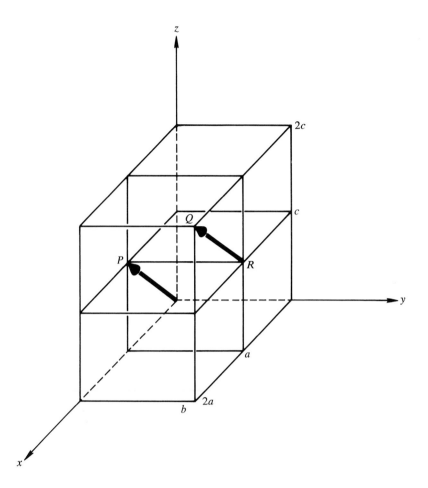

FIG. 3-11   Illustration of the equivalence of space directions through different points. The Miller indices of direction of the two vectors are identical.

HEXAGONAL INDICES   With hexagonal systems, the *Miller–Bravais* indices of planes are sometimes used, where the form (*hkil*) is calculated from the intercepts of a given plane with a set of hexagonal axes x_1, x_2, x_3, and z, in the same way as the Miller indices. Because of the geometry of the hexagonal system, the algebraic sum of the first three indices h, k, and i is zero [8].

The relation between the Miller–Bravais and the Miller indices is very simple for the hexagonal system. If one defines a set of axes using only the x_1, x_2, and z axes but not the x_3 axis, then the Miller indices are (*hkl*) with the index i omitted. The third coplanar index only serves to emphasize the hexagonal symmetry of the system; it offers no new information.

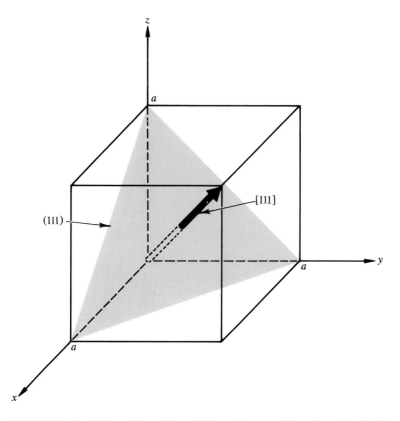

FIG. 3-12   Illustration of the relation of the vector normal to a given plane in the cubic system. In general, [*hkl*] is normal to (*hkl*).

If the only differences between the two systems of indices were those above, there would be little reason to prefer one or the other unless one was concerned with preserving the uniform usage of the Miller indices in other systems. However, derivation of the Miller–Bravais indices of direction presents a cogent reason for abandoning the hexagonal system. Miller–Bravais direction indices are complicated to derive and somewhat confusing in their results; they bear no simple relationship to the Miller indices of direction. This is a result of the requirement of projecting a vector onto three coplanar axes instead of two, as illustrated by the line parallel to the positive x_1 axis in the hexagonal system shown in Fig. 3-13. This line has the same set of indices as the x_1 axis. The *bc* segment, cut out by the intersection of $\overline{BC}$ with the x_3 and $-x_2$ axes, would have Miller indices [100].

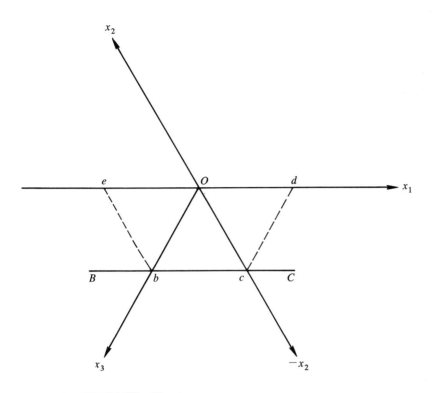

FIG. 3-13   Diagram illustrating derivation of the Miller–Bravais indices of direction in a hexagonal lattice.

When bc is projected parallel to the x_3 axis, it intersects the negative x_2 axis at c and the positive x_1 axis at d. When this same segment is projected parallel to the x_2 axis, it intersects the x_3 axis at b and the negative x_1 axis at e. The algebraic signs given these projections will depend on the sign of the projected vector. The total projected distance on the x_1 axis is $eO + Od = 2a$; on both the x_2 and x_3 axes it is $-a$ because the direction of the projection is from b to O, lying in the negative direction along the axis. Thus, the Miller–Bravais indices for this direction are $[2\bar{1}\bar{1}0]$.

The principal value of the Miller–Bravais directional indices is that equivalent directions have similar indices. For example, the positive axes of the hexagonal system have the indices $[2\bar{1}\bar{1}0]$, $[\bar{1}2\bar{1}0]$, and $[\bar{1}\bar{1}20]$, respectively, whereas the corresponding Miller indices are $[100]$, $[010]$, and $[110]$.

Using the Miller rather than Miller–Bravais indices of direction for hexagonal systems somewhat obscures the underlying symmetry, but the greater ease in deriving the Miller indices more than offsets this. Therefore, Miller indices are frequently used to describe crystallographic directions in hexagonal systems. Because the use of Miller–Bravais indices for planes and Miller indices for directions would be inconsistent, only Miller indices will be used in any subsequent discussions of hexagonal crystal systems.

INTERPLANAR SPACINGS Crystal interplanar spacings are important and can be calculated for perfect crystals from the indices of a given plane by the methods of analytic geometry. Since the Miller indices (*hkl*) are proportional to the reciprocals of the intercepts of the plane with the reference axes, one can calculate for the cubic system from analytic geometry that the interplanar spacing distance *d* is given by

$$d = \left[\frac{a^2}{h^2 + k^2 + l^2} \right]^{1/2} \tag{3.7}$$

where *a* is the lattice parameter and (*hkl*) are the Miller indices of the plane. The results of similar calculations made for other crystal systems are listed [12] in Table 3-6.

TABLE 3-6

Interplanar Spacing of Simple Lattices*

LATTICE	SPACING, d
Cubic	$\left[\dfrac{a^2}{h^2 + k^2 + l^2} \right]^{0.5}$
Tetragonal	$\left[\dfrac{a^2}{h^2 + k^2} + \dfrac{c^2}{l^2} \right]^{0.5}$
Orthorhombic	$\left[\dfrac{a^2}{h^2} + \dfrac{b^2}{k^2} + \dfrac{c^2}{l^2} \right]^{0.5}$
Monoclinic	$\left[\dfrac{a^2 \sin^2 \beta}{h^2} + \dfrac{b^2}{k^2} + \dfrac{c^2 \sin^2 \beta}{l^2} - \dfrac{ac \sin^2 \beta}{2hl \cos \beta} \right]^{0.5}$
Rhombohedral	$\left[\dfrac{a^2(1 - 3\cos^2 \alpha + 2\cos^3 \alpha)}{(h^2 + k^2 + l^2)\sin^2 \alpha + 2(hk + kl + hl)(\cos^2 \alpha - \cos \alpha)} \right]^{0.5}$
Hexagonal	$\left[\dfrac{3}{4} \cdot \dfrac{a^2}{h^2 + hk + k^2} + \dfrac{c^2}{l^2} \right]^{0.5}$

*From *The Solid State for Engineers*, by M. J. Sinnott, copyright 1961, John Wiley and Sons, Inc. Used by permission.

EXAMPLE 3.5

Compute the spacing between the (111) planes in a tetragonal cell for which $a = b = 3.4\,\text{Å}$ and $c = 2a$.

Solution:

From Table 3-6,

$$d = \left[\frac{a^2}{h^2 + k^2} + \frac{c^2}{l^2}\right]^{1/2} = (3.4)\left[\frac{1}{1 + 1} + \frac{4}{1}\right]^{1/2}\,\text{Å}$$

$$= 7.6\,\text{Å}$$

Experimental values of interplanar spacings for real crystals are either determined by x-ray or by electron diffraction. The first technique is more universally applicable to the study of crystal structures due to the shorter wavelengths and greater penetrating power of the x-rays. (X-rays used in crystallographic work usually range from 0.7 Å to 3.0 Å.) Electrons, which penetrate less than x-rays, are useful for the study of crystal surface phenomena.

Figure 3-14 schematically illustrates two layers of atoms in a crystal structure, separated by the distance *d*. Two incident waves strike the

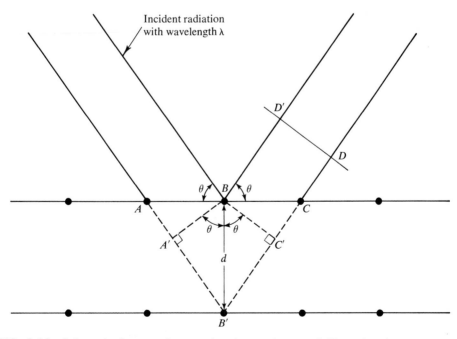

FIG. 3-14 Schematic diagram of two surface layers of a crystal, illustrating the geometry necessary to derive Bragg's diffraction equation.

crystal at points A and B. The wave which contacts the first layer at point A is either reflected or transmitted onto point B' in the second layer, and there this process is repeated. The wave reflected from B interferes constructively at $\overline{DD'}$ with the wave which is transmitted from A to B' and subsequently reflected, providing $\overline{A'B'C'}$ is an integral multiple of the wavelength λ of the incident radiation. From geometry, $\overline{A'B'} = \overline{B'C'} = d \sin \theta$. The condition of constructive interference, known as *Bragg's diffraction law*, is

$$n\lambda = 2d \sin \theta \tag{3.8}$$

where θ is the incidence angle of the radiation. For the principal diffraction line, $n = 1$. Since normal interplanar spacings are of the order of a few anstroms and the maximum value of $\sin \theta$ is unity, radiation must be used with wavelengths of the order of a few angstroms or less which fall in the x-ray region of the spectrum.

Experimentally, a powdered sample of the crystalline material is placed at the center of a circular strip of photographically sensitive film and bombarded by a collimated beam of x-rays [13], as shown

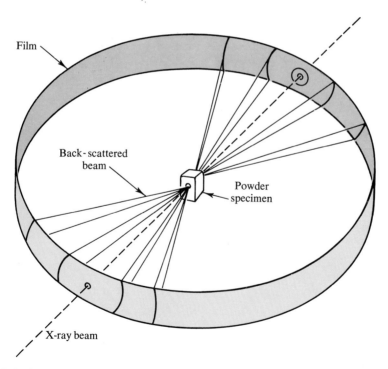

FIG. 3-15 Schematic representation of experimental arrangement used for powder x-ray crystallographic studies.

schematically in Fig. 3-15. Since the diffraction occurs from planes in the crystals that are more or less randomly oriented relative to the source, the diffraction pattern will be spherically distributed so that the circular arrangement of the film also intercepts the back diffraction patterns. This technique is very sensitive to small changes in interplanar spacings, and with ordinary care can measure crystal lattice parameters to precisions of the order of 0.001 Å in 2–5 Å.

Theoretical Crystal Strengths

A number of the concepts already developed can now be employed in making theoretical estimates of some of the mechanical strength properties of perfect crystals. Attention here will be restricted to ceramic or ionic crystalline solids because of the simplicity of the mathematical representations of their operative attractive and repulsive forces. The results of our calculations, although not quantitatively precise for other bonding types, will nonetheless provide useful qualitative insights in subsequent chapters.

YOUNG'S MODULUS   In response to externally applied mechanical forces, the elastic restoring force of a solid arises from the bonds which hold the molecules in their respective lattice sites. If deformations are small and only minor displacements of the molecules from their equilibrium lattice sites occur, then the restoring force is approximately proportional to the magnitude of the displacements. This situation can be analyzed approximately for ceramic crystals as follows.

According to both Eq. (2.15) and Eq. (2.17), the total energy of an ionic crystal can be written as

$$V(R) = V_0 \left\{ \frac{1}{\xi} - \frac{\xi^{-n}}{n} \right\} \tag{3.9}$$

where $V_0 = -Ae^2Z^2/R_0$ and $\xi = R/R_0$. For ionic crystals of the Na^+Cl^- type, where each ion has six nearest neighbors of the opposite charge, the net energy per ion pair is $V(R)/6$. Therefore, the interionic force $f(R)$ acting between the ion pair *in the crystal* is

$$f(R) = \frac{1}{R_0} \frac{dV(R)}{d\xi} = \frac{V_0}{6R_0\xi^2} \{\xi^{1-n} - 1\} \tag{3.10}$$

When $R = R_0$ ($\xi = 1$), Eq. (3.10) indicates that $f(R_0) = 0$. For values of $R = R_0 + \Delta R$, where $\Delta R/R_0 = \Delta \xi \ll 1$, we may write

$$f(R_0 + \Delta R) \doteq f(R_0) + \frac{df}{dR}\bigg|_{R_0} \Delta R \tag{3.11}$$

which is valid as long as ΔR is a small quantity. (Only the first order or linear term of a Taylor series expansion of $f(R)$ about the equilibrium point R_0 is retained here, so that this analysis is valid only for small values of $\Delta \xi$.)

Since $f(R_0) = 0$, Eq. (3.11) reduces to the approximate result

$$f(R_0 + \Delta R) \doteq \frac{(n - 1)Ae^2Z^2}{6R_0^3} \Delta R \tag{3.12}$$

Now that attention has been focused on a crystal ion pair, it can be seen that this force acts on an area essentially R_0 on a side to cause a separation ΔR of the pair. Thus, if $\Delta R / R_0 \ll 1$, an approximation may be written $\Delta R / R_0 = \epsilon$ and $f(R_0 + \Delta R)/R_0^2 = \sigma$. Consequently, Eq. (3.12) becomes

$$\sigma = \left[\frac{(n - 1)Ae^2Z^2}{6R_0^4} \right] \epsilon \tag{3.13}$$

Reference to Eq. (1.1) clearly shows that the term enclosed in brackets in Eq. (3.13) is an approximate expression for Young's modulus. That is,

$$E \doteq \frac{(n - 1)Ae^2Z^2}{6R_0^4} \tag{3.14}$$

EXAMPLE 3.6

Using the data of Table 2-4 and Problem 2.19, estimate Young's modulus for NaCl crystals in tension.

$$A \doteq 1.748 \qquad n = 8 \qquad R_0 = 2.81 \text{ Å}$$

$$e = 4.774(10^{-10})(\text{erg cm})^{1/2}$$

$$E = \frac{(8 - 1)(1.748)(4.774)^2(10^{-10})^2(1)^2}{(6)(2.81)^4(10^{-8})^4} \frac{\text{dyne}}{\text{cm}^2}$$

$$= 7.45(10^{11}) \frac{\text{dyne}}{\text{cm}^2} = 1.08(10^7) \text{ psi}$$

The experimental value of E for NaCl crystals in tension, as computed from the data of Hunter and Siegel [14], is 7.89 (10^6) psi—somewhat less than the value estimated from Eq. (3.14). The reason for this discrepancy will be discussed in Chapter 7.

SHEAR STRENGTH The key process in plastic deformation, the shearing or slipping of one plane of atoms past another, will also be described in Chapter 7. This process is nearly always observed on

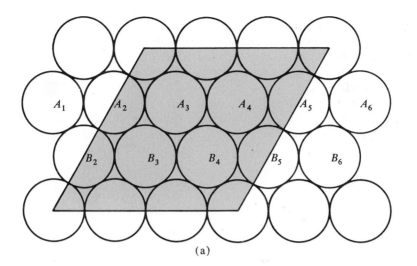

(a)

Shear force

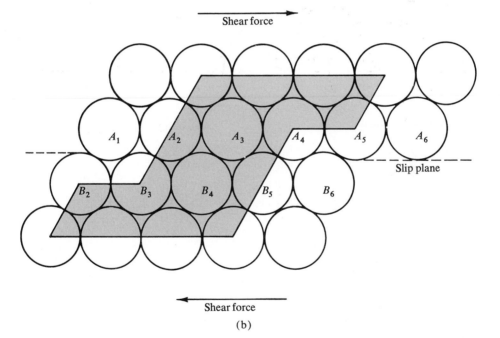

(b)

Shear force

FIG. 3-16 Schematic illustration of the slip process in perfect close-packed crystal. Atoms of row A under the action of the shear force couple slip one lattice distance R_0 to the right past the atoms of row B. Solid lines delineate a region in (a) the unslipped crystal and (b) the crystal after the unit slip to the right.

planes of densest atomic packing ((111) in FCC or face-centered cubic crystals), and in a direction of densest linear packing on the plane ([110] in FCC crystals). Using the model of this slip process schematically illustrated in Fig. 3-16, a theoretical estimate of the maximum shearing strength of perfect crystalline solids can be made.

In order for the layer of atoms marked A in Fig. 3-16 to slip one lattice distance R_0 past the atoms in the row marked B, it is necessary that they achieve a position directly above the atoms of the lower row.

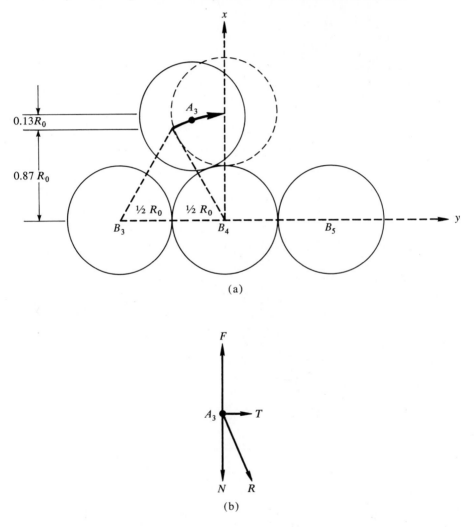

(a)

(b)

FIG. 3-17   Schematic representation of forces acting on a molecule during shear. (a) Molecular arrangement during transfer process. (Center of A_3 follows curved arrow; dashed circle shows A_3 midway through transfer.) (b) Vector representation of forces acting on A_3.

To do this, work has to be done *on each of the atoms* in order to raise its center of gravity the appropriate amount.

Figure 3-17 is a schematic representation of the process of molecular transfer in an intermediate position, showing the forces acting on the molecules. The force component T is applied to the atom in a tangential direction to cause shear. The total work done by this force on the molecule is

$$W = \tfrac{1}{2} R_0 \bar{T} \tag{3.15}$$

where $\bar{T}$ is the average value of the force.

In Fig. 3-17(b), the forces acting on the center of gravity of the upper atom A_3 are indicated. N is the intermolecular attractive force of bonding which holds A in place, and R is the resultant of N and T. From Fig. 3-17(a), it is evident that R does no work because it always acts in a direction normal to the motion of A.

Therefore, the work W calculated in Eq. (3.15) is also that which would have to be done by an external force F opposing N in order to raise the center of gravity of atom A_3 from its equilibrium position $0.87R_0$ to the maximum position R_0 possible above the line of B atom centers. That is

$$\tfrac{1}{2} R_0 \bar{T} = \int_{0.87R_0}^{R_0} F(x)\, dx \tag{3.16}$$

For small displacements, Hooke's law is obeyed and $F(x) = kx$. Since $F = S\sigma$, with S the cross-sectional area of atom A_3 normal to F, and σ the externally applied tensile stress, we can write

$$F(x) = \frac{SEx}{L_0} \tag{3.17}$$

where $L_0 = 0.87R_0$ and $\epsilon = x/L_0$ for small x. Thus we have

$$\tfrac{1}{2} R_0 \bar{T} = \frac{1}{2} \frac{SE}{L_0} (0.13R_0)^2 \tag{3.18}$$

or

$$\bar{\tau} \equiv \frac{\bar{T}}{S} \doteq (0.13)^2 \frac{E}{0.87} \doteq 0.02E \tag{3.19}$$

Since $\epsilon = 0.13/0.87$ is not small when compared with unity, the assumption of Hooke's law is probably not valid and a nonlinear relation should have been used for $F(x)$. However, Fig. 1-1 clearly shows that the proper nonlinear relation rises more slowly than Hooke's law, making Eq. (3.19) an upper bound on $\bar{\tau}$. Another estimate of this value may be obtained by the following method.

Suppose that the total energy of the ion pair is to be given by $\frac{1}{6}$ of Eq. (2.15). Then, for an adiabatic shear operation, the first law of thermodynamics gives us $-w = \Delta V$. Now, $\Delta V = V(1.13) - V(1)$, where $V(\xi)$ is evaluated at the two limits corresponding to $\xi = 1$ at $0.87R_0$ and $\xi = 1.13$ at R_0. Making this calculation for a typical ionic crystal (Na^+Cl^-) with $n = 8$, we obtain $w = 0.005R_0^3$ and $E = 0.5R_0\bar{T}$. This method gives us as an estimate $\bar{\tau} = 0.01E$, but this is probably low because there is some heat generated by friction in the process. Therefore, a rough estimate of the order of magnitude of $\bar{\tau}_{cr}$ for a perfect ionic crystal should be approximately

$$\bar{\tau}_{cr} \text{ (perfect crystal)} \doteq 0.015E \qquad (3.20)$$

Experimental values of $\bar{\tau}_{cr}$ are generally about a factor of 10^3 smaller than this value—a fact which will be explained in Chapter 7.

References Cited

[1] Present, R. D., *Kinetic Theory of Gases*, McGraw-Hill Book Co., Inc., New York, 1958.

[2] Glasstone, S., K. J. Laidler, and H. Eyring, *The Theory of Rate Processes*, McGraw-Hill Book Co., Inc., New York, 1941.

[3] Frenkel, J., *Kinetic Theory of Liquids*, Dover Publications, Inc., New York, 1955.

[4] Warren, B. E., *J. Appl. Phys.*, **8**, p. 643 (1937).

[5] Moffatt, W. G., G. W. Pearsall, and J. Wulff, *The Structure and Properties of Materials, Volume I, Structure*, John Wiley and Sons, Inc., New York, 1964, p. 112*ff*.

[6] Azaroff, L. V., *Introduction to Solids*, McGraw-Hill Book Co., Inc., New York, 1960, p. 429*ff*.

[7] Buerger, M. J., *Elementary Cystallography*, John Wiley and Sons, Inc., New York, 1963, Chapter 10.

[8] Smith, M. C., *Principles of Physical Metallurgy*, Harper and Brothers, New York, 1956, Chapter 2.

[9] Azaroff, *op. cit.*, Chapter 2.

[10] Sinnott, M. J., *The Solid State for Engineers*, John Wiley and Sons, Inc., New York, 1961, p. 23.

[11] deJong, W. F., *General Crystallography, A Brief Compendium*, W. H. Freeman, and Co., San Francisco, 1959, p. 7.

[12] Sinnott, M. J., *op. cit.*, p. 45.

[13] Azaroff, L. V., and M. J. Buerger, *The Powder Method in X-Ray Crystallography*, McGraw-Hill Book Co., Inc., New York, 1958, pp. 8–9.

[14] Hunter, L., and S. Siegel, *Phys. Rev.*, **61**, p. 84 (1942).

PROBLEMS

3.1 Derive the Miller–Bravais indices of the negative x_3 axis.

3.2 Calculate the mean thermal speeds of helium, argon, and xenon.

3.3 Calculate the internal thermal energy of an ideal gas at 300°K.

3.4 Numerous correlations have been developed which relate thermodynamic properties of substances to critical properties. Thus, the normal boiling point is related to the
critical temperature by $T_b = 0.67\ T_c$. Calculate the normal boiling points of Cl_2, CS_2,
and CH_4 by this method and compare them with the experimental values.

3.5 The *effective* molecular diameter of a gas molecule may be estimated by the Flynn
and Thodos method (*A.I.Ch.E. Journal*, **8**, p. 362 (1962)) from critical constants as
$\bar{\sigma} = 0.561\ (V_c^{1/3})^{5/4}$, where $V_c = 1/\rho_c$ is the critical specific volume of the gas. Compute
the molecular diameters of methane, oxygen, and xenon.

3.6 By drawing several unit cells, show that the end-centered cubic and the tetragonal
unit cells are equal.

3.7 Compute the Miller indices of the planes which pass through parallel base diagonals
of the body-centered tetragonal cell. (HINT: there are two such planes.)

3.8 Compute the planar atomic density on the (111) plane of Cu (FCC). (HINT: count
only those atoms whose centers lie in the plane.)

3.9 Compute the planar atomic density on the (111) plane of Fe (BCC).

3.10 Compute the planar atomic density on the (110) plane of Cu (FCC). Determine
the Miller indices of all planes crystallographically equivalent to this one.

3.11 What are the Miller indices of the normal to the (110) plane in a body-centered
orthorhombic structure?

3.12 What are the Miller indices of the line of intersection of the (110) and ($\bar{1}$10) planes
in Cu (FCC)? Compute the linear atomic density on this intersection line. (HINT: count
only those atoms whose centers lie on the line.)

3.13 Compute the atomic planar density on the (011) plane in Al (FCC) and the linear
atomic density of the normal direction to this plane.

3.14 Repeat Problem 3.13 for Fe (BCC).

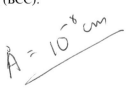

$\overset{\circ}{A} = 10^{-8}\ cm$

4

THE BEHAVIOR
OF REAL
SOLID MATERIALS

The concepts of geometrical or ideal crystallography introduced in Chapter 3, although very useful in understanding the behavior of solid materials, only serve as guidelines when defining the real solid materials an engineer must deal with in his daily work. The properties of real solid materials are usually determined by the imperfections and disruptions of their theoretical ideal structures rather than by the ideal geometrical arrays of spheres visualized in Chapter 3.

The study of the behavior of real materials will be divided between two main subjects: first we will see how atoms pack together in the solid state and examine the properties of these actual atomic arrays; then the defects and imperfections which exist in these structures and how they govern the mechanical responses of the real materials will be considered.

Atomic Coordination

Ligancy - what it means

The *coordination number* or *ligancy* is the number of atoms in a crystal which surround a particular atom as its *nearest neighbor*. Nearest

107

neighbors are defined as those atoms or ions whose effective spheres of radii equal to their ionic or other appropriate radii are just tangent to the effective sphere of the ion or atom in question.

The ligancy of an ion is largely determined by the geometrical problem of packing other ions around it as nearest neighbors, commensurate with the condition of electrical neutrality; thus it depends on the relative sizes of its surrounding neighbors. The only possible ligancies in a regular three-dimensional array are 1 (a trivial case), 2, 3, 4, 6, 8, and 12. The range of the relative sizes of ions for which various ligancies are expected to be stable can be estimated in terms of the following simple rules:

1. Cations and anions occupy their equilibrium spacings so that the spheres of the ionic radii of nearest neighbors are just tangent.
2. Anions do not overlap; their centers maintain a minimum distance of separation equal to twice their ionic radii.
3. Each cation is surrounded by the largest possible number of anions as nearest neighbors. Thus, as the difference between ionic radii decreases, higher ligancies are possible.

According to the above conditions, a given ligancy should be stable between that *radius ratio* (or ratio of cation to anion radii) at which the anions and the central cation are mutually tangent (called the *critical* radius ratio) and the radius ratio at which the next highest ligancy becomes possible. Below the critical radius ratio the cation-anion distance exceeds the equilibrium interionic distance, and a new ligancy will probably be stable.

The calculation of the range of radius ratios for which various ligancies are stable is a simple geometrical problem. Joining the centers of the coordinated anions by straight lines generates a regular poly-

TABLE 4-1

Ligancies as a Function of Ionic or Atomic Radius Ratios

LIGANCY	r/R*	POLYHEDRON	PACKING
2	0–0.155	line	linear
3	0.155–0.225	triangle	triangular
4	0.225–0.414	tetrahedron	tetrahedral
6	0.414–0.732	octahedron	octahedral
8	0.732–1.0	cube	cubic
12	1.0	(see Fig. 4-2)	HCP or FCC†

*r is the radius of the smaller ion or atom; R is the radius of the larger.
†Hexagonal close-packed or face-centered cubic arrangements.

hedron, called the *anion polyhedron*. Table 4-1 contains a tabulation of the range of radius ratios corresponding to stable ligancies, together with the coordination polyhedra and packing types calculated from these considerations. The calculation of the radius ratio for triangular coordination is illustrated in Fig. 4-1.

Spheres uniform in diameter can be packed with a coordination number of 12 by either the hexagonal close-packed (HCP) or the face-centered cubic (FCC) arrangement, both illustrated [1] in Fig. 4-2. In the HCP structure, the two triangular arrays of atoms are similarly vertically oriented above and below the plane of the hexagonal array in the center (see Fig. 4–2(a)), while in the FCC structure (Fig. 4-2(b))

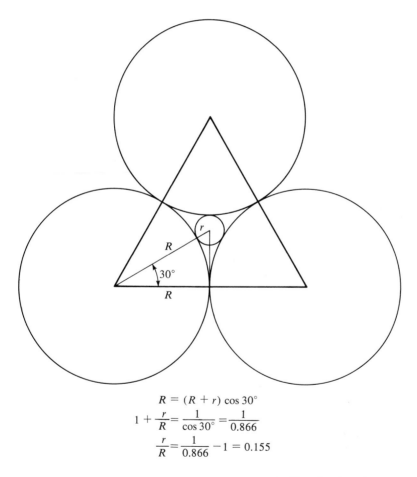

$$R = (R + r) \cos 30°$$

$$1 + \frac{r}{R} = \frac{1}{\cos 30°} = \frac{1}{0.866}$$

$$\frac{r}{R} = \frac{1}{0.866} - 1 = 0.155$$

FIG. 4-1   Illustration of the geometrical arrangement of ions in triangular coordination, leading to the critical ligancy radius ratio of 0.155.

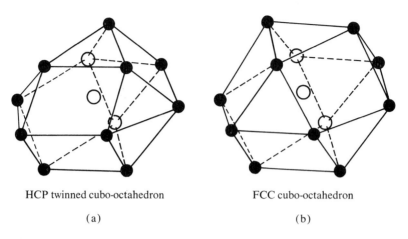

HCP twinned cubo-octahedron   FCC cubo-octahedron

(a)   (b)

FIG. 4-2 Schematic representation of coordination polyhedra for ligancy 12: (a) HCP or hexagonal close packing and (b) FCC or cubic close packing. Note that the lower part of each polyhedron is the same: In (a) the upper triangle mirrors the lower one, while in (b) the upper triangle is rotated by 60° relative to the lower one.

the upper triangular set is rotated 60° with respect to the lower triangle. In either case, twelve atoms surround the central atom as nearest neighbors. The coordination polyhedron for the FCC structure can be viewed as a cube with all its corners octahedrally planed off, standing on one of these octahedral planes. The HCP polyhedron, however, is a twinned cubo-octahedron in which the upper half of the polyhedron is the mirror image or twin of the lower half, thus making the central hexagonal plane a symmetry plane. The close relationship between these two most highly coordinated structures will become very important when the intermixing of the crystal structures of two different substances in a solution is considered in Chapter 5.

A comparison of the predicted and observed ligancies for several compounds with different ionic radius ratios [1] in Table 4-2 shows that, although many substances exhibit the expected ligancy, a significant number of compounds show a ligancy pattern different from that predicted on the basis of their ionic radius ratios. Since very few solids contain purely ionic or metallic bonds, the degree of covalency or partial ionic nature of the bonds evidently affects the observed ligancy of the ions. Also, observed values were obtained from the analysis of crystalline samples which consisted essentially of large aggregates of coordination polyhedra stacked together. In such arrays, second nearest neighbors also influence the ligancy. It is interesting that so many real crystalline compounds do, in fact, follow the simple ligancy stability rules.

TABLE **4-2**

Predicted and Observed Ligancies for Different Ionic Radius Ratios*

COMPOUND	r/R	PREDICTED LIGANCY	OBSERVED LIGANCY
B_2O_3	0.14	2	3
BeS	0.17	3	4
BeO	0.23	3–4	4
SiO_2	0.29	4	4
LiBr	0.31	4	6
MgO	0.47	6	6
MgF_2	0.48	6	6
TiO_2	0.49	6	6
NaCl	0.53	6	6
CaO	0.71	6	6
KCl	0.73	6–8	6
CaF_2	0.73	6–8	8
CsCl	0.93	8	8
BCC metals†	1	8–12	8
FCC metals	1	8–12	12
HCP metals	1	8–12	12

*From *Structure and Properties of Materials, Vol. I, Structure,* by W. G. Moffatt, G. W. Pearsall, and J. Wulff, copyright 1964, John Wiley and Sons, Inc. Used by permission.
†Body-centered cubic arrangements

Atomic Packing

The efficiency of atomic packing within a crystal cell is measured in terms of the *atomic packing factor,* which is the ratio of the volume occupied by the atoms (that is, spheres having the given atomic or ionic radius) to the volume of the unit cell. The difference between the packing factor and unity is the *void fraction,* or fraction of void space in the structure.

Polymorphism

The term *polymorphism* means that a given substance exists in more than one stable crystalline phase. The element iron is an example of a polymorphic substance. At room temperature, the stable crystalline form of iron is a body-centered cubic (BCC) crystal with a ligancy of 8. At a temperature of 910°C, iron transforms reversibly into an FCC structure with the ligancy 12. At 2550°F, it again transforms reversibly to a BCC lattice and retains this structure until it melts. Several other materials also exhibit polymorphism, and some, such as SiC, show as many as 20 polymorphic modifications.

Polymorphs of crystalline substances differ in density as well as in other physical properties. Such differences between the polymorphic varieties of iron alloys make possible the heat treatment processes that modify the mechanical and other properties of steel. (These processes will be examined in greater detail in Chapter 6.)

Several types of polymorphic transformations can occur, and these are usually classified as order-disorder changes, packing rearrangements, or structural group rotations. The order-disorder type of transformation is observed [2] in the compound AuCu, which exists at high temperatures as an FCC crystal with the Au and Cu atoms interspersed, but transforms at low temperatures to alternate layers of Au and Cu arranged in a tetragonal structure with a ratio c/a close to unity.

The order-disorder transformations occur either as a first-order or second-order transition. If, as in the case of Cu_3Au, the *alternation order*[1] decreases gradually up to a certain temperature T_0 and then decreases precipitously, the transition is quite similar to ordinary melting. Such first-order transitions occur with the absorption of a latent heat.

In the CuAu composition, the decrease in the alternation order is continuous with an ever increasing rate as the transition temperature is approached where the order will vanish completely. The transition is accompanied by an abnormal rise of the specific heat C_p to a sharp peak at the transition temperature, followed by a very rapid drop to its normal value as the transition temperature is passed. This is a transformation of the second order.

The FCC-BCC transition in iron mentioned earlier is a packing rearrangement type of polymorphism. The BCC (alpha) iron unit cell contains only two net atoms, while the FCC (gamma) iron cell contains four. The alpha to gamma transition (the (110) planes in alpha-iron becoming the (111) planes in gamma-iron) occurs with little strain since the lattice parameters of the two cells are different and the cubic axes are not parallel. Similarly the [111] direction in the alpha-iron cell becomes the [110] direction in the gamma-iron unit, as illustrated schematically in Fig. 4-3. The respective shifts involved are small and the transformations occur quite rapidly.

Cubic Crystal Properties

Because iron has two polymorphic cubic forms and is of considerable industrial importance, the properties of cubic structures are of

---

[1]The alternation order of a crystalline material is the variation of the different constituent atoms. In a perfect binary AB crystal, this order would be ABABABA in three-dimensional symmetry.

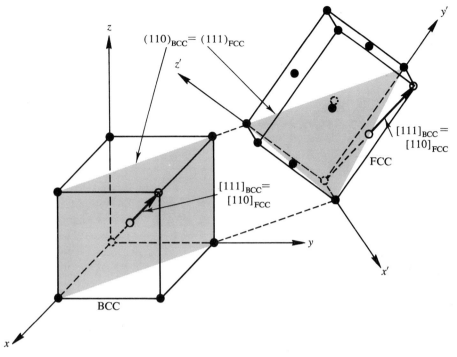

FIG. 4-3 Schematic illustration of BCC-FCC transformation in iron. Note how $[111]_{BCC} \rightarrow [110]_{FCC}$, and $(110)_{BCC} \rightarrow (111)_{FCC}$. Because this transformation requires very small atom movements, it can occur quite rapidly.

special interest to the engineer. The cubic system includes three basic lattices: simple cubic, body-centered cubic, and face-centered cubic.

The simple cubic (SC) lattice, illustrated previously in Fig. 3-5, p. 82, contains only one net atom per unit cell; every corner atom in each unit cell of an assemblage participates in the corner position of seven other unit cells. Therefore, each of the eight corner atoms in the unit cell contributes only one octant of its effective sphere, resulting in one net atom per unit cell.

The BCC cell illustrated schematically in Fig. 3-5 has two atoms per unit cell due to the inclusion of the body-centered atom. Metals do not crystallize in the SC structure because this atomic packing is too inefficient.

In terms of the effective atomic radius (illustrated in Fig. 4-4), the calculation of the lattice parameter for the BCC cell gives the relation as

$$a_{\mathrm{BCC}} = \frac{4r}{\sqrt{3}} \tag{4.1}$$

The third member of the cubic family is the FCC lattice, also illustrated earlier in Fig. 3-5. Clearly there are $8(\frac{1}{8})$ atoms at each corner of the cell, plus $6(\frac{1}{2})$ in the center of each face—a total of four net atoms per unit cell. The calculation of the relation between the lattice constant and the atomic radius of the FCC structure, as illustrated in Fig. 4-5, gives

$$a_{\mathrm{FCC}} = \frac{4r}{\sqrt{2}} \tag{4.2}$$

The atomic packing factor (PF) for each of these cubic lattices can now be calculated easily. For example, the PF for the FCC structure is

$$\mathrm{PF}_{\mathrm{FCC}} = \frac{(4)(4\pi r^3/3)}{a^3} = \frac{(16)(\pi r^3)(2\sqrt{2})}{(64r^3)(3)} = 0.74 \tag{4.3}$$

It is evident from Eq. (4.3) that the packing factor is independent of the radius of the spheres being packed. The order of packing efficiency is SC (PF $=$ 0.52), BCC (0.68), FCC (0.74), accounting for the absence of simple cubic structures in materials which occur naturally.

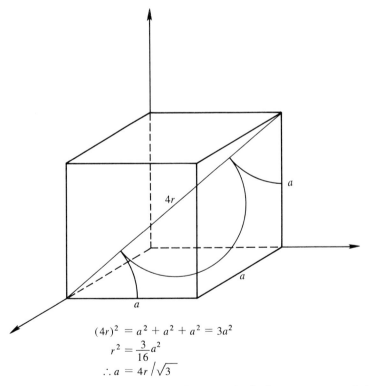

$$(4r)^2 = a^2 + a^2 + a^2 = 3a^2$$
$$r^2 = \frac{3}{16}a^2$$
$$\therefore a = 4r/\sqrt{3}$$

FIG. 4-4   Calculation of the relation between the lattice parameter and the atomic radius for uniform spheres in BCC packing.

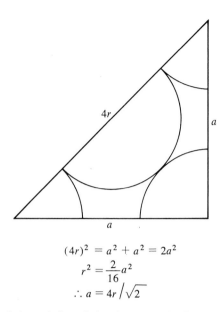

$$(4r)^2 = a^2 + a^2 = 2a^2$$
$$r^2 = \frac{2}{16}a^2$$
$$\therefore a = 4r\big/\sqrt{2}$$

FIG. 4-5 Calculation of the relation between the lattice parameter and the atomic radius for uniform spheres in FCC packing.

If the crystalline lattice type and atomic radius of a material are known, its maximum theoretical density can be calculated by

$$\rho = \frac{nA}{N_0 V} \tag{4.4}$$

where ρ is the density, n is the net number of atoms per unit cell, A is the atomic or molecular weight of the element or compound, N_0 is Avogadro's number, and V is the unit cell volume (a^3 for cubic cells). The lattice parameter for a cubic crystal may be calculated from its atomic radius by one of the preceding equations, or from similar expressions for other types of crystal symmetry. Equation (4.4), for example, is valid for any type of crystal and is not restricted merely to the cubic system.

Structure Types

IONIC SOLIDS   The lattice energy calculations discussed in Chapter 2 can be used in determining the relative stability of ionic polymorphs, but the calculations involved are strictly valid only at the temperature of absolute zero. Because the effect of temperature on atomic spacing can be appreciable and the differences in polymorphic lattice energies are frequently small, relative stability predictions are somewhat questionable. However, if the computed energy of formation is positive for a given compound, it is likely to be unstable.

Numerous ionic solid structures are known, but a complete discussion of their individual characteristics is beyond the scope of this book. One very common ionic crystal of the A^+X^- type is the sodium chloride structure shown in Fig. 4-6. This consists of two interpenetrating FCC structures, and should not be confused with the simple cubic lattice. The former is designated as the NaCl structure, although many other ionic compounds such as MgO, NaBr, RbCl, KBr, and LiI have this same crystal structure. The lattice constants are measured between the Na^+ ions or between the Cl^- ions, but not from the Na^+ to the Cl^- ions. Each ion is coordinated by six nearest neighbors of the other type.

Some important common variations of the A^+X^- structure are shown in Figs. 4-7, 4-8, and 4-9 for the CsCl, polymorphic ZnS zincblende, and wurtzite structures, respectively. The CsCl structure is basically BCC, with one type of ion at each corner and the other at each center. Compounds which exhibit this type of crystal structure are CsCl, CsBr, CuZn, and TlCl. As in the ordinary BCC metallic crystal, the ligancy of each ion is 8.

In the zincblende structure illustrated (Fig. 4-8) the basic arrangement of the ions is that of the carbon atoms in diamond. However,

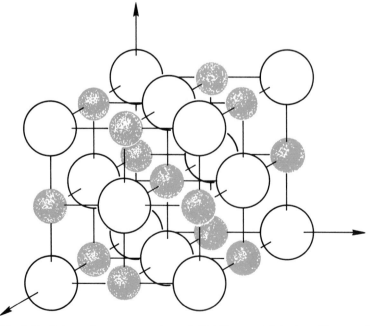

FIG. 4-6 Schematic representation of Na^+Cl^- crystal lattice. The smaller spheres represent the Na^+ ions; the larger spheres are the Cl^- ions.

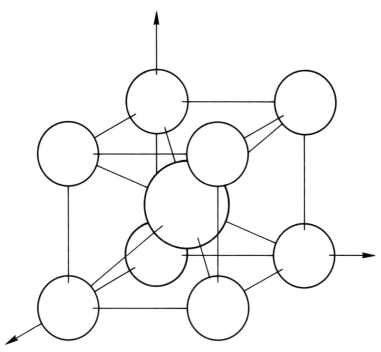

FIG. 4-7   Schematic illustration of the atomic arrangement in CsCl crystal. The large, central ion is Cl^- and the smaller, corner ions are Cs^+.

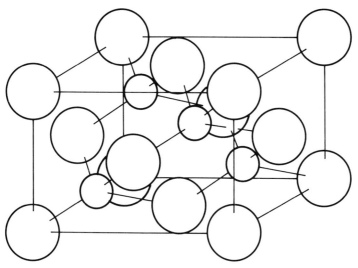

FIG. 4-8   Schematic arrangement of ions in the zincblende polymorph of ZnS. The larger spheres arranged in the FCC pattern are the S^{--} ions; the tetrahedrally arranged, smaller spheres within the crystal are the Zn^{++} ions.

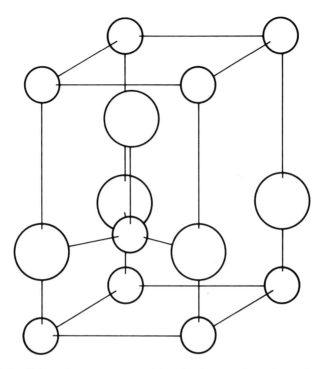

FIG. 4-9 Schematic arrangement of ions in the wurtzite polymorph of ZnS. The large spheres are the S^{--} ions; the small spheres are the Zn^{++} ions.

the S atoms form an FCC structure while the Zn atoms intersperse tetrahedrally at appropriate points within the lattice.

Upon structural rearrangement of the ZnS zincblende crystal, the polymorphic wurtzite structure shown in Fig. 4-9 results. The compounds HgS (β), CuCl, BeS, and CuBr form typical zincblende structures, while NH_4F, BeO, AgI, and ZnO form typical wurtzite structures.

Another common set of structures arises from the AX_2 compounds where coordination is 2:1. Two common examples are cuprite (Cu_2O) and rutile (TiO_2), shown in Figs. 4-10 and 4-11, respectively. Cuprite has a BCC structure of O^{--} ions with four tetrahedrally arranged Cu^+ ions contained in the interior of the cell, resulting in a 4:2 coordination.

In rutile, the Ti^{4+} ions form a body-centered tetragonal structure with the O^{--} ions lying in (001) planes. Many oxides and fluorides such as VO_2, MoO_2, NiF_2, and MgF_2 are so structured.

METALLIC SOLIDS Because metals form more or less undirected bonds, their structures tend to contain the highest possible atomic coordinations. Nearly 70 per cent of the metallic elements crystallize

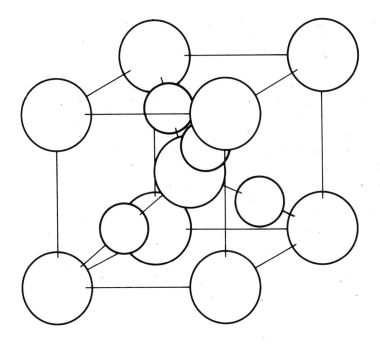

FIG. 4-10   Schematic arrangement of ions in cuprite, Cu_2O. The O^{--} ions (large spheres) form a BCC structure with the Cu^+ ions (small spheres), tetrahedrally coordinating the body-centered O^{--} ion.

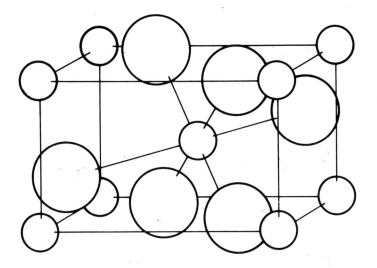

FIG. 4-11   Schematic arrangement of ions in rutile, TiO_2. The O^{--} ions (large spheres) lie in parallel planes and the Ti^{4+} ions (small spheres) form a body-centered, tetragonal cell structure.

in either the BCC, FCC, or HCP crystal structures; the remaining elements crystallize in the simple hexagonal, rhombohedral, or ortho-rhombic structures (see Appendix IV).

COVALENT SOLIDS   Carbon and silicon are the main constituents of the third important group of real solid crystalline materials—covalently bonded solids. The hybridized tetrahedral bond orbitals of carbon discussed in Chapter 2 almost exclusively result in the formation of covalent bonds. The two familiar forms of solid carbon are graphite, a sheetlike structure illustrated in Fig. 4-12, and diamond, a very hard, rigid structure illustrated in Fig. 4-13. In the graphite structure, carbon atoms are arranged in parallel sheets of regular hexagons with C–C bond distances of 1.42 Å in the hexagons and 3.41 Å between the layers. The overall structure of graphite is that of giant, two-dimensional planar molecules stacked one on top of another and held together by weak, secondary van der Waals forces. The facility with which these sheets can slide past each other imparts a greasy feeling to graphite, making it suitable for use as a dry lubricant.

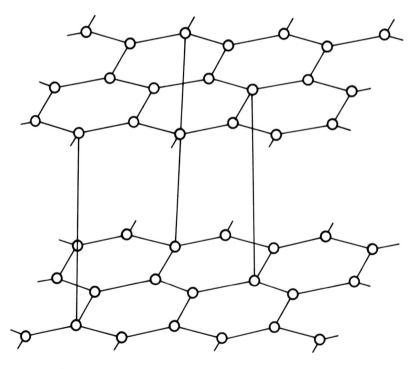

FIG. 4-12 Schematic representation of carbon atoms in the graphite sheet structure.

POLYMORPHS OF CARBON

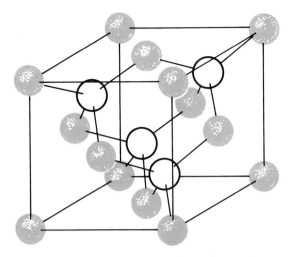

FIG. 4-13   Schematic illustration of the diamond crystal structure. Shaded spheres form FCC structure; the four unshaded spheres are interspersed in tetrahedral locations, as shown.

In the diamond, carbon atoms are tetrahedrally bonded together in a rigid network of carbon-to-carbon covalent bonds, making the structure extremely hard and brittle.

A third form of solid carbon, amorphous carbon, is commonly encountered in the forms of carbon black, lamp-black, coal, coke, and charcoal. These materials are microcrystalline in structure and have the basic crystalline form of graphite. All forms of solid carbon are stable at ordinary temperatures and pressures, but revert to graphite at high temperatures. It is possible, therefore, to convert graphite into low grade industrial diamond under sufficiently high pressure [3].

A fourth polymorph of carbon, the hexagonal diamond [4], is a recent result of laboratory experiments and has been discovered in certain meteorites. This polymorph, with a density of 3.51 g/cm^3, forms a hexagonal cell which has a wurtzitelike structure containing four atoms per cell and lattice parameters of $a = 2.52$ Å and $c = 4.12$ Å. This polymorph of carbon does not occur in natural minerals [4].

Silicon also forms important, covalent solid crystalline phases. In many respects, silicon is chemically similar to carbon, but there are several significant differences between them as well. The oxides of carbon are gaseous, while those of silicon are solid. The chlorides, fluorides, and sulfides of carbon are stable; those of silicon are decomposed by water. Hydrocarbons are extremely stable, but silicon hydrides decompose readily. In reacting with oxygen, the silicon atom preserves the tetrahedral arrangement of its bonds due to the divalence of the oxygen atoms. Consequently, many of the solid silicon compounds, although called silicates, are almost entirely oxides.

The basic unit of crystalline silica (SiO_2) is tetrahedral, with the silicon atom at the center coordinated tetrahedrally by four oxygen atoms, as illustrated in Fig. 4-14. Depending upon the spatial arrangement of these tetrahedra, different structures can be obtained. If the corners of the tetrahedra are shared so that every oxygen atom is common to two tetrahedra, an electrically neutral structure, called silica, results where the silicon-oxygen ratio is 1:2. The silica structure, an open, three-dimensional network with strong primary bonds between the silicon and oxygen atoms, is the lowest density form of the silicate compounds. Mechanically, silica is very strong and has a relatively high melting point (3110°F). It has three crystalline polymorphs: quartz, tridymite, and cristobalite, as well as an amorphous form variously referred to as silica gel, diatomite, or vitreous quartz.

In other combinations, several chain- and sheet-type structures can be formed by the silica tetrahedra. A simple, single unit chain structure is obtained by the common sharing of the tetrahedral corners. A large group of minerals, known as pyroxenes, are formed by chains of tetrahedra of varying lengths. A series of minerals with double chain structures, called amphiboles, results from the formation of the basic ring

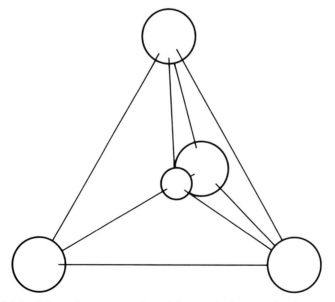

FIG. 4-14   Schematic representation of the tetrahedral coordination of silicon by four oxygen atoms in the silica crystal. The small sphere represents the Si^{++}, and the four large spheres are the O^{--}.

structure illustrated in Fig. 4-15. Other sheet structures produce a large class of substances, which includes kaolinite, talc, and pyrophyllite. These materials are very similar to graphite in that the bonds in the plane of the sheets are strongly formed, but the adjacent sheets are only loosely bound by van der Waals forces. Mica, another important sheet structure silicate, is harder than either kaolinite or talc, but because its basic layered structure is the same it splits easily into very thin sheets, making it suitable for use as an electrical insulator.

Another large class of molecular solids is the high molecular weight organic polymers. These polymeric solid materials consist of a large, more or less amorphous mass of giant *macromolecules*, entangled somewhat like the strands of spaghetti in a bowl. These macromolecules consist of thousands of carbon atoms linked together in a chainlike structure with hydrogen or another atom surrounding them. Macromolecules behave much differently in forming crystals than do the single spherical atoms of the ceramic and metallic materials already mentioned. The mechanisms by which these giant molecules are formed will be discussed in Chapter 12.

The randomness of the amorphous polymeric mass described above is periodically disrupted by crystalline regions which yield definite x-ray diffraction patterns [5]. These crystalline regions were originally interpreted as small, relatively perfect crystals embedded in an amor-

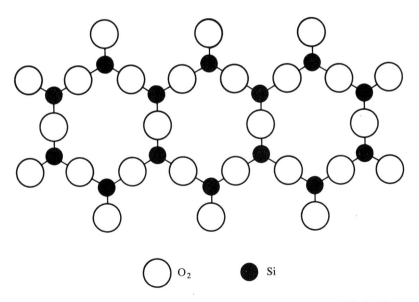

O_2 Si

FIG. 4-15   Schematic arrangement of Si and O atoms in the amphibole sheet structure. Large circles are O^{--} ions; small circles represent the Si^{4+} ions.

phous matrix; this has come to be known as the *fringed-micelle model* [5].

The fringed-micelle model (illustrated in Fig. 4-16) was supposed to consist of an amorphous mixture of long polymer molecules in which regions of relative order or coordination of adjacent molecular sections dictate crystalline behavior. The existence of these *crystallites* was believed to be the result of the increased secondary bonding caused by more efficient packing when the molecular sections aligned properly [6, 7]. Because of the large linear dimension of the polymer molecules, one molecule was supposed to wander through several alternating regions of amorphous and crystalline arrangements, providing *links* between the crystallites and thus accounting for the mechanical strength of such polymers as polyethylene.

Recent studies have revealed that the fringed-micelle model is probably fundamentally incorrect [5]. Careful examination of single polymer crystals grown in laboratory solutions shows that they are really monolayers or *lamallae* of the order of 100 Å thick and several microns in lateral dimensions [5]. More complex polymer crystals often contain numerous lamallae originating from a common nucleus, as illustrated [5] in Figs. 4-17 and 4-18.

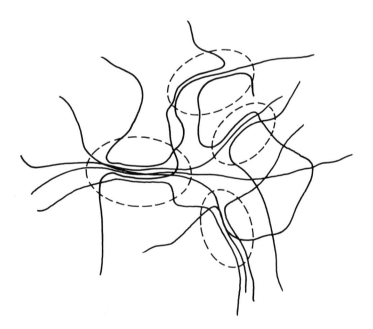

FIG. 4-16 Schematic representation of the fringed-micelle model of a polymer structure. Dashed ovals enclose crystallites formed by local coordination of the molecular sections.

FIG. 4-17 Electron micrograph of spiral crystals of polyethylene (×23,000). (Courtesy of E. S. Clark, Senior Research Chemist, DuPont Company, Plastics Department, Wilmington, Delaware. Used by permission.)

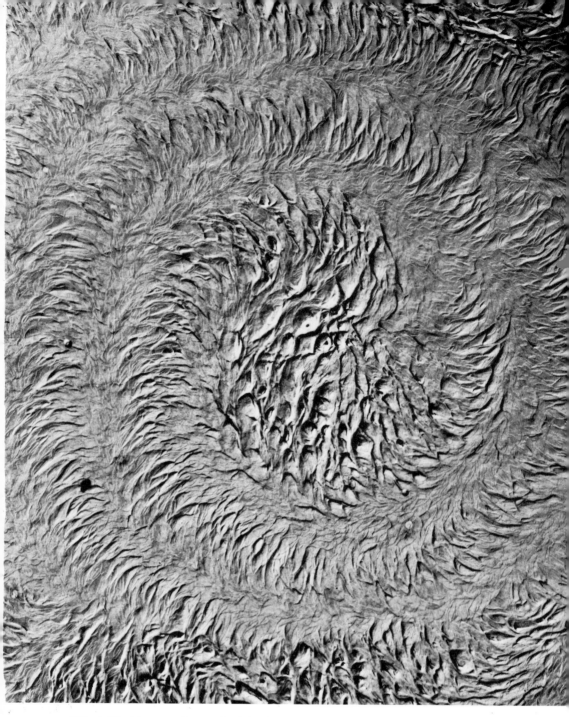

FIG. 4-18 Electron micrograph of the ringed spherulite growth of polyethylene ($\times$8,000). (Courtesy of E. S. Clark, Senior Research Chemist, DuPont Company, Plastics Department, Wilmington, Delaware. From *Polymer Single Crystals,* by P. H. Geil, copyright 1963, John Wiley and Sons, Inc., Interscience Publishers. Used by permission.)

The current interpretation of these lamellar crystals formed from melts or dilute solutions involves three models: the adjacent re-entry regular, the adjacent re-entry irregular, and the nonadjacent re-entry or switchboard model, all illustrated in Fig. 4-19. The molecules are visualized as axially folded along their skeletons, like undulating corrugated paper. The tops of these corrugated folds lie in a plane which is microscopically observable. In the case of polyethylene, these folded planar sheets form a symmetrical, diamond-shaped pattern which appears as a hollow pyramid in solution [5]. When these hollow, pyramidal crystals are sedimented onto a glass surface for observation, as in Fig. 4-20, the diamond-shaped residue looks very much like a collapsed tent.

When polymers crystallize from the melt, more than one lamella frequently originate from the same nucleation site. These multiple lamallae twist and turn as they grow outward from the nucleus, developing splits in their growth faces. The result is a spherically symmetric, grossly uniform growth composed of radially oriented lamallae, called a *spherulite* [5].

In addition to pyramidal and spherulitic structures, polymers may also crystallize in a dendritic or branching, tree-like formation, as well as numerous other crystalline types [5]. One particularly interesting

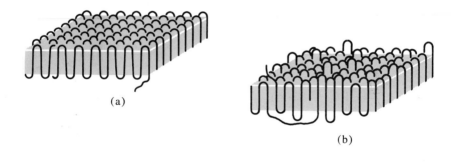

(a)

(b)

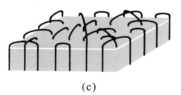

(c)

FIG. 4-19   Three models of lamellar polymer crystals: (a) regular adjacent re-entry, (b) irregular adjacent re-entry, and (c) nonadjacent re-entry or switchboard model. (From "Polymer Morphology," by P. H. Geil, *Chem. & Engr. News,* **43,** p. 72 (Aug. 16, 1965). Used by permission.)

phenomenon is the observed existence [8] of intercrystalline links in such polymers as polyethylene. (These links are illustrated in Figs. 4-21, and 4-22 for a linear polyethylene [8]. From these photomicrographs, it is easy to understand the origin of the fringed-miscelle model.) Such links are due to the existence of *tie molecules* which participate in more than one spherulitic structure, thereby substantially reinforcing the crys-

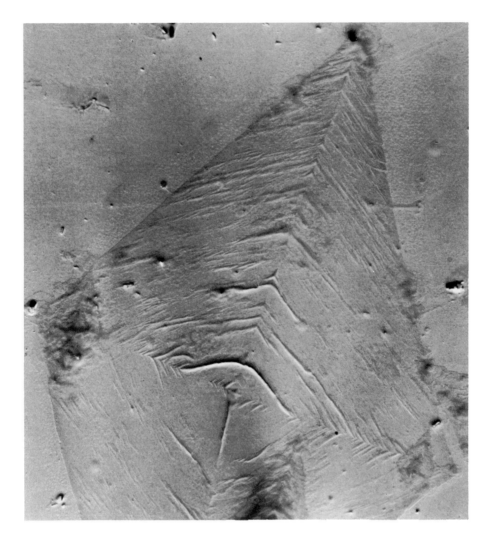

FIG. 4-20 Electron micrograph of a collapsed, pyramidal, single crystal of polyethylene (×5,500). (Courtesy of E. S. Clark, Senior Research Chemist, DuPont Company, Plastics Department, Wilmington, Delaware. From *Polymer Single Crystals,* by P. H. Geil, copyright 1963, John Wiley and Sons, Inc., Interscience Publishers. Used by permission.)

FIG. 4-21 Electron micrograph of bulk crystallized polyethylene, showing intercrystalline tie molecules. (Courtesy of H. D. Keith, Bell Telephone Laboratories, Murray Hill, New Jersey. From "Intercrystalline Links in Bulk Polyethylene" by H. D. Keith, F. J. Padden, Jr., and R. G. Vadimsky, *Science*, **150**, p. 1026 (Nov. 19, 1965), and copyright 1965 by the American Association for the Advancement of Science. Used by permission.)

talline polymer structure. The ductility and strength of polyethylene is much greater than would be expected from only van der Waals forces.

The constituent elements of the polymer skeletal chain are important in determining the degree to which a polymer is susceptible to crystallization. Polymers with long-repeating units or low symmetry do not crystallize readily. Specific arrangements of the elements of the skeletal chain of such a polymer can be produced by using special catalysts, called stereospecific because they act only for specific orientations of molecules. Thus, usually amorphous polystyrene can be made highly crystalline by using stereospecific catalysts which effectively influence the symmetric spacing of its benzene ring side units about the skeletal carbon chain of its polymer molecule. If these rings are consistently spiraled around the main chain at regular angular displacements of 120° from one another, the molecule has a high degree of symmetry and considerable crystallinity is possible. Such a structure is called *isotactic*. If the ring arrangement is random, the structure is *atactic*; if the rings alternate on both sides of the chain, the structure is *syndiotactic*. Each of these three types of side group arrangements is illustrated schematically in Fig. 4-23 for polypropylene, which has $-CH_3$ side units.

The degree of crystallinity possible in a substance is important in determining its physical properties. In polyethylene, for example, spherulitic crystallites interconnected by tie molecules refract light randomly so that bulk polyethylene appears opaque. Highly crystalline, isotactic polystyrene, however, is crystal clear because of the high regularity of its structure.

It should be clear from this discussion that the simple notions of unit cells, although useful in identifying ceramics and metals, are inadequate when dealing with polymer crystals.

FIG. 4-22   Enlarged electron micrograph of bulk crystallized polyethylene, showing more detail of spherulites linked by tie molecules. (Courtesy of H. D. Keith, Bell Telephone Laboratories, Murray Hill, New Jersey.) From "Intercrystalline Links in Bulk Polyethylene" by H. D. Keith, F. J. Padden, Jr., and R. G. Vadimsky, *Science*, **150**, p. 1026 (Nov. 19, 1965), and copyright 1965 by the American Association for the Advancement of Science. Used by permission.)

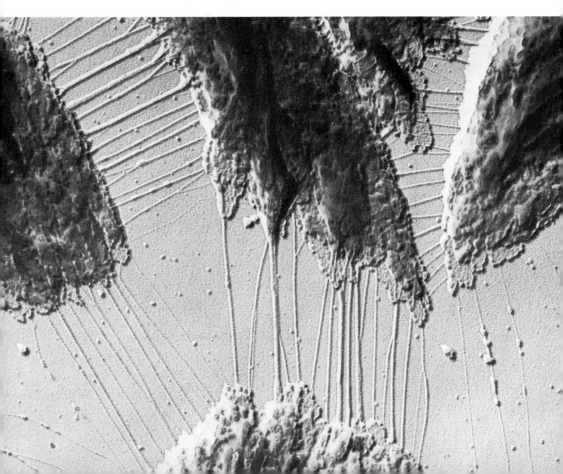

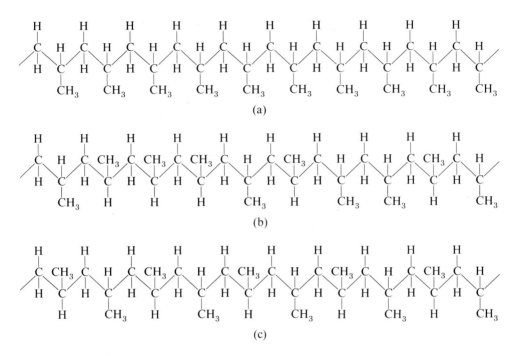

FIG. 4-23 Schematic illustration of stereotactic arrangements of −CH₃ groups of polypropylene: (a) isotactic, (b) atactic, and (c) syndiotactic.

Crystal Imperfections

All real crystalline materials possess defects in their lattices. These imperfections may occur in many forms [9], varying from point defects to line defects and stacking faults.

Point Defects

The simplest and most easily understood defects are those which are localized to the region of a single lattice site, called *point defects*. One of the most important of these defects is the simple *vacancy* or absence of an atom at a lattice site. Because of the improper coordination of the remaining atoms resulting from a vacancy, the electronic bonding fields which surround it are disrupted from the ideal pattern. Effective radii of those atoms adjacent to the vacancy change slightly, and the crystal is weakened. In fact, if sufficient vacancies are generated in the crystal structure, it loses its long-range order and symmetry, becoming a liquid. (These vacancies are the holes mentioned in Chapter 3 when the properties of the liquid state were introduced. Since a liquid normally results from sufficient heating of a solid, it is evident that the breaking bonds by increasing the thermal energy of a crystal is one mechanism for introducing vacancies into a solid structure.

If the vacancy occurs as the absence of a cation-anion pair in an ionic or ceramic crystal, it is called a *Schottky defect* [10] (illustrated schematically in Fig. 4-24(a)). In this type of point defect, the ionic crystal retains its overall electrical charge balance, but large, local charge imbalances do occur in the immediate neighborhood of the vacancies, creating a local region of higher energy.

Another type of point defect observed in ceramic materials is the *Frenkel defect* [10] (Fig. 4-24(b)). Here, either an anion or a cation is somehow dislodged from its proper lattice site and relocated to an interstitial location nearby. This simultaneously creates two types of disruption in the normal crystal structure: a vacancy at the original site, and a strained interstitial ion. The effect of these defects is increased by the electrical charge imbalances created locally by their formation. As in the case of the Schottky defect above, the overall electrical neutrality of the entire crystal is preserved in this defect.

A common defect observed [9] in metallic crystals is the *intersticialcy* illustrated schematically in Fig. 4-24(c). This differs from the Frenkel defect of ceramic materials in that it is caused by the presence of a foreign atom located at an interstitial location in the metallic crystal. This foreign atom may be that of a specifically added alloying agent or it may simply be an impurity.

All of the point defects introduced here produce local variations in the order of atomic arrangement in a crystal and consequently affect the bond structure and energy distributions in crystal formations. The local strains thus induced are very important in determining the response of the material to various external stimuli.

Line Defects

DISLOCATIONS   When the dimensions of an imperfection become large by atomic dimensions but are still small when compared with visible macrocrystals, the imperfection is called a *dislocation*. Dislocations are linear arrays of atoms where the coordination characteristics

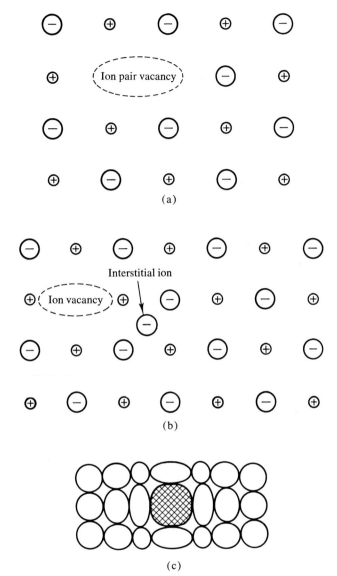

FIG. 4-24   Schematic representation of local or point defects: (a) Schottky defect, (b) Frenkel defect, and (c) interstialcy.

differ systematically from the ideal. Depending upon the type of coordination mismatch which occurs at the dislocation line, we define an edge, a screw, or a mixed dislocation.

The concept of dislocations was originally proposed [11] as an hypothesis to account for certain aspects of the plastic deformation of metallic crystals. Direct observations of dislocations have since been made, using techniques such as high resolution electron microscopy. Figures 4-25 and 4-26 show photomicrographs of edge and screw dislocations, respectively.

The *edge dislocation* (also called a *Taylor* or *Taylor–Orowan* dislocation) is a configuration of continuous rows of atoms, each with one

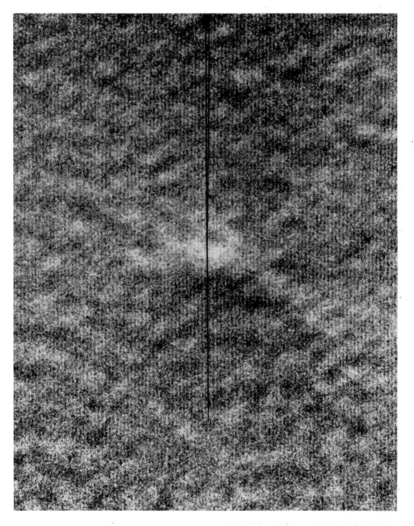

FIG. 4-25 Electron photomicrograph showing an edge dislocation in an aluminum crystal. (Courtesy of J. R. Parsons, Atomic Energy of Canada, Ltd. From "Crystal-lattice Images of End-on Dislocations in Deformed Aluminum," by J. R. Parsons and C. W. Hoelke, *J. Appl. Phys.*, **40**, No. 2, p. 867 (1969). Used by permission.)

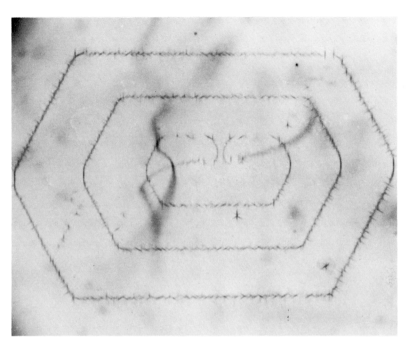

FIG. 4-26 Photomicrograph showing ringed growth due to screw dislocations in a Si crystal. (Courtesy of The General Electric Research and Development Center. From the cover of *J. Appl. Phys.,* by Dr. William Carter Dash (November 1956). Used by permission.)

less atom coordinating it than the crystal habit (or the ideal crystal structure into which the material should form) requires. When a row of insufficiently coordinated atoms is viewed end-on, it appears as the terminus of an extra plane of atoms extending only part way into the crystal structure (see Fig. 4-27). *The student should understand that the dislocation is not an extra plane of atoms* [11], *but a configuration of improperly coordinated sites in the crystal structure.*

The atoms immediately above the *slip plane* (the plane separating the last layer with an extra atom from the planes below it) are in a state of compression, while those immediately below the slip plane are under tension. Thus, the edge dislocation is surrounded by a stress field, roughly inversely proportional in magnitude to the distance from the dislocation [11].

The *screw* or *Burgers dislocation* is a configuration in which the line of atoms defining the dislocation has the *proper number* of coordinating neighbors, but the *coordination polyhedron is distorted*. The screw dislocation can be visualized as originating from the partial slipping of a section of a crystal plane, with the slip displacement terminating

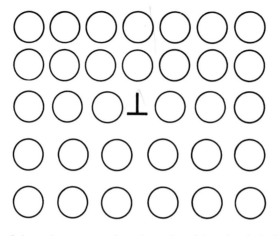

FIG. 4-27 Schematic representation of an edge dislocation (⊥). At this point in the structure an atom is missing. The rows below the dislocation close together again, so that it appears as if a plane of atoms terminates at ⊥.

internally to the crystal and on the dislocation line which will lie in the slip plane. This is illustrated schematically in isometric projection in Fig. 4-28. The same crystal is illustrated in frontal elevation in Fig. 4-29, which shows the distortion of the coordination polyhedron. Here, the atoms on either side of the dislocation line are in proper registry above and below the plane of slip on the right side. (In Fig. 4-29, the small, dark circles represent the atomic plane closest to the viewer; the larger, open circles represent the plane immediately behind it.) The normal coordination octahedron is indicated in the figure by the solid lines which join the four planar coordinating neighbors on the left side of the dislocation in the unslipped region. The polyhedron is completed by atoms above and below the central atom. The same coordination octahedron crossing the line of the dislocation is no longer regular but severely distorted by the partial slip which has occurred on the right side of the dislocation line.

In general the dislocation line through a crystal will be curved [11], as in Fig. 4-30(a) and (b). The portion of this curved dislocation line near the bottom of the figure is in the *screw orientation*, while that portion near the top right hand part of Fig. 4-30(b) is in the *edge orientation*. Clearly, the portion between these two extremes must be a combination of the two *pure* types. Such a combination dislocation, called a *mixed dislocation*, arises from the partial slip of the crystal planes in the lower right hand portion of the figure. In general, a dislocation line cannot terminate internally in the crystal, but must either close on itself or terminate in the crystal surface [11].

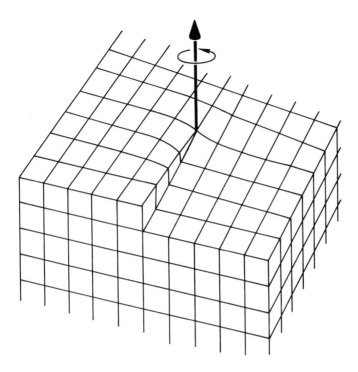

FIG. 4-28 Schematic representation of a screw dislocation, showing spiral nature of dislocation. Arrow shows axis with direction of rotation indicated.

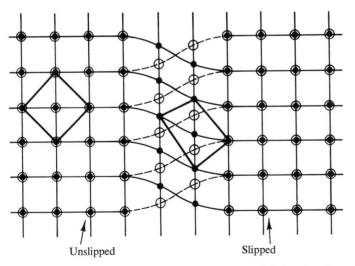

Unslipped   Slipped

FIG. 4-29 Frontal elevation of a screw dislocation model, showing distortion of coordination polyhedron (see dark solid lines).

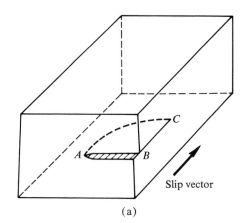

(a)

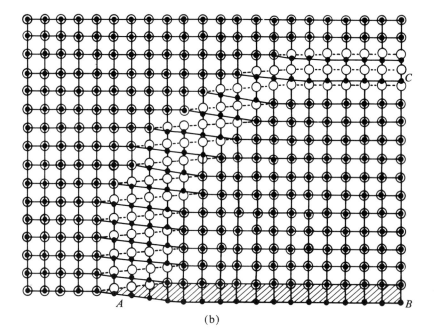

(b)

FIG. 4-30 Schematic representation of a mixed dislocation. (From *Dislocation in Crystals*, by W. T. Read, Jr., copyright 1953, McGraw-Hill Book Co., Inc. Used by permission.)

BURGERS VECTORS The *Burgers circuit*, or the closed path through a crystal lattice, is used to differentiate between types of dislocations. This circuit is composed of equal numbers of lattice parameter translations taken sequentially in positive and negative directional senses, illustrated schematically in Fig. 4-31 for an edge dislocation. If the

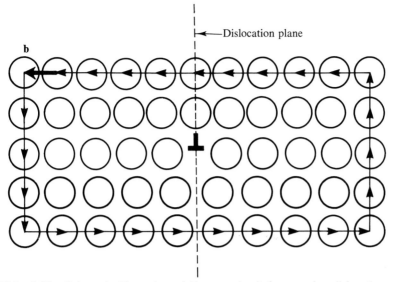

FIG. 4-31 Schematic illustration of Burgers circuit for an edge dislocation, showing Burgers vector **b**. Note that **b** is normal to the plane containing ⊥.

Burgers circuit encloses only perfect crystalline material, it will close on itself. If the loop encloses a dislocation, however, the circuit will fail to close in the manner characteristic of that particular dislocation. From Fig. 4-31, it is evident that the closure failure for an edge dislocation is described by a vector with a magnitude equal to a unit lattice parameter *normal to the plane of the edge dislocation*. This closure vector, called the *Burgers vector*, is characteristic of a given dislocation [12] and is *constant for the entire region* of the crystal enclosed by a dislocation loop. It is, therefore, constant along the entire extent of this dislocation loop [11], changing from edge to mixed to screw orientation as the direction of the dislocation loop changes.

Figure 4-32 illustrates the Burgers circuit for a pure screw dislocation. From it, it is clear that the Burgers vector is parallel to the line of dislocation in the screw orientation. Figure 4-33 illustrates schematically a region (shaded) in a crystal enclosed by a single, mixed dislocation line which has a Burgers vector **b**. That **b** is constant throughout the slipped region is easily shown [12] by considering the *PQRS* region in Fig. 4-34. Assume that part of the region has a Burgers vector $\mathbf{b}_1$, and the other part has a Burgers vector $\mathbf{b}_2$. Because dislocation lines must be closed curves, the two regions with differing Burgers vectors in *PQRS* must themselves be separated by a dislocation (indicated by the dashed *RTP* curve which must have a Burgers vector $\mathbf{b}_3 = \pm(\mathbf{b}_1 - \mathbf{b}_2)$. However, if *PQRS* is to be a single, unintersected dislocation, as postulated, it follows that $\mathbf{b}_3 = 0$ and, consequently, $\mathbf{b}_1 = \mathbf{b}_2$.

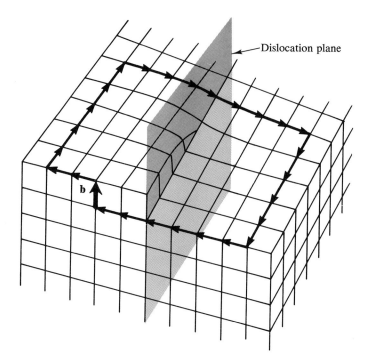

FIG. 4-32 Schematic illustration of Burgers circuit for a screw dislocation, showing Burgers vector **b**. Note that **b** is parallel to the plane containing the screw dislocation.

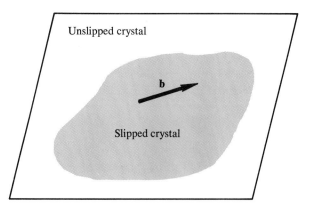

FIG. 4-33 Schematic illustration of slipped region (shaded) in a crystal, showing Burgers vector **b**.

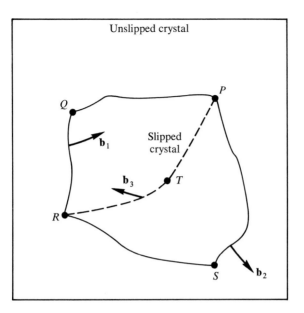

FIG. 4-34   Illustration of proof that Burgers vector is constant along any given dislocation.

ORIGIN OF DISLOCATIONS   Dislocations arise from two fundamental sources: growth mechanisms and plastic deformation. We shall defer the discussion of plastic deformation to Chapter 7, where the response of materials to mechanical stresses will be examined.

An analysis of the mechanisms by which crystals grow and a discussion of the free energy differences that must exist between the two phases for this growth to occur will appear in Chapter 6. In studying the phenomenon of the growth of iodine crystals from the vapor phase, Volmer and Schultze observed [13] that appreciable growth occurred when the free energy difference between the phases (as measured by a vapor pressure[2] difference) was considerably less than the theoretical value for perfect crystals. In order to explain such observations, Frank and his colleagues [14] suggested that in reality crystals contain dislocations which insure the continual presence of steps on their faces. These steps, in turn, provide sites where new vapor phase molecules can affix themselves to the growing crystal with less than the free energy difference normally required.

---

[2]The vapor pressure is the pressure of vapor molecules above the solid or liquid phase at any given temperature. When this pressure exceeds that of the ambient atmosphere, vapor condenses. When it is less than atmospheric, the solid or liquid vaporizes.

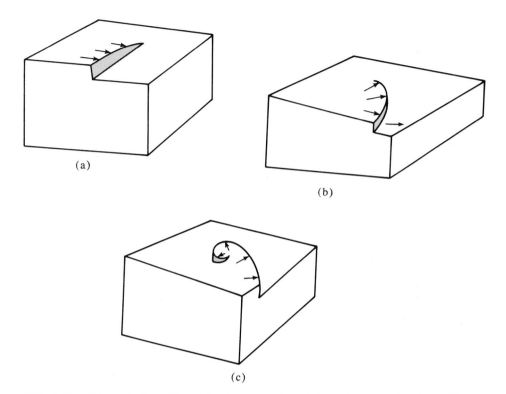

(a)

(b)

(c)

FIG. 4-35 Schematic illustration of the formation of a spiral step in a crystal surface which has a screw dislocation terminating in it. As growth proceeds at a uniform rate, the outer end of the dislocation falls behind the center, generating a spiral step.

If a screw dislocation exists in the crystal structure and terminates in the face, atoms from the vapor can attach themselves to the terminal edge of the dislocation and cause it to pivot continuously about its axis, providing a perpetual step in the surface. However, the velocity of the addition of atoms to the step is uniform all along the step. Consequently, the central regions of the step will outdistance[3] the outer portions, and a *spiral step* will be formed (see Fig. 4-35). The actuality of this postulated mechanism was first observed by Griffin [15] in the form of growth rings on the surface of a beryl crystal. Numerous examples have since been noted; another example is illustrated in Fig. 4-36.

---

[3]It is a well known fact that in uniform circular motion the speed of particles must increase linearly with the radial distance from the center of rotation. Therefore, if the speed of addition is constant, particles farther from the center must continually fall behind, forming a spiral pattern.

FIG. 4-36 Electron micrograph showing spiral growth pattern in a crystal of polyoxymethylene (×14,000). (Courtesy of E. S. Clark, Senior Research Chemist, DuPont Company, Plastics Department, Wilmington, Delaware. From *Polymer Single Crystals,* by P. H. Geil, copyright 1963, John Wiley and Sons, Inc., Interscience Publishers. Used by permission.)

Several possible sources have been suggested [16] as the origins of crystal dislocations:

1. Nucleation in regions of locally high supersaturation.[4]
2. Plastic buckling of very thin and fragile nuclei.
3. Nucleation on an imperfect substrate or foreign particle.
4. Buckling under large strains created by composition gradients.
5. The collapse of sheetlike cavities formed by the aggregation of vacant cites in the structure.
6. Nuclei with dissimilar, crystallographic orientation growing together into a single crystal.

The more rapidly a crystal is grown, the more active the above mechanisms and the more numerous the dislocations and imperfections it will contain. Thus, the process of growing nearly perfect crystals is a slow and tedious one, requiring extreme care on the part of the experimenter.

References Cited

[1] Moffatt, W. G., G. W. Pearsall, and J. Wulff, *The Structure and Properties of Materials, Volume I, Structure*, John Wiley and Sons, Inc., New York, 1964, p. 39*ff.*

[2] Frenkel, J., *Kinetic Theory of Liquids*, Dover Publications, Inc., New York, 1955, p. 54*ff.*

[3] Hall, H. T., *US Patent No.* 2,947,608 (Aug. 2, 1960); *J. Chem. Ed.*, **38,** p. 484 (1961).

[4] Hanneman, R. E., H. M. Strong, and F. P. Bundy, *Science*, **155,** p. 995 (Feb. 24, 1967).

[5] Geil, P. H., *Chem. & Engr. News*, **43,** p. 72 (Aug. 16, 1965).

[6] Jastrzebski, Z. D., *Engineering Materials*, John Wiley and Sons, Inc., New York, 1959, Chapter 2.

[7] Van Vlack, L. H., *Elements of Materials Science, 2nd ed.*, Addison-Wesley, Reading, Mass., 1964, Chapters 3, 7.

[8] Keith, H. D., F. J. Padden, Jr., and R. G. Vadimsky, *Science*, **150,** No. 3699, p. 1026 (Nov. 19, 1965).

[9] Van Bueren, H. G., *Imperfections in Crystals, 2nd ed.*, North-Holland Publishing Company, Amsterdam, 1961, Chapter 2.

---

[4]Nucleation is the process of forming tiny nuclei or *seed crystals* which can subsequently grow into full crystals. Supersaturation is the excess vapor pressure required for small crystal nuclei, a process to be discussed in detail in Chapter 6.

[10] Azaroff, L. V., *Introduction to Solids*, McGraw-Hill Book Co., Inc., New York, 1960, Chapter 5.

[11] Read, W. T., Jr., *Dislocations in Crystals*, McGraw-Hill Book Co., Inc., New York, 1953.

[12] Cottrell, A. H., *Dislocations and Plastic Flow in Crystals*, Oxford University Press, 1956, Chapter 1.

[13] Volmer, M., and W. Schultze, *Zeit. Phys. Chem.*, (A) **156**, p. 1 (1931).

[14] Frank, F. C., N. Cabrera, and W. K. Burton, *Nature*, **163**, p. 398 (1949).

[15] Griffin, L. J., *Phil. Mag.*, **47**(Ser. 7), p. 196 (1952).

[16] Frank, F. C., *Advances in Physics*, **1**(1), p. 91 (1952).

PROBLEMS

4.1   The element Zn normally crystallizes in an HCP crystal cell. Compute its theoretical density.

4.2   Titanium transforms from BCC ($a = 3.32$ Å) to HCP ($a = 2.95$ Å, $c = 4.683$ Å) at 880°C. Does the material expand or contract during this transformation? Compute the per cent of change in the theoretical density of Ti undergoing this transformation.

4.3   The element Fe shows the BCC-FCC allotropic transformation at 910°C. Compute the theoretical change in density accompanying this transformation.

4.4   Compute the packing factor of an HCP system.

4.5   NaCl crystallizes in a unit cell (shown in Fig. 4-6) which is FCC with respect to its Cl^- ions. If the equilibrium, interionic Na^+Cl^- distance is 2.814 Å, compute the density of NaCl.

4.6   Compute the planar density of Cl^- ions on the (111) plane in Fig. 4-6, using the data of Problem 4.5.

4.7   Compute the planar density of ions on a $(1\bar{1}0)$ plane in NaCl.

4.8   Compute the linear density of ions in the [111] direction of NaCl.

4.9   Compute the packing factor of the NaCl crystal as a function of the ratio r/R, where r is the radius of the Na^+ ion and R is the radius of the Cl^- ion. Using the data of Table 4-2 and Problem 4.5, compute the packing factor and the individual ionic radii of Na^+ and Cl^-.

4.10   Prove the validity of the entries $(r/R)_c = 0.414$ and 0.732 for ligancies of 4 and 6, respectively, in Table 4-1.

4.11   The mean C–C bond distance in diamond is 1.542 Å. Using this value and Fig. 4-13 as guides, compute the theoretical density and the packing factor of carbon in the diamond crystal. (The atoms are tetrahedrally arranged and the tetrahedral bond angle is 109° 28′.)

4.12   Using the data in Problem 4.11, compute the planar atomic density on the $(\bar{1}11)$ plane of diamond.

4.13   Using the $R(L12)$ data in Table 2-12, compute the theoretical densities of Pb(FCC), Cu(FCC), and Ni(FCC).

4.14   Using the data for $R(L12)$ in Table 2-12, compute the theoretical density of Se which crystallizes in a hexagonal unit cell.

4.15   An edge dislocation has occurred so that the slip plane is the most densely packed plane in a crystal cell and the Burgers vector lies in the most densely packed direction on that plane. Find the magnitude and direction (in terms of the Miller indices) of the Burgers vector for Ni(FCC) for unit slip (one full interatomic distance in the slip direction). Is your result unique?

4.16   Repeat Problem 4.15 for Fe(BCC).

4.17   Repeat Problem 4.15 for Zn(HCP).

5

PHASE
EQUILIBRIUM

Thermodynamic Background

Thus far our discussion of material properties has only been generally applied to the relationships between atoms in a single state of aggregation. Since most important materials are multiphase, it now becomes essential to investigate the interrelations that exist between several phases of matter.

The entire structure of phase relation analysis relies upon the fundamental laws of thermodynamics. Although the advanced concepts and formalism of this discipline will not be discussed, an understanding of certain terms and basic concepts in thermodynamics will be necessary before preceding.

The energy content of a phase is fundamental to the study of phase relationships. Energy contents may be expressed in a number of ways, but the most familiar and easily understood energy content of matter is the internal energy E. This comprises the thermal energy of the body, together with electronic energy, vibrational energy, and all other forms of energy which are stored internally in the molecular structure of the substance. When the product of pressure p and volume V is added to the internal energy, we obtain the enthalpy H. The first law of thermo-dynamics deals with the conservation of the internal energy or enthalpy, depending upon how it is applied. When dealing with an isolated mass of substance in what is called a *closed system*, we speak only in terms of the internal energy. However, when the quantity of mass under consideration is permitted to vary, and to enter or leave the boundaries of this system, the enthalpy must be considered.

The second law of thermodynamics deals with the conservation of another thermodynamic quantity, the entropy S. Entropy is related to the degree of randomness of organization in an assemblage of particles. If more order exists, the entropy is lower, and conversely. Spontaneous processes always occur with an increase in the total entropy of the system. Because entropy has the physical dimensions of energy divided by temperature, the product TS of the temperature T and the entropy S has dimensions of energy. When this product is subtracted from the enthalpy, the free energy $G = H - TS$ is obtained. It is this thermo-dynamic quantity which becomes all important in determining the behavior of phase relationships. Since free energy incorporates both the thermal energy content of a material through H and the degree of organization in a material through S, it characterizes the nature of the material completely.

A *phase* is a particular aggregation of molecules which is physically distinct and, therefore, distinguishable from its surrounding neighbors. The phase generally possesses different physical properties from neigh-boring phases, and is separated from them by a continuous and identifiable boundary, called a *phase boundary*. A given state of ag-gregation of matter may be multiphase; for example, two liquid phases are evident when an oil film is spread on the surface of water. The variable thickness of the oil film refracts light as does a prism, resulting in a rainbow effect which clearly delineates the oil phase from the water phase.

Many solid materials are multiphase, for example, steel-reinforced concrete. We shall discover later that each of the obvious phases in this particular example, steel and concrete, are themselves multiphase on a microscopic scale.

Everyone is familiar with the simultaneous existence of different phases of the same substance, as in the case of ice in water. It is a common experience to see ice form on a body of cold water and melt when the sun supplies sufficient thermal energy to it. The two phases are somehow related to each other in relative amount by the degree of heat energy which is supplied or removed. Our goal in this chapter is to discover the qualitative and quantitative relationships that exist between different phases of matter when they are in equilibrium with one another.

Phase Equilibrium in Pure Substances

Concept of Equilibrium

Before the complex systems of interest to the engineer can be explored, a discussion of the fundamental concepts of single-component phase equilibrium is necessary. A simple familiar example will illustrate the concept well enough for present purposes.

COOLING CURVES   In Chapter 2, the phenomenon of vapor to liquid condensation was discussed in terms of van der Waals forces. As a pure vapor is cooled, the average kinetic energy of the molecules decreases and the temperature falls. Eventually, an average energy content is reached at which the van der Waals forces can sufficiently overcome the kinetic motion of most of the molecules. Some of the molecules with less than average energy stick together or *condense*, forming the liquid phase. The temperature at which this first condensation occurs is the dew point [1] of the vapor. Further removal of energy results in the vapor molecules giving up their kinetic energies of motion and condensing in increasingly greater fractions. The energy withdrawal proceeds without further decrease in temperature, until nearly all the vapor molecules have condensed. Some of the molecules possess energies too great to condense, and remain as vapor molecules. However, when the majority[1] of the molecules have condensed, further removal of heat again results in a temperature decrease in the liquid phase and further condensation of the energetic vapor molecules. This phenomenon is illustrated schematically in Fig. 5-1 by the *cooling curve*.

The process just described is reversible. If the liquid were heated, the reverse order of phenomena would occur and the first bubbles of vapor would be observed at the temperature indicated as T_B in Fig. 5-1. Further addition of heat would occur isothermally until all the liquid had been vaporized; the vapor temperature would then rise. (The temperature at which the first bubbles of vapor form is called the *bubble point* [1] of a liquid. For a pure substance, such as the one in Fig. 5-1, the dew point and bubble point temperatures are identical and are called the *boiling point*.)

If heat addition or removal is arrested at the temperature T_B, the relative amounts of the two phases will not change. This is obvious since sufficient energy to overcome the van der Waals forces and free the molecule must be added in order to vaporize more liquid, but in

---

[1]The determining condition is that the pressure exerted by the remaining vapor molecules must be equal to the prevailing ambient pressure.

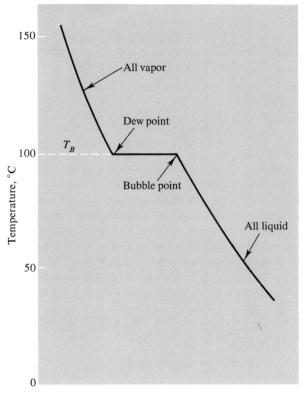

FIG. 5-1   Schematic cooling curve, illustrating the vapor-liquid condensation for water. In pure substances, the dew and bubble points are identical, and are called the boiling point.

order to condense more vapor molecules, their energy of translation must be removed for the van der Waals forces to bind them together. When the relative amounts of the two phases are in this stationary condition, a dynamic process is actually occurring. The higher energy vapor molecules give up their kinetic energy and condense upon striking the liquid surface. This excess energy cannot be removed from the system, so it is distributed among the neighboring liquid molecules which are already sufficiently energetic, and these very slight additions of energy free some of them from the van der Waals forces. Consequently, liquid molecules are vaporized by the addition of energy provided by the condensation of higher energy vapor molecules. Therefore, in the macroscopically stationary system, a microscopic, dynamic balance exists between the condensing vapor molecules and the vaporizing liquid molecules. Since the relative amounts of the two phases remain constant in such a situation, the rates of vaporization and condensation are balanced. *This condition of balanced rates is the condition of phase equilibrium.* The vapor phase is in *equilibrium* with the liquid phase; thermodynamically speaking, the free energies of the two phases are identical. This dynamic balance corresponds to a balance between changes in enthalpy and entropy.

If the pressure of this equilibrium system were changed, the balance of rates would be destroyed. The enthalpy would change and, in turn, change the relative amounts of the two phases present until a new condition of equilibrium was established or one phase vanished. At equilibrium, all environmental variables such as pressure and temperature must be identical in each of the phases, so that no free energy difference driving forces will enhance the rate of change of either phase.

VAPOR PRESSURE Similar equilibria exist between the vapor and liquid phases at temperatures other than the boiling point. At any temperature below the boiling point, a certain vapor pressure exists above the liquid because the vapor molecules are in dynamic equilibrium with it. If the total pressure to which the liquid phase is subjected is decreased to the vapor pressure at a given temperature, the liquid will vaporize or boil. The boiling condition which occurs under the normal atmospheric pressure (760 mm Hg or 14.696 psia) is called the *normal boiling point* of the liquid.

Phase Diagrams

The vapor pressure of water plotted as a function of temperature illustrates a *P-T* or phase diagram, as shown in Fig. 5-2. The region of the diagram lying above the vaporization curve represents liquids under pressures in excess of the vapor pressure. The region below the vaporization curve represents vapors at pressures below the liquid vapor pressure. To be condensed at constant pressure, these super-heated vapors must be cooled through a temperature difference equivalent to the horizontal distance from the point of interest to the vaporization curve on the diagram. The amount of energy which must be removed is called the *vapor superheat.*

Similar phase equilibria exist between vapors and solids. Vapors can condense directly into solids under certain conditions, as in the case of solid CO_2 or dry ice which passes from the solid to the vapor state without going through an intermediate liquid phase at ordinary atmospheric pressure. This phenomenon, called *sublimation*, is dependent on pressure and temperature.

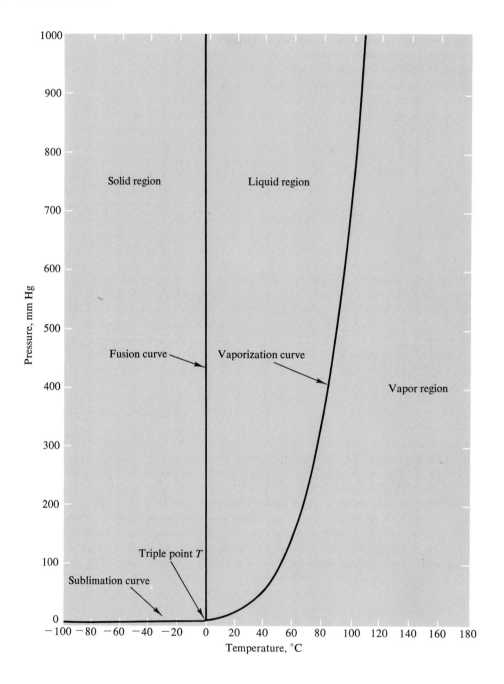

FIG. 5-2 *P-T* phase diagram for water.

The phenomena of sublimation and vaporization become the same at conditions which correspond to the intersection of the sublimation and vaporization curves (at point T in Fig. 5-2).

Triple Points

Phase equilibrium between solids and liquids is represented by the fusion curve illustrated in Fig. 5-2. The various pairs of phase transition equilibrium curves intersect at a single point T, called the *triple point. At this unique point, all three phases (vapor, liquid, and solid) are simultaneously in equilibrium.* Triple point data [2] for several substances are listed in Table 5-1.

TABLE **5-1**

Triple Point Data for Selected Substances

SUBSTANCE	TEMPERATURE (°C)	PRESSURE (MM HG)
Water	0.0098	4.579
Sulfur dioxide	−72.7	16.3
Carbon dioxide	−56.6	3880
Oxygen	−218.4	2.0
Argon	−189.4	516
Nitrogen	−209.8	96.4
Hydrogen	−259.2	54.1

Allotropic Solid Phases

Water is interesting because it is transformed through six different stable solid phases [2] as pressure is raised to relatively high values. Consequently, equilibria exist between these different solid phases, in addition to the phase equilibria already discussed above. Many materials in the solid state exhibit *allotropy* or *polymorphism,* which is the simultaneous existence of multiple solid phases that are thermodynamically stable under appropriate conditions. At the temperature and pressure conditions appropriate to the polymorphic transformation, solid-solid phase equilibrium exists.

An atmospheric pressure cooling curve for iron illustrating its allotropic transformations is shown in Fig. 5-3. The thermal arrest at 1539°C corresponds to the liquid-solid transition, or the freezing of liquid Fe. The thermal arrests at 1400°C and 910°C correspond to allotropic transformations. Crystallographic examination reveals that

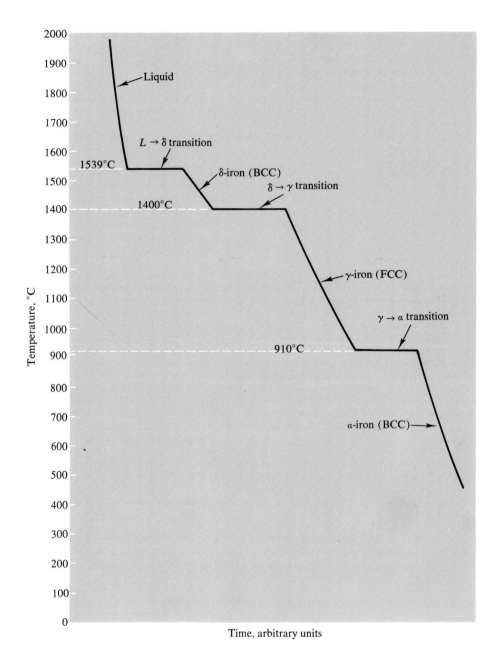

FIG. 5-3 Schematic cooling curve for iron at atmospheric pressure, showing allotropic solid phase transitions.

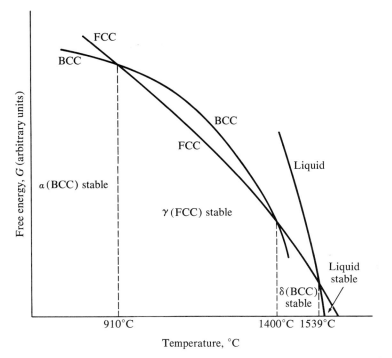

FIG. 5-4 Schematic representation of the free energy-temperature curves for the allotropic phases of iron.

a BCC solid phase, called delta iron or the δ-phase, exists between 1400°C and 1539°C. γ-iron or *austenite*, a stable FCC structure, exists between 910°C and 1400°C, while below 910°C the stable form of the iron is again BCC, called α-iron or *ferrite*.

Small differences in the BCC and FCC crystal free energies cause these transformations to occur. The atomic arrangement in a crystalline solid is determined by the requirement that the crystal free energy be minimal. The free energy-temperature curves of the two allotropes of Fe are illustrated schematically in Fig. 5-4. The lowest curve at any temperature represents the stable phase. The radius of the curvature of a curve is given by its second derivative. The curvature of the G vs T curve is therefore given [2] by

$$\left[\frac{\partial^2 G}{\partial T^2}\right]_p = \frac{-C_p}{T} \tag{5.1}$$

where C_p is the heat capacity at constant pressure. The value of C_p for BCC structures exceeds that for FCC structures; consequently, the radius of curvature for the BCC curve is smaller than that for the FCC curve, and the two will intersect in two places, as illustrated.

Although the phase relations of pure substances are of occasional theoretical interest to the engineer, multicomponent systems are the materials of preemptory significance to him. These multicomponent systems take several forms: gaseous mixtures are of importance in many applications, such as the production of liquid air and the separation of uranium isotopes by gaseous diffusion; vapor-liquid equilibria are necessary in many separation or purification processes in the chemical industry; liquid-liquid mixtures are often encountered in the chemical and nuclear industries; and metallurgists are frequently concerned with the liquid-solid and solid-solid systems. In order to discuss multicomponent phase equilibria, the concepts of solubility and the composition of solutions must first be considered.

Solutions and Solubility

Solution Concept

A clear distinction must be made between solutions and mixtures. The latter term is often used interchangably with the former, although the two are not strictly equivalent. *A solution is a homogeneous mixture of chemically distinct substances in which the atoms or molecules of one substance are uniformly and randomly dispersed throughout the other.* The substance which is present in major proportions is called the *solvent* while the substances or substance present in minor proportions are called *solutes.* Thus, air is a multicomponent solution of oxygen, water vapor, carbon dioxide, and various other trace elements in nitrogen.

In a *mixture* of salt and sugar crystals, two crystalline phases are distinguishable which, with sufficient care and patience, can be picked apart with tweezers. In this case, the mixture is neither homogeneous nor random. Placed in water, this mixture of crystals would dissolve in dilute concentration. Thus, salt and sugar form a mixture as solids, but a solution when both are placed in water.

Air is an example of many gases dissolved in another gas. Water and alcohol are liquids which form solutions in each other. When copper and nickel are melted together they form a liquid solution, and microscopic examination reveals that, after solidification, the solution of the two metals persists into the solid state, forming a *solid solution.*

Two types of solid solutions are identifiable: *substitutional* and *interstitial.* In a substitutional solid solution the solute atoms replace some of the solvent atoms in the regular crystal matrix of the solvent. In an interstitial solid solution the solute atoms are interspersed among the solvent atoms at interstitial positions. This is one source of the interstitialcy defects already mentioned in Chapter 4.

Solubility Curves

In the example of air mentioned above, only a single phase is present. Furthermore, if the oxygen concentration is systematically varied through the entire range of its possible values, all possible combinations of oxygen-nitrogen solutions result in a uniform gaseous mixture or solution. Consequently, we say that these substances are soluble or *miscible* in all proportions, or that each gas is infinitely soluble in the other. This is true of gases in general. With liquid and solid systems, however, complete miscibility is not always observable.

COMPOSITION VARIABLES   Composition is important in the discussion of phase equilibria and solubility. For gaseous solutions, a convenient composition variable is the *mole fraction* $x_i = N_i/N$, where N_i is the number of moles of a given gas in the mixture and $N = \sum_i N_i$ is the total number of moles. (In this and subsequent expressions, the notation $\sum_i N_i$ means $N_1 + N_2 + \cdots N_n$, or a sum over all members of the set.) From the definition of x_i it follows that

$$\sum_i x_i = \frac{\sum_i N_i}{N} = \frac{N}{N} = 1 \tag{5.2}$$

If the gases are ideal (obeying the simple gas law of Eq. (3.5), compositions are frequently given as *partial pressures*

$$P_i = x_i P, \qquad \sum_i P_i = P \tag{5.3}$$

where P is the total pressure. Partial pressures are meaningless for liquid or solid systems.

A composition variable often used for liquids and solids is the *weight fraction* $w_i = m_i/m$, where m_i is the mass of the constituent substance and the total mass is $m = \sum_i m_i$. As with mole fractions,

$$\sum_i w_i = 1 \tag{5.4}$$

Mole and weight fractions are related by the equations

$$x_i = \frac{(w_i/M_i)}{\sum_j (w_j/M_j)} \tag{5.5}$$

$$w_i = \frac{(x_i M_i)}{\sum_j (x_j M_j)} \tag{5.6}$$

where M_i is the molecular weight of the ith species.

A composition variable occasionally used in solid solution work is the *atomic fraction* a_i; this fraction is defined in the same manner as the weight fraction, except that the relative *numbers* of atoms of the ith species and the total number of atoms are used. The number of atoms of the ith species is $m_i N_0/A_i$, where N_0 is Avogadro's number and A_i is the atomic mass of the element. Thus,

$$a_i = \frac{m_i N_0/(\mathbf{m} A_i)}{\sum_j [m_j N_0/(\mathbf{m} A_j)]} = \frac{w_i/A_i}{\sum_j (w_j/A_j)} \tag{5.7}$$

When $A_i = M_i$, Eq. (5.7) reduces to Eq. (5.5), and $a_i = x_i$.

EXAMPLE 5.1

A solution contains 2.5 lb benzene (C_6H_6, $\rho = 0.879$ g/cm³) and 3.4 lb toluene (C_7H_8, $\rho = 0.867$ g/cm³). Compute the composition in terms of weight fraction, weight per cent ($w/0$), mole fraction, and mole per cent ($m/0$) benzene. Assuming these liquids form an ideal[2] solution (no volume change on mixing), compute the volume per cent ($v/0$) benzene present.

Solution:

$$m = \sum_i m_i = 2.5 + 3.4 = 5.9 \text{ lb}$$

$$w_B = \frac{2.5}{5.9} = 0.424$$

$$(w/0)_B = 100 w_B = 42.4\%$$

$$x_B = \frac{0.424/78.11}{(0.424/78.11) + (0.576/92.13)} = 0.465$$

$$(m/0)_B = 100 x_B = 46.5\%$$

$$(v/0)_B = \frac{100 v_B}{\sum_i v_i} = \frac{100}{\rho_B} \bigg/ \sum_i \frac{1}{\rho_i} = \frac{100}{0.879} \bigg/ \left(\frac{1}{0.879} + \frac{1}{0.867}\right) = 49.7\%$$

When the molecules of two different substances are interspersed, the entropy of the overall system is increased and the spontaneous unmixing of the constituents is highly unlikely. Noticeable changes relative to the unmixed constituents can also occur in the enthalpy of the solution, as when NaOH is dissolved in water and generates large quantities of heat. Enthalpy changes may be positive, negative, or zero, depending upon the nature of the materials mixed. When no enthalpy change results from the mixing process, the solution is called *ideal*.

The cumulative effects of both of the above phenomena are ex-

---

[2]In ideal solutions, the volume and energy are equal to the sum of the corresponding properties of the components.

pressed by the free energy function G. If the free energy of a solution is decreased [2] relative to the sum of the free energies of the separate constituents, the solution is stable. If the reverse occurs, the solution is unstable and separates into two or more stable, immiscible fractions, each with lower free energies than the overall total. The latter situation, illustrated schematically in Fig. 5-5, is strongly dependent on temperature. The effect of the enthalpy change on mixing is relatively insensitive to temperature changes, while that of the entropy increase on mixing is very temperature dependent. Consequently, as the temperature level is changed, a given stable mixture may become unstable.

The unstable binary solution of composition z, illustrated in Fig. 5-5, separates into two stable solutions, having compositions w_1 and w_2 which correspond to the minima in the free energy curve.

Any solution of overall composition z' in the shaded region of Fig. 5-5 is unstable, while solutions in the unshaded portions are stable. Solutions with $z < w_1$ have B as solvent and A as solute; in solutions

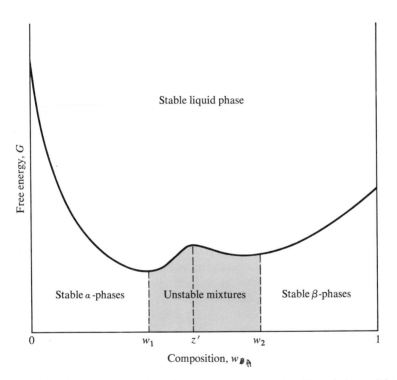

FIG. 5-5 Schematic variation of free energy G as a function of the weight fraction W_B in a binary liquid mixture. Solutions of overall composition z are unstable, and sepatate into two stable solutions of the compositions w_1 and w_2. Any solution of composition z' in the range $w_1 < z' < w_2$ is unstable.

with composition $z > w_2$, A is the solvent and B is the solute. If these two types of stable solutions are designated α and β, respectively, then the shaded region of Fig. 5-5 represents those mixtures which separate into definite proportions of immiscible mixtures of type α and β solutions, having respective compositions of w_1 and w_2. Therefore, regions of $z < w_1$ or $w_2 < z$ are single-phase regions, while the shaded region is two-phase.

If curves such as the schematic one in Fig. 5-5 are obtained at several different temperatures, a great deal of information can be noted about the phase-concentration-temperature behavior of the particular system in question. This information is usually presented as a plot of temperature vs composition, or a *phase diagram*, which contains the composition curves of the stable phases at each given temperature.

The information presented in phase diagrams can be obtained by other, more convenient means than the measurement of free energies in solution. The methods most commonly employed [2] to determine this phase data involve actual phase examination at various temperatures and compositions.

Multicomponent Phase Diagrams

Figure 5-3 illustrates the thermal behavior of a pure substance in terms of a cooling curve. Cooling curves obtained for various mixtures of differing overall composition provide valuable information concerning multicomponent phase relations at various temperatures [3].

In the cooling curve for the pure vapor shown in Fig. 5-1, the vapor to liquid phase transition defines the dew point temperature of the vapor, while the reverse transformation defines the bubble point temperature of the liquid. Similarly, the cooling curve for the liquid to solid phase transition defines the *freezing temperature* [3] of the liquid melt as the temperature at which the first crystals of the solid phase appear in the liquid, and the *melting temperature* [3] of the solid as the temperature at which the first drops of liquid appear upon heating. In the case of a pure substance, these two temperatures are identical. (See Fig. 5-3, where a thermal arrest occurs at 1539°C.) This is not the case, however, with a multicomponent mixture where the dew and bubble points, or the freezing and melting temperatures, are different from one another.

In order to understand this separation of the phase transition temperatures in a multicomponent system, a quantitative relationship between the thermodynamic variables which describe the system will be necessary. This quantitative relation of interest is known as the *phase rule* [3, 4].

THE PHASE RULE   The phase rule defines the relationship which exists between the number of pertinent thermodynamic variables, the number of restrictive relations possible among these variables, and the number of variables which are free for arbitrary specification in a multiphase, multicomponent system.

Consider a region of space filled by a single phase of matter, composed of C components with concentrations expressed in weight fraction units. In order to completely specify the thermodynamic state of the phase, we must define $(C - 1)$ weight fractions (the remaining one being fixed by Eq. (5.4)), and E variables of environmental condition (temperature, pressure, magnetic field, etc.). If we call this number of variables N_{vi}, then

$$N_{vi} = C - 1 + E \qquad \text{(5.8)}$$

Now suppose the region contains P phases, each of the above type. The total number of variables N_v which must be specified in order to fix the thermodynamic state of the multiphase system is evidently

$$N_v = P(C - 1 + E) \qquad \text{(5.9)}$$

If the P phases are in a condition of mutual equilibrium, these various phases have the same free energies. Therefore, for the E environmental variables, $E(P - 1)$ independent conditions of equilibrium must be specified.[3] A similar consideration exists for the concentrations of the C components, which are assumed to be distributed among all phases. Thus, for P phases with C compositions, $C(P - 1)$ composition equilibrium relations must be specified.

The number of free variables, or *degrees of freedom* N_f, of the system is therefore the total number of variables minus the number of restrictive conditions

$$N_f = P(C - 1 + E) - (C + E)(P - 1) = C - P + E \quad \text{(5.10)}$$

In order to use the phase rule (Eq. (5.10)) properly, certain definitions of the meanings of C, P, and E are necessary.

P is simply the total number of physically distinguishable phases within a system. Some examples of a multiphase system are: ice and water $(P = 2)$; water at the triple point $(P = 3)$; a solution of alcohol and water $(P = 1)$; and a mixture of two different solid solutions of iron and carbon $(P = 2)$.

E is the total number of physical variables which are external to a system and characterize the surrounding environment, such as temperature, pressure, magnetic field intensity and direction, etc.

---

[3]The factor $P - 1$ is easily inferred by starting with two phases (one equilibrium relation) and then progressing sequentially to three phases (two, independent, equilibrium relations), four phases (three relations), etc.

The term C, or the number of components in a system, most frequently confuses the student. In the absence of chemical reactions or ionic dissociations, this term represents the total number of separate, identifiable chemical species present in a system. In the case of ionic dissociation in an aqueous solution or dissociation due to a chemical reaction, the correct evaluation of C may be somewhat more subtle [4]. In the example systems used to illustrate P above, $C = 1$ in the cases of ice and water or of water at the triple point. This is so because the only identifiable chemical substance present in these instances is water, H_2O.

Ignoring the slight dissociation of water, in the case of alcohol and water $C = 2$. In the case of the mixture of the two solid solutions of carbon in iron, again only two components are present, C and Fe.

EXAMPLE 5.2

Determine the number of degrees of freedom for: (1) a mixture of ice and water at equilibrium; (2) water at the triple point; (3) the benzene-toluene solutions of Example 5.1; (4) steel at its melting temperature.

Solution:

(1) $C = 1, P = 2, E = 2$ (temperature and pressure), $N_v = 1 - 2 + 2 = 1$
(This means we can fix one variable. At atmospheric pressure, therefore, the equilibrium temperature is fixed.)

(2) $C = 1, P = 3, E = 2, N_v = 1 - 3 + 2 = 0$
(This point is invariant, occurring only at a unique value of temperature and pressure which cannot be specified arbitrarily.)

(3) $C = 2, P = 1, E = 2, N_v = 2 - 1 + 2 = 3$
(If temperature, pressure, and composition were varied, a liquid solution would still result.)

(4) $C = 2$ (Fe and C), $P = 2, E = 2, N_v = 2 - 2 + 2 = 2$

COOLING CURVES   The number of degrees of freedom calculated from the phase rule corresponds to the different types of information on a phase diagram. For example, the invariant triple point in water corresponds to the single point T on the P-T phase diagram in Fig. 5-2, while the univariant ice-water equilibrium corresponds to the freezing point curve also shown in Fig. 5-2. The trivariant benzene-toluene solution is represented by an open area on a phase diagram where not only the pressure and temperature, but also the composition of the solution can be changed within limits without destroying the phase structure of the system.

Although a P-T diagram is a valid phase diagram, many of the phase equilibrium data with which the engineer must deal are presented in the form of a temperature-composition diagram. Much of the information used in this form is determined experimentally at atmospheric pressure, particularly for metallurgical systems or solid systems in general. (This means that $E = 1$ in Eq. (5.10).)

Figure 5-6 shows a set of hypothetical cooling curves for several solutions of Cu in Ni. The thermal arrests at 1455°C and 1083°C in the curves for $w = 0$ and 1, respectively, correspond to the melting temperatures of the pure components. The curves for the intermediate compositions of Cu do not exhibit thermal arrests. From the phase rule, consideration of the melting condition at constant pressure gives $N_v = 2 - 2 + 1 = 1$. Therefore, heat is removed during the freezing equilibrium so that the relative phase amounts and their compositions change, and the one degree of freedom is used up causing the temperature to drop. The change of slope observed in a cooling curve is due to the fact that some of the energy is removed as latent heat, while some is removed as sensible heat of the phases. Hence, the temperatures of freezing and melting are different for a binary solution. The solution freezing and melting points vary systematically from 1083°C to 1455°C in Fig. 5-6.

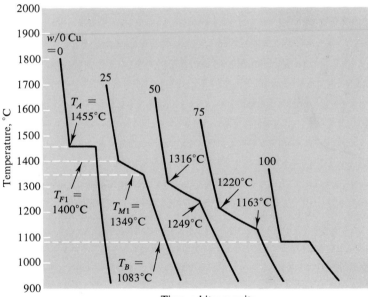

FIG. 5-6 Schematic binary cooling curves for completely miscible solutions of Cu and Ni. Note the separation of the freezing and melting temperatures for binary solutions.

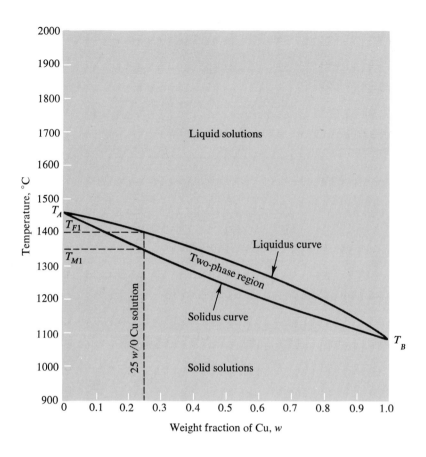

FIG. 5-7   Phase diagram for the Cu-Ni system. The liquidus curve is the locus of all freezing points, and the solidus curve is the locus of all melting points. Details of these curves can be deduced from cooling curves such as those in Fig. 5-6.

The solution freezing points are plotted as ordinates and the corresponding solution compositions as abscissae in Fig. 5-7; the locus of all possible freezing points, called the *liquidus curve*, is thus obtained. A similar plot of the melting points of the various solution compositions is shown in Fig. 5-7 as the *solidus curve*.

Figure 5-7 corresponds to a binary pair, miscible or soluble in all proportions in the solid phase. The bivariant region ($N_v = 2$ from the phase rule) of the diagram above the liquidus curve corresponds to stable liquid solutions, and the bivariant region below the solidus curve corresponds to stable solid solutions. The monovariant region between the two curves represents a two-phase region where liquid and solid phases are in equilibrium.

In Fig. 5-6, the freezing and melting points of binary solutions were separated. *Therefore, it follows that the melting temperature of a solution of one composition corresponds to the freezing temperature of a solution of a different composition.*

THE LEVER RULE Suppose that F lbs of a solution of overall composition z have been cooled to the temperature T, as illustrated in the schematic phase diagram in Fig. 5-8. The isotherm (the solid line at constant temperature T) drawn through the two-phase region is called a *tie line* because it connects the two phase compositions which are in equilibrium at this temperature. That is, T is the *freezing point* of a *liquid* whose composition is x and the *melting point* of a *solid* whose composition is y. *Therefore, since the temperatures between the two phases in equilibrium must be equal, these two phases must be the ones which are in equilibrium at this temperature.* This is true, in general, for any pair of liquid-solid compositions obtained from the intersection of an isotherm with the liquidus and solidus curves.

At temperature T, the total solution of composition z is unstable as a single phase (Fig. 5-5) and separates into the two stable phases

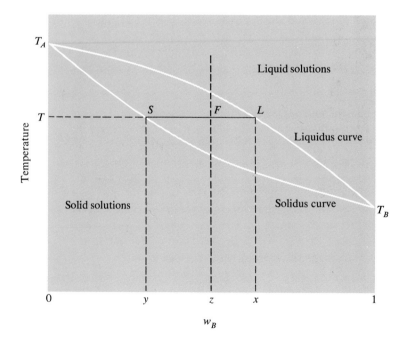

FIG. 5-8 Schematic binary phase diagram for a completely miscible pair, illustrating the derivation of the lever rule.

indicated by the ends of the tie line in Fig. 5-8. The *amount* of liquid phase L of composition x and the *amount* of solid phase S of composition y are related to the total amount of material F by

$$F = L + S \tag{5.11}$$

Similarly, the solute is distributed according to the mass balance relation

$$Fz = Lx + Sy \tag{5.12}$$

Equations (5.11) and (5.12) are the only independent mass balances because the solvent distribution can be obtained by differences.

Substitution of $L = F - S$ from Eq. (5.11) into Eq. (5.12) gives the ratio S/F of the material present in solid form as

$$\frac{S}{F} = \frac{x - z}{x - y} \tag{5.13}$$

Similarly, solving Eq. (5.11) for $S = F - L$ and substituting this into Eq. (5.12), gives the ratio L/F of the material present in liquid form as

$$\frac{L}{F} = \frac{z - y}{x - y} \tag{5.14}$$

A third alternative would have been to substitute Eq. (5.11) directly into Eq. (5.12), and then to solve for the ratio L/S of the liquid to solid phase amounts as

$$\frac{L}{S} = \frac{z - y}{x - z} \tag{5.15}$$

These mass balance relations have simple geometrical interpretations in terms of the phase diagram. The composition differences $x - z$, $z - y$, and $x - y$ are equal to the line segments $\overline{FL}$, $\overline{SF}$, and $\overline{SL}$ in Fig. 5-8, respectively. Thus, Eq. (5.13) is equivalent to the ratio of line segments $\overline{FL}/\overline{SL}$, and Eq. (5.14) is equivalent to the ratio of line segments $\overline{SF}/\overline{SL}$. The following useful rule can, therefore, be observed: *the relative fraction of material present in a particular phase is equal to the ratio of the tie line portion which lies between the overall composition and the phase boundary curve, away from the phase of interest to the total tie line length.* This is the statement of the *lever rule*, to which Eqs. (5.13)–(5.15) are all equivalent. From Eq. (5.15), we obtain $L(x - z) = S(z - y)$, which states that the amount of liquid L times its lever arm $x - z$ equals the amount of solid S times its lever arm $z - y$ and balances about the overall composition point z like the two halves of a teeter-totter. Although the term lever rule is most meaningful when applied to Eq. (5.15), it can also be applied to both Eqs. (5.13) and (5.14), which are actually the more useful computational forms.

Nonideal Binary Solutions

The phase diagrams considered thus far have been valid only for the special cases of ideal or nearly ideal binary solutions in which the components are completely miscible. However, many nonideal solutions are of interest to engineers. When the simple linear additivity of component free energies is not observable, the deviations are reflected in the phase equilibrium behavior of the solutions. For example, positive deviations from ideality, occurring when the solution enthalpy exceeds the linear sum of the component enthalpies, cause an unstable solution situation such as the one depicted in Fig. 5-5.

AZEOTROPIC LIQUID SOLUTIONS   The enthalpy concentration curve for a liquid solution exhibiting positive deviations from ideality is illustrated schematically in Fig. 5-9. An azeotropic solution boils from

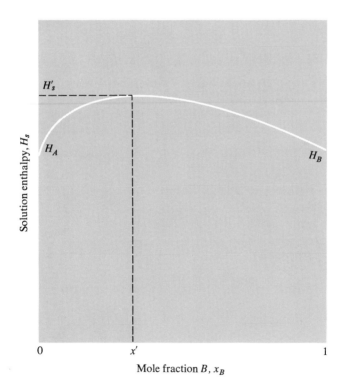

FIG. 5-9 Schematic enthalpy-concentration diagram for a binary liquid solution exhibiting positive deviations from ideality. The enthalpy H'_s corresponds to composition x', and represents a maximum boiling point condition.

the liquid to the vapor phase without changing composition (see Fig. 5-10).

When isobutanol-water solutions are boiled, the composition of each phase tends separately toward the azeotropic composition. Thus, by ordinary distillation procedures, isobutanol-lean solutions can only be enriched to the azeotropic composition and isobutanol-rich solutions can only be stripped to the azeotropic composition. In order to overcome this limitation, a third component must be added to disturb the azeotropic enthalpy relation. This process is known as *azeotropic distillation* [5].

Many liquid systems possess minimum boiling azeotropes corresponding to negative deviations from solution ideality. Azeotropic liquid systems are of considerable importance in the chemical and petroleum industries where much separative treatment of solutions occurs.

EUTECTIC SOLID SOLUTIONS   Solid phase deviations from solution ideality result in limited solubility or immiscibility, with the formation of two or more crystallographically distinct equilibrium phases each having different compositions.

Figure 5-5 could equally well represent an unstable solid solution situation. The stable phase free energy curve at any temperature is the *envelope* of the lowest free energy curves for the possible phases present. This envelope varies as a function of temperature, as illustrated sche-

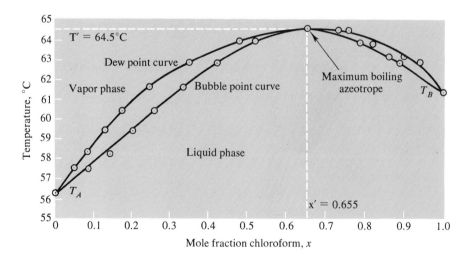

FIG. 5-10   Phase diagram for maximum boiling, nonideal liquid solutions of chloroform and acetone at 760 mm Hg pressure. (From *Chemical Engineer's Handbook, 4th ed.*, by R. H. Perry, *et. al.*, copyright 1963, McGraw-Hill Book Co., Inc. Used by permission.)

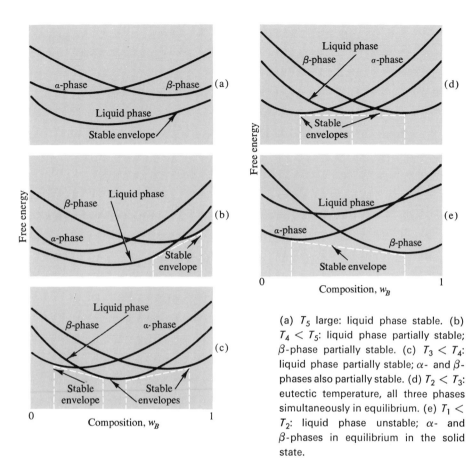

(a) T_5 large: liquid phase stable. (b) $T_4 < T_5$: liquid phase partially stable; β-phase partially stable. (c) $T_3 < T_4$: liquid phase partially stable; α- and β- phases also partially stable. (d) $T_2 < T_3$: eutectic temperature, all three phases simultaneously in equilibrium. (e) $T_1 < T_2$: liquid phase unstable; α- and β-phases in equilibrium in the solid state.

FIG. 5-11 Schematic representation of free energy curves for the nonideal solid solution system possessing a eutectic point.

matically in Fig. 5-11 for a binary system with positive, solid phase deviations from ideality. The corresponding phase diagram is shown in Fig. 5-12, with the five temperatures indicated as isotherms. At temperature T_5, the liquid phase is stable and the system is completely miscible, forming only a single phase. At T_2, the liquid free energy curve is exactly tangent to the dashed straight line which, in turn, is tangent to the minima of the solid α- and β-phase free energy curves. This unique situation corresponds to the simultaneous equilibrium of the three phases. Application of the phase rule at this temperature results in $N_v = 0$, corresponding to the invariant point E on the phase diagram which represents the intersection of the two separate liquidus curves. At this temperature, the transformation reaction $L_E = S_\alpha + S_\beta$ occurs. At T_1, the completely unstable liquid phase no longer exists.

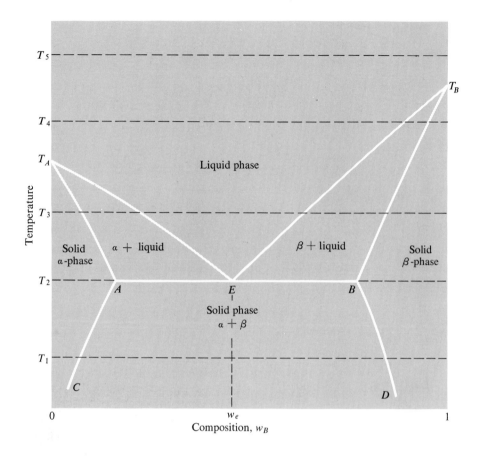

FIG. 5-12 Schematic phase diagram for a binary pair, showing a simple eutectic point. Temperatures T_1–T_5 correspond to the free energy curves in Fig. 5-11.

In Fig. 5-12, the $\overline{AEB}$ isotherm at T_2 connects the three phases A, E, and B which are in equilibrium. The $\overline{AE}$ segment represents the *bottom* tie line of the L-α two-phase region, and $\overline{AB}$ represents the *top* tie line of the α-β two-phase region. The eutectic liquid of composition w_e has a unique freezing and melting point T_2. The cooling curve for a eutectic liquid would have a single thermal arrest at the eutectic temperature T_2, and would appear identical to that of a pure substance. As the eutectic liquid freezes, both α- and β-solid solutions crystallize simultaneously, resulting in a solid structure composed of interspersed crystals of the two phases in alternating sequence. This *lammelar structure*, characteristic of eutectic solid mixtures, is composed of platelike layers that can easily be distinguished under the microscope (see Fig. 5-13).

The student must realize that *this characteristic lammelar structure is not a separate phase in itself, but an intimate physical mixture of two distinct phases.* With sufficient patience, the crystals of one phase could be picked apart from those of the other. The relative amounts of the two phases co-crystallized during the eutectic reaction are given by applying the lever rule to the tie line $\overline{AB}$, using the shorter tie lines $\overline{AE}$ and $\overline{EB}$ in Eqs. (5.13) and (5.14), respectively. With decreasing temperature, the relative amounts of α and β in equilibrium change as the solubility curves $\overline{AC}$ and $\overline{BD}$ change, but the microscopic physical appearance of the eutectic mixture does not alter significantly. That is, once the lamellar structure is formed, it retains its characteristic appearance even though the distribution of the two phases may vary.

EUTECTOID SOLID SOLUTIONS   Although the sequence of free energy curves shown in Fig. 5-11 illustrates liquid-solid transformations, similar curves apply for an all solid system in which one solid phase becomes unstable in favor of two others. This phenomenon is observed in many systems which form interstitial solid solutions. A typical phase diagram

FIG. 5-13   Photomicrograph of a eutectic mixture for the Fe-C system ($\times$150). (Courtesy of W. F. Kindle, United States Steel Corporation. Used by permission.)

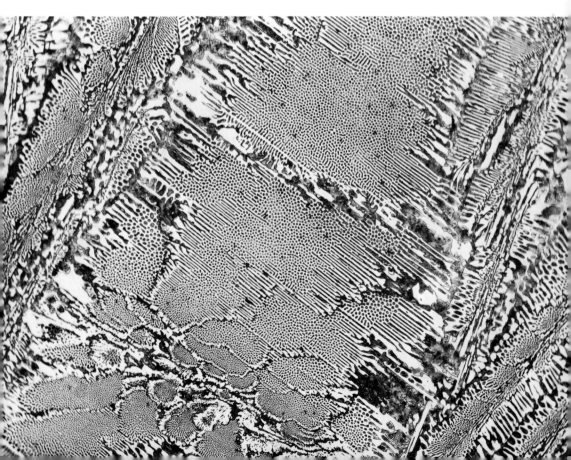

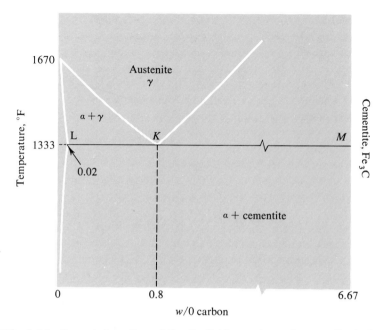

FIG. 5-14   Expanded portion of the Fe-C binary phase diagram, illustrating eutectoid pearlite transformation.

illustrating such a solid-state transformation reaction for the binary Fe-C pair appears in Fig. 5-14. Such a reaction, known as a *eutectoid reaction*, is $S_A = S_B + S_C$. Again, a characteristic lammelar structure is obtained, illustrated for this system in Fig. 5-15. The lammelar solid formed when this particular system undergoes a eutectic reaction has been given the name *pearlite* because of its resemblance to mother-of-pearl under the microscope. Pearlite is an important constituent of steel, and its influence on the properties of this material will be discussed more fully in Chapter 6.

In the case of the pearlite transformation, the phase which disappears at 1333°F is a FCC crystalline solid solution of 0.8 $w/0$ C in Fe called austenite. This transforms into two solid solutions, ferrite (α-iron) and *cementite* (iron carbide). Ferrite is a BCC phase of nearly pure iron, with a maximum of 0.025 $w/0$ C content and is very soft and ductile in comparison with the austenite from which it is formed or the cementite with which it co-crystallizes. The cementite phase[4] is

---

[4]Strictly speaking, cementite is a metastable phase which will decompose to the stable phases of ferrite and graphite under suitable conditions. If silicon is present in an Fe-C mixture in appreciable amounts and is allowed sufficient time at high temperatures, for example, the cementite will decompose or *graphitize*. Normally, however, cementite retains its integrity, and it will be treated here as a stable phase.

an intermetallic compound with an orthorhombic unit cell in which the Fe:C ratio is 3:1. This extremely hard and brittle material is responsible for strengthening the softer ferrite matrix in pearlite. A lever rule calculation shows that a pearlite composition steel has 88 per cent ferrite and twelve per cent cementite at room temperature. This ferrite-cementite distribution for noneutectoid compositions, although different, can easily be determined by applying the lever rule.

Figure 5-16 is a photomicrograph of a steel which has less of a carbon content than pearlite. If a vertical line is drawn at the composition z of this solution in the Fe-C phase diagram, it intersects the α-γ two-phase region, to the left of point K in Fig. 5-14. According to the lever rule, the amount of pearlite composition austenite at the pearlite transition temperature is only $(z - L)/(K - L)$ times the original quantity of steel. The remainder is in the form of α-phase crystals that have already developed because of the sloped solubility curve $\overline{HK}$ which marks the lower limit of stability in the austenite region. These ferrite crystals will not undergo the pearlite transformation since they will already have formed by the time the steel cools to the pearlite

FIG. 5-15 Photomicrograph of pearlite microstructure ($\times$1,000). (Courtesy of W. F. Kindle, United States Steel Corporation. Used by permission.)

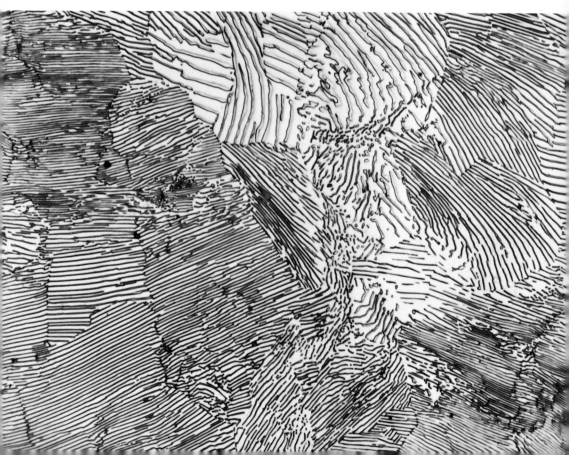

FIG. 5-16 Photomicrograph of a low carbon steel of less than pearlite carbon content ($\times$1,000). White areas are primary regions of ferrite crystals; darker areas are pearlite mixtures of ferrite and cementite. (Courtesy of W. F. Kindle, United States Steel Corporation. Used by permission.)

transition temperature and, therefore, cannot possess the pearlite composition. They remain in their original form as the *primary crystals* of ferrite (the white areas visible in Fig. 5-16). The total amount of ferrite present is thus distributed between these primary crystals and the portion of it which serves as the matrix for the pearlite crystals. (Again it must be emphasized that pearlite itself is *not* a phase, but is a composition of the two phases ferrite and cementite.) Below 1333°F only these two phases of ferrite and cementite exist, but their *physical* distribution may be rather complex depending upon how quickly the solution is cooled from the melt.

Many other solid, binary solution systems undergo eutectoid transformations. Phase diagrams for a number of binary systems are included in Appendix III. The Fe-C system is included here because, although relatively simple, it illustrates the eutectoid reaction very well and is industrially important to the engineer since it forms the basis of the steelmaking industry.

PERITECTIC SOLID TRANSFORMATIONS   In the eutectic and eutectoid transformations, one phase becomes unstable and transforms into two stable phases. Another interesting phase transformation often observed involves the inverse of this reaction, that is, two phases becoming unstable and combining to produce a single phase. This inverted eutectic or eutectoid reaction is illustrated in Fig. 5-17 for the Fe-C system.

This particular reaction is not especially important to this system, since it occurs very near the melting point of the material and is crowded into the very low carbon content portion of the composition scale. However, it is included in the discussion of the Fe-C system here for illustrative purposes as well as for the sake of completeness. Many other systems undergo such transformations in much lower ranges of temperature.

At temperature T', the solid BCC crystalline phase δ-iron of composition D is in equilibrium with the liquid solution of composition B in Fig. 5-17. As heat is removed, both phases become unstable, favoring the peritectic austenite of composition P. Depending upon the overall composition of the solution as this phase forms, one or the other (or both, if $z = P$) of the phases D or B disappear, leaving a two-phase mixture of $\delta + \gamma$ or $\gamma + L$ (or all γ in the special case $z = P$). The peritectic solid which forms is a *single-phase* FCC crystal structure with no lamellar appearance.

THE FE-C SYSTEM   The iron-carbon binary illustrates all three of the nonideal liquid and solid state solution characteristics we have discussed; a complete Fe-C phase diagram is shown in Fig. 5-18. The liquid phase exhibits a eutectic point at E with the composition 4.3 $w/0$ C in Fig. 5-18. The lamellar solid formed from this eutectic liquid is composed of cementite and austenite of the composition

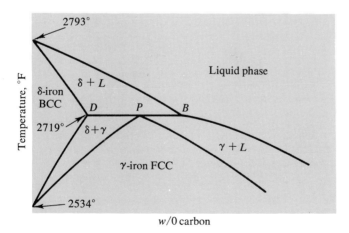

FIG. 5-17 Enlarged portion of the Fe-C binary phase diagram, showing peritectic transformation. (This is a schematic diagram, and is not drawn to scale.)

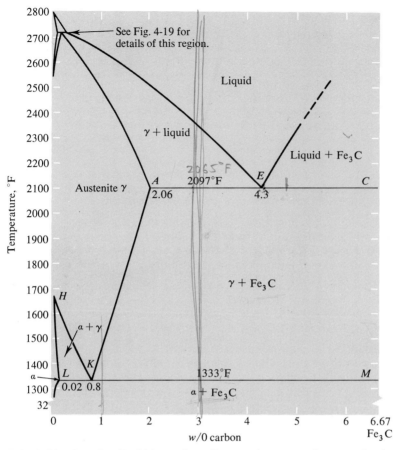

FIG. 5-18 Complete Fe-C binary phase diagram, drawn to scale, except for the peritectic region and point *L*. These two BCC regions have been purposely distorted to make them observable here.

2.0 *w/0* C at point *A*. This eutectic solid has been given the name *ledeburite*. The system also exhibits a peritectic point at *P* and a eutectoid point at *K*.

Because the solubility curve $\overline{KA}$ for the bottom of the austenite region is sloped, the relative distribution of the ledeburite phases changes as the temperature decreases, until the pearlite transition temperature is reached. Then all austenite in the system has eutectoid composition and transforms to pearlite. Because of this transformation, the ledeburite, which still maintains its lamellar structure below the eutectoid temperature, is now called *transformed ledeburite*.

Whether a solution is called steel or cast iron is determined by its carbon content. All solutions of an overall composition of less than

2.0 $w/0$ C are steels, while solutions of any overall composition greater than this are considered cast irons. In general, steels are more ductile and tougher than cast irons which are brittle as a result of a higher proportion of the brittle cementite phase in their microstructure. The physical distribution of the phases at room temperature is extremely important in determining the physical properties of the steel or cast iron. The factors which contribute to this distribution will be the subject of a large portion of Chapter 6.

Nonideal Ternary Solutions

LIQUID-LIQUID EQUILIBRIA   Many important industrial processes rely upon three-component solution nonideality, particularly in the liquid phase. In the chemical and nuclear industries, a great number of difficult separations are accomplished by means of liquid-liquid extraction or azeotropic distillation. These separations are based on the fact that a third component, when introduced into a nonideal, partially miscible, binary liquid system, will generally distribute itself nonuniformly between the two phases that form. Thus, if we wish to obtain the third component and it is in solution with one of the immiscible liquids, adding the second immiscible liquid to the solution will cause the formation of a two-phase system. The soluble component is thus *extracted* from the original solution.

Ternary equilibria are usually presented in terms of an isothermal[5] triangular phase diagram, such as the one in Fig. 5-19 for the water-acetone-benzene system. In this particular example, water and benzene are immiscible, as is evident from the dome-shaped region in the two-phase region at the base of the diagram. The acetone is the distributed component.

The diagram is plotted as an equilateral triangle, with the compositions of the three possible binary pairs running from 0 to 1 along all three sides. This use of the equilateral triangle is based on its geometric property that the sum of the perpendicular distances from any point to the three sides is a constant. If this constant is defined as 100 per cent, the sum of the three perpendicular distances is the total composition of the solution, and each perpendicular distance can represent $w/0$ or $m/0$ of the component that lies at the apex of the triangle, opposite the side to which the perpendicular distance is measured.

The nonparallel, sloped lines running through the two phase region are tie lines which connect solutions in equilibrium at this temperature.

---

[5]It should be emphasized that the ternary phase equilibria discussed are isothermal. Figure 5-19 is an *isothermal* section of a ternary phase diagram. Its appearance would change if the temperature were changed.

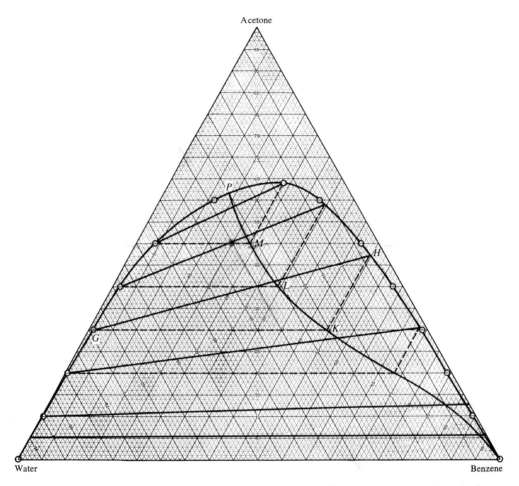

FIG. 5-19 Ternary phase diagram for the water-acetone-benzene system at atmospheric pressure. (From "Effect of Temperature on Liquid-Liquid Equilibrium," by S. W. Briggs and E. W. Commings, *Ind. Engr. Chem.*, **35**, p. 411 (1943).

It is evident that these tie lines converge to the single point P somewhat off center at the top of the two-phase region. This convergence point, called the *plait point*, is related to the tie lines in a unique manner. Consider the $\overline{GH}$ tie line in Fig. 5-19. If lines are drawn through G and H parallel to the $\overline{WB}$ and $\overline{WA}$ sides of the triangle, respectively, they intersect at point K in the figure. Similar points labeled L and M can be plotted for the other, experimentally determined tie lines. The smooth curve, or *conjugate curve*, drawn through these points and the plait point P, defines the locus of all possible tie lines and serves as interpolation between the experimental tie lines.

Many other ternary systems are of industrial importance [4, 5], but further discussion of this interesting topic is beyond the scope of this book.

SOLID-SOLID EQUILIBRIA   Many industrially important metallurgical systems are at least ternary and often higher in multiplicity of components. A particularly important ternary system is Fe-C-Cr, which is the basis of stainless steel. The addition of Cr to the Fe-C binary greatly distorts the austenitic region of the phase diagram of the steel, particularly the eutectoid reaction. The additional Cr drastically modifies the phase structure of the material, making the formation of austenitic stainless steels which are stable at room temperature possible without the occurrence of a eutectoid transformation. Stainless steel exhibits considerable resistance to corrosive chemical attack.

References Cited

[1]   Smith, B. D., *Design of Equilibrium Stage Processes*, McGraw-Hill Book Co., Inc., New York, 1963, pp. 97–98.

[2]   Zemansky, M. W., *Heat and Thermodynamics, 3rd ed.*, McGraw-Hill Book Co., Inc., New York, 1951.

[3]   Findlay, A., *The Phase Rule and Its Applications, 9th ed.* (revised by A. N. Campbell and N. O. Smith), Dover Publications, Inc., New York, 1951, p. 471*f.*

[4]   Ricci, J. E., *The Phase Rule and Heterogeneous Equilibrium*, D. Van Nostrand Company, Princeton, N. J., 1951.

[5]   Smith, *op. cit.*, Chapter 11.

Additional References

Gordon, P., *Principles of Phase Diagrams in Materials Systems*, McGraw-Hill Book Co., Inc., New York, 1968.

Prince, A., *Alloy Phase Equilibria*, American Elsevier Publishing Co., Inc., New York, 1966.

PROBLEMS

(NOTE: Phase diagrams can be found in Appendix III.)

5.1   Use the phase rule to compute the number of degrees of freedom for each of the following cases: (All are equilibrium situations.)
   (a)   An alcohol-water solution with ice-cubes in it.
   (b)   A Cu-Ni-Cr-Mo-C-Fe steel at the eutectoid temperature.

(c)   A thirteen component hydrocarbon oil being boiled in a fractionating column at an oil refinery.

(d)   Water at its triple point (that is, ice, water, and steam simultaneously in equilibrium).

5.2   A certain Fe-C steel mixture of eutectoid composition solidifies in the presence of a magnetic field. Assuming that the field has an effect on the phase equilibrium, compute the number of degrees of freedom when the material is in the austenite phase.

5.3   25 lb of a 1.4 $w/0$ C steel are cooled slowly from 2000°F to room temperature. Compute the lbs of carbide in the pearlite which forms.

5.4   Draw a schematic cooling curve from 1800°F to room temperature for a 60 $w/0$ Cu-40 $w/0$ Ag alloy. Using the phase rule, compute the number of degrees of freedom for each segment of the curve. Draw schematic microstructures for each segment of the curve, showing the phases present.

5.5   For the alloy in Problem 5.4, determine the phases present at 779.4°C after all the material has frozen. For a 75 g sample, compute the amounts of each phase present and the phase distribution.

5.6   For a 60 g sample of a 30 $w/0$ Ag alloy with copper, compute the amount of Ag present in the primary crystals of the β-phase at room temperature.

5.7   Using the Fe-C phase diagram, compute the following for a 75 g sample of a 97.0 $w/0$ Fe alloy at room temperature:
(a)   The weight of ledeburite present.
(b)   The weight of pearlite present in the form of primary crystals.
(c)   The total weight of cementite present.
(d)   The weight of carbon in the ferrite present in the total pearlite.

5.8   Using the Be-Cu phase diagram, compute the following for a 55 g sample of a 3.5 $w/0$ Be alloy at room temperature:
(a)   The total weight of α-phase present.
(b)   The weight of α-phase present in the eutectoid solid.
(c)   The weight of β'-phase which precipitates from the *primary* grains of the α-phase when it cools from eutectoid to room temperature.

5.9   Draw a schematic cooling curve from 3000°F to room temperature for a 1.7 $w/0$ C steel. Using the phase rule, compute the number of degrees of freedom for each segment of the curve. Draw schematic microstructures for each segment of the curve, showing the phases present.

5.10   For a 150 g sample of the alloy in Problem 5.9, compute the weight of pearlite formed and the weight of carbide crystals in the form of primary grains at room temperature.

6

NONEQUILIBRIUM
PHASE
TRANSFORMATIONS

Transformation Rates: Theory

Phase relationships between the multicomponent, nonideal solutions discussed in Chapter 5 apply only when these solutions are in *equilibrium*. That is, the transformations indicated by the phase diagrams for such solutions are the stable reactions observable when a *sufficiently long time* is allowed for equilibration to occur. Allowing sufficiently long time eliminates time as a variable from consideration. However, because time can be a very important variable in many vital industrial processes, we will concern ourselves here with nonequilibrium phase transformations rather than equilibrium phase relations. This means that the concept of *rates of transformation* becomes important.

Nucleation

A phase boundary on a phase diagram indicates the existence of a transformation reaction and implies that the unstable phase transforms into the stable phase at this boundary. In reality, however, the temperature must be reduced somewhat below the equilibrium transformation temperature before the phase transformation can occur at a measurable rate. This decrease in temperature, called *supercooling*, is a very necessary part of the transformation process.

181

Supercooling precedes phase transformation because the process of *nucleation* must occur before a phase can change. Nucleation requires the driving force of a free energy difference in order to occur. Such a force is usually the result of a temperature difference.

VAPOR PHASE NUCLEATION   A nucleus is essentially a small assemblage of molecules capable of sustaining further growth [1]. Although nucleation can be most easily visualized in the vapor phase, many of the following principles can also be applied to liquid and solid nucleation.

As the temperature of the vapor is reduced, the thermal energy of the molecules on the average is decreased, and certain of the lower energy molecules can occasionally be held in contact after they collide by van der Waals forces. If their kinetic energy of translation can be temporarily absorbed by the walls of the container, an impurity particle, or other vapor molecules, a small cluster of molecules can coalesce as a roughly spherical droplet of liquid. If the pressure of the vapor phase is higher than the equilibrium vapor pressure of the liquid phase, there is a greater probability that this initial coalescence will occur. The quantitative measure of this excess pressure is the *supersaturation*

$$S = \frac{p - p^*}{p^*} \tag{6.1}$$

where p is the vapor pressure and p^* is the saturation or equilibrium vapor pressure.

As nuclei form, they may be considered [2] as liquid drops with the same liquid-vapor energy per area and the same free energy per volume relative to the vapor phase that a macroscopic liquid phase would have. Hence, the free energy change ΔG which accompanies the formation of the spherical droplet is

$$\Delta G = 4\pi r^2 \gamma - \tfrac{4}{3}\pi r^3 |\Delta G_v| \tag{6.2}$$

The first term on the right is the energy required to create the interface or surface; the second term is the energy released by the molecules as they condense. The terms γ and $|\Delta G_v|$ represent the free energy of the droplet per unit area and per unit volume, respectively. These energies vary with the size of the nucleus, as illustrated schematically in Fig. 6-1.

Here it is clear that, at first, the positive, surface energy term dominates the free energy change accompanying nucleation; however, as the radius of the nuclei increases, [1] the energy release due to the volumetric association of the molecules predominates, causing a downturn

---

[1] Increasing the radius of the nuclei corresponds to an increase in supersaturation, resulting in a decrease of the temperature below the equilibrium phase transformation value.

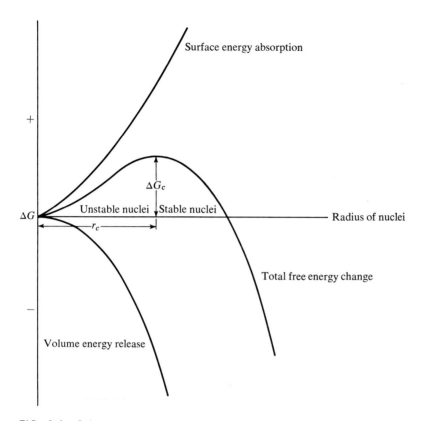

FIG. 6-1 Schematic representation of the dependence of vapor phase nucleation free energies upon nucleus size.

of the overall curve. The radius at which $d\,\Delta G/dr = 0$ is called the *critical radius of nucleation*. Nuclei with less than critical radii are unstable and are broken up by colliding with energetic vapor molecules, while those with radii equal to or greater than the critical value can grow by absorbing impinging vapor molecules.

The critical radius of nucleation is found from Eq. (6.2) by equating $d\,\Delta G/dr$ to zero; thus

$$r_c = \frac{2\gamma}{|\Delta G_v|} \tag{6.3}$$

Substituting Eq. (6.3) into Eq. (6.2) gives ΔG_c as the critical [2] free energy change:

$$\Delta G_c = \frac{16\pi\gamma^3}{3(\Delta G_v)^2} \tag{6.4}$$

The nucleation process can occur at numerous locations in the vapor, resulting in condensation in the form of many small droplets which will eventually coalesce into larger pools of liquid.

LIQUID PHASE NUCLEATION   While the mechanism of the formation of crystal from a melt or solution is not as well understood as is vapor phase condensation [1], the process of nucleation still plays an essential role. The formation of a nucleus in vapor phase nucleation essentially corresponds to a relative increase in vapor density and a simultaneous decrease in thermal energy content. Since it is virtually impossible to increase the density in the liquid phase, the number of intermolecular collisions cannot increase.

The removal of the latent heat of solidification is the predominant factor in the formation of nuclei in a liquid melt. This energy must be absorbed either by the walls of the crucible, by the supercooled melt, or by an impurity particle which may be present. In order for this energy to be transmitted away from the nucleation site, a temperature differential must exist between the nucleus and the melt or the other portions of the system.

Two methods for removing this latent heat of crystallization are commonly encountered in metallurgical operations and characteristically lead to the formation of single crystals or polycrystalline materials. The Bridgman–Stockbarger method [1], used to produce large single crystals, consists of lowering a crucible of molten metal slowly through a temperature gradient. As the bottom of the crucible attains a temperature slightly below the equilibrium freezing point, nucleation occurs and freezing begins with the formation of a crystalline interface between the solid and liquid phases. This interface is advanced by a continuous extraction of heat from the solid through the bottom of the crucible as it is still lowered slowly. The temperature gradient for this system is illustrated schematically in Fig. 6-2. According to Eq. (1.10), the temperature gradient for a fixed heat flow will be determined by the respective thermal conductivities in the two phases, so that the slopes of the two portions of the temperature profile will differ.

In the second method of crystallization, the bulk of the liquid is cooled quickly and a significant amount of supercooling occurs, resulting in the temperature gradient illustrated schematically in Fig. 6-3. The flow of heat away from the interface in both directions produces the dendritic or treelike crystals, illustrated in Fig. 6-4. If any portion of the crystal surface advances ahead of its neighboring regions, it encounters a region of greater supercooling where it grows faster than the remainder of the crystal, forming a spike which extends into the liquid [3]. However, the rapid growth of this spike will result in a large, local release of latent heat, temporarily raising the melt temperature around the spike and terminating its growth. The base of the spike is

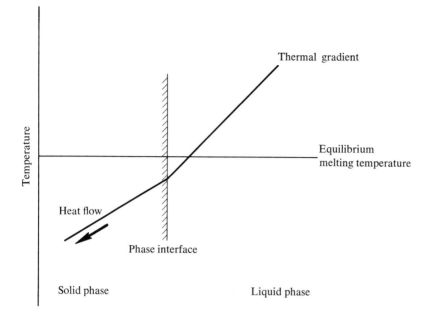

FIG. 6-2 Schematic representation of the temperature gradient for the Bridgman–Stockbarger method of growing single crystals.

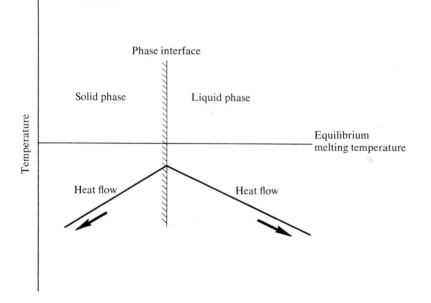

FIG. 6-3 Schematic representation of the temperature gradients in a highly supercooled melt. This process leads to the formation of dendrites.

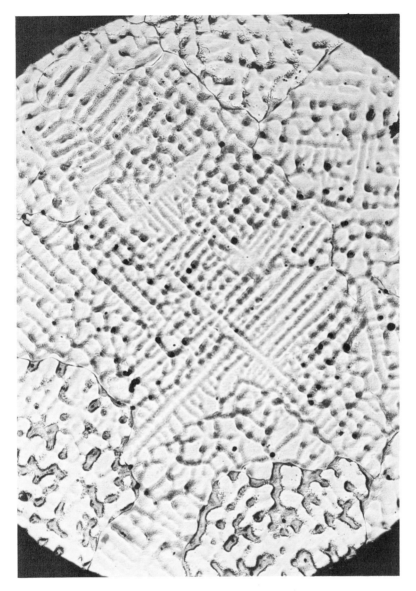

FIG. 6-4 Photomicrograph of a 70% Cu-30% Ni cast alloy, showing dendritic crystal growth ($\times$60). (Courtesy of R. A. Kozlik, The International Nickel Company. Used by permission.)

then in an even more supercooled region, a lateral spike begins to grow from it; this process eventually gives a branched appearance to the crystals. This severe supercooling of the melt causes widespread nucleation and results in the formation of polycrystalline solids.

ARRHENIUS EQN.

Solid Phase Nucleation

As the temperature is lowered in the solid phase, the solubility limitations of certain crystal structures require the formation of new phases. Solid phase nucleation is governed by many of the same principles encountered in the liquid phase, but some important new factors also arise.

The free energy difference needed to drive the phase transformation requires supercooling for its generation and is composed of several terms.

$$\Delta G = \Delta G_s + \Delta G_v + \Delta G_\sigma \qquad (6.5)$$

where, as before, ΔG_s and ΔG_v are the free energy changes associated with the creation of the new surface and the volumetric term.

The term ΔG_σ arises from the strains introduced into the host structure as the solute atoms agglomerate to form the new phase (illustrated in Fig. 6-5). As the solute atoms which were uniformly dispersed in Fig. 6-5(a) congregate to form an incipient nucleus in Fig. 6-5(b), they introduce severe distortion and local strain into the crystal. (ΔG_σ was also negligibly present in liquid phase nucleation.)

Rate Equations

In order to grow, a nucleated phase must be fed additional molecules from the parent phase. The rate at which this growth takes place is governed by many factors, including temperature level, rate of energy extraction, and rate of diffusion of molecules to the reaction site; it is, therefore, considerably important in determining the physical properties of the final material.

THE ARRHENIUS EQUATION   The basic empirical equation observed in describing most rate-type processes is the *Arrhenius equation*

$$r = Ae^{-Q/RT} \qquad (6.6)$$

in which r is the reaction rate, A is a constant sometimes called the frequency factor, Q is an energy of activation, R is the universal gas constant, and T is the *absolute temperature*. The activation[2] energy Q is usually expressed in units of calories per mole or kcal/mole (1 kcal = 1000 cal), and R has the value 1.987 cal/mole °K. The constant A has the same units as r; both vary according to the process being represented.

Because Eq. (6.6) is exponential, its graphical representation is

---

[2]The significance of activation will be discussed more fully in the next section.

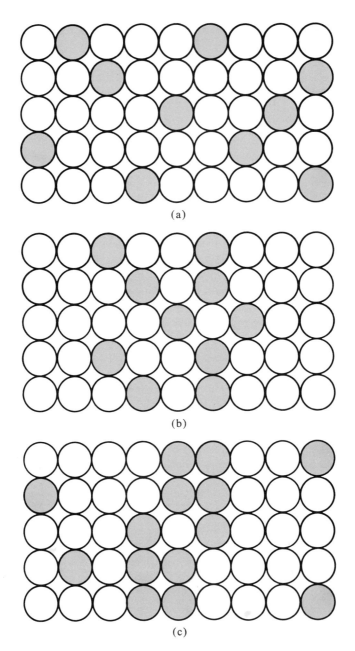

FIG. 6-5 Schematic illustration of the development of ΔG_σ in the solid state nucleation phenomenon: (a) Unnucleated α-phase, with unshaded atoms as its matrix. (b) Nucleated α-phase, with shaded atoms diffusing more closely together to form new β-phase. Here the α-phase is strained, but the β-phase is not yet clearly formed. (c) Separated α- and β-phases. β-phase has shaded atoms as its matrix.

linear if the logarithm of r is plotted as ordinate, with $1/T$ as abscissa. This is evident when the logarithms of both sides of Eq. (6.6) are taken:

$$\log r = \log A - k\left(\frac{Q}{R}\right)\left(\frac{1}{T}\right) \qquad (6.7)$$

Here the factor k is introduced to call attention to the base of logarithms used. If the base e of *natural* logarithms is chosen, $k = 1$; if *common* logarithms are chosen, $k = 1/2.303$. Thus, a plot of $\ln (r)$ vs $1/T$ is linear, with slope[3] $-Q/R$ and intercept $\ln (A)$ at $1/T = 0$. (The Arrhenius equation will be encountered many times in subsequent chapters in connection with the temperature dependence of rate processes. Although not universal for all rate processes, it is very widely observed.)

ACTIVATION ENERGIES In dealing with rate processes, it is essential for the student to obtain a conceptual understanding of the activation energy Q which appears in the exponent of the Arrhenius equation. This particular quantity is of great importance in determining the magnitude of the reaction rate r.

As we saw in previously considering the process of nucleation, it is necessary for a certain amount of supercooling to occur before a nucleus can form. This corresponds to the free energy of the material decreasing below the equilibrium value of transformation. Until the free energies of a sufficient number of these molecules have been reduced below the equilibrium value, nucleation is not possible and no reaction can occur. That is, the rate of reaction is zero and the system is behaving as if some sort of barrier existed that prohibited this reaction from occurring.

The existence of this energy barrier is not unique in nucleation, but is characteristic in all rate processes. Until enough of the reactant molecules acquire the requisite amount of energy to become activated, no measurable reaction occurs. The energy which must be acquired, called the *activation energy*, is illustrated schematically in Fig. 6-6, where the molecular free energy is plotted against a generalized *reaction coordinate*. The latter could represent the interatomic distance (in the case of diffusion), or viscous flow (discussed in Chapter 8), or some characteristic dimension of a molecular group (in the case of a chemical reaction).

The first plateau in Fig. 6-6 represents the free energy of the reactants or starting substances; the second one corresponds to the free energy of the products, or the final state of the system. If the reaction is to proceed at all, the energy of the initial state must be greater than

---

[3]If $\log_{10}(r)$ is plotted or if commercially ruled, semi-logarithmic paper is used, the slope is $-Q/2.303R$.

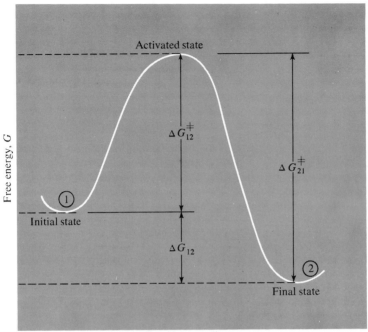

Free energy, G

Reaction coordinate

FIG. 6-6 Schematic representation of reaction rate theory in terms of free energy coordinates. $\Delta G_{12}^{\ddagger}$ is the activation energy for the reaction: state 1 $\rightarrow$ state 2. $\Delta G_{21}^{\ddagger}$ is the activation energy for the reverse reaction.

that of the final state. Before the molecules in the initial state can react, however, they must acquire an amount of energy $\Delta G_{12}^{\ddagger}$ and be activated. This intermediately activated condition is indicated by a peak in the energy curve, called a *barrier*, that separates the initial and final states; it is a highly unstable state from which the molecules rapidly move in one direction or the other as energy is released. Because the energy release is greater in the forward direction here, the probability of a forward reaction naturally exceeds that of a reverse reaction.

This notion of probability may be used in interpreting the energy term Q in the Arrhenius equation. In calculating the distribution of molecular velocities in a gas, Boltzmann found that the probability P of a molecule having the internal energy $E = \overline{E} + \Delta E$, where $\overline{E}$ is the mean thermal energy at the particular temperature T, is

$$P = e^{-\Delta E/kT} \tag{6.8}$$

where k is Boltzmann's constant (1.38×10^{-16} erg/°K). The similarity between this probability factor and the Arrhenius factor is evident. If Q could be interpreted as the excess free energy ΔG acquired by a reacting molecule as it becomes activated (in the supercooling process, for example), then the exponential factor in the Arrhenius equation could be interpreted as the probability factor for that particular reaction.

Since thermal energies are nonuniformly distributed, some molecules possess sufficient energy to overcome the energy barrier in the reverse direction. The net rate of reaction is the difference between the forward and the reverse rates

$$r = n_1 e^{-\Delta G_{12}^{\ddagger}/RT} - n_2 e^{-\Delta G_{21}^{\ddagger}/RT} \qquad \textbf{(6.9)}$$

where n_1 is the population of the initial state, n_2 is the population of the final state, $\Delta G_{12}^{\ddagger}$ is the free energy of activation for the forward reaction, and $\Delta G_{21}^{\ddagger}$ is the free energy of activation for the reverse reaction. The equilibrium condition here is one in which a dynamic balance exists between forward and reverse rates of change. This corresponds to the net rate $r = 0$ in Eq. (6.9), giving

$$\frac{n_2}{n_1} = \exp\left(\frac{\Delta G_{21}^{\ddagger} - \Delta G_{12}^{\ddagger}}{RT}\right) = \exp\left(\frac{-\Delta G_{12}}{RT}\right) \qquad \textbf{(6.10)}$$

where ΔG_{12} is the net difference between the final and initial states in Fig. 6-6. Equation (6.10) expresses [4] the thermodynamic equilibrium constant and shows that the relative populations of two states are given by a Boltzmann probability factor. (These rate equation concepts are applicable in numerous situations and will be referred to frequently.)

Diffusional Phenomena

hnow 191-195

MOLECULAR MECHANISMS   Diffusion is the molecular process of transporting mass from one part of a material to another under the influence of differences in concentration. This is the limiting process in many rate phenomena.

The stable state of a system corresponds to minimum, overall free energy; any deviation from this state results in a shift of the system toward the equilibrium state at a rate proportional to the degree of displacement therefrom. Nonuniformities in the concentrations of multicomponent systems represent free energy deviations and serve as driving forces for mass flow. Ease of diffusion is dependent upon the relative freedom with which the molecules are allowed to move under the influence of a concentration imbalance. Consequently, gaseous diffusion is several orders of magnitude faster than liquid diffusion, which, in turn, is much faster than solid state diffusion.

FLUXES: FICK'S FIRST LAW   Experimental investigations of diffusion processes have shown that the *flux* of matter, rather than the simple mass or molar rate of flow, is the important quantity which must be related to the gradient of concentration. Flux is defined as the rate of flow, divided by the magnitude of the transport area normal to the direction of the flow. The *concentration gradient* $\partial c/\partial x$ represents the spatial rate of change of the concentration of the diffusing species. The relation between the flux J_x and the concentration gradient is given by *Fick's first law of diffusion*

$$J_x = -D \frac{\partial c}{\partial x} \qquad \text{KNOW UNITS} \qquad (6.11)$$

where D is a coefficient of proportionality called the *diffusion coefficient*. (Equation (6.11) is only written for one-directional fluxes (those in the x direction, for example). If diffusion occurs in more than one direction, terms in $\partial c/\partial y$ and/or $\partial c/\partial z$ would have to be included.)

The similarity between Eq. (6.11) and Eqs. (1.10) and (1.12) is obvious. Each relates the flux of a transported quantity to the potential gradient which caused the flux to occur. Consequently, Fourier's law of heat conduction (Eq. (1.10)), Ohm's law of electrical conduction (Eq. (1.12)), and Fick's first law of mass diffusion are mathematically equivalent. This means that if a flux problem can be solved in one type of system, the mathematical solution with the symbols altered can be used in the corresponding flux problem for a different system. This is the basis for the electrical analog computation of many heat and mass flow problems.

FICK'S SECOND LAW OF DIFFUSION   Equation (6.11) alone is useful in calculating fluxes from known concentration gradients, but not in determining either concentration distributions within diffusing systems or computing diffusion coefficients from experimentally measured concentration distribution data. In order to perform such calculations, an expression for the law of conservation of matter in a diffusing system must first be obtained.

In Fig. 6-7, a difference in concentration in the x direction is causing mass to diffuse toward the right in the channel with cross-sectional area S. A mass balance (or statement of the law of conservation of mass) around the volume $dV = S\,dx$, bounded by the walls of the channel and the imaginary planes at x and $x + dx$, takes the form (rate of inflow of mass) $-$ (rate of outflow of mass) $=$ (rate of accumulation of mass in volume). The rate of inflow of mass is $J(x)S$ and the rate of outflow is $J(x + dx)S = J(x)S + (\partial J/\partial x)S\,dx$, because the bounding walls are assumed to be impervious to mass flow. The amount of mass within the volume dV at any instant of time is $c\,dxS$, where c is the concentration. Thus, if there are no chemical reactions taking place,

$$J(x)S - \left[J(x)S + \left(\frac{\partial J}{\partial x} \right) S\,dx \right] = \left(\frac{\partial c}{\partial t} \right) S\,dx \qquad (6.12)$$

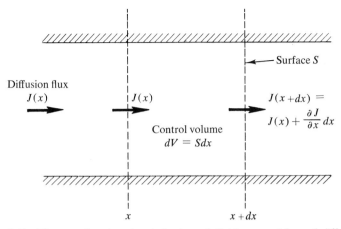

FIG. 6-7 Diagram showing the derivation of Fick's second law of diffusion. It is assumed that the bounding walls are impervious to mass transport and that no chemical reactions occur within the control volume.

which reduces to

$$\frac{-\partial J}{\partial x} = \frac{\partial c}{\partial t} \tag{6.13}$$

because the term $dV = S\,dx$ is independent of time and cancels from the equation. If the one-dimensional form of Eq. (6.11) is introduced into Eq. (6.13),

$$\frac{\partial}{\partial x}\left(\frac{D\,\partial c}{\partial x}\right) = \frac{\partial c}{\partial t} \tag{6.14}$$

which is known as *Fick's second law of diffusion.*

In Eq. (6.14) D is retained inside the differential operator to include the often observed possibility that it is a function of composition. In the special case where D is independent of composition, Eq. (6.14) reduces to

$$D\frac{\partial^2 c}{\partial x^2} = \frac{\partial c}{\partial t} \tag{6.15}$$

which is the one-dimensional form of Fick's second law more commonly encountered.

DIFFUSION COEFFICIENTS   In Eq. (6.11), if the flux J_x is expressed in units of g/cm^2 sec, and the concentration gradient $\partial c/\partial x$ is expressed in $(g/cm^3)/cm$, then the *diffusion coefficient* D has units of cm^2/sec. (In general the physical dimensions of D will be $(length)^2/time$.) The diffusion coefficient represents the magnitude of the flux caused by a concentration gradient of unit magnitude. In Eq. (6.11), only the symbol D is used to indicate the diffusion coefficient. If the diffusion problem is considered in more depth [5], however, it soon becomes apparent that D is not only strongly dependent upon the type of species present, but on the number of components in the solution as well. That is, the diffusion coefficient for material A in itself is different from the diffusion coefficient for A in an AB binary and this, in turn, differs from the diffusion coefficient for the same material in an ABC ternary solution, and so on. This fact makes it complicated to calculate D from theoretical considerations.

The simplest situation which can be considered is the diffusion of a material through itself, called *self-diffusion*. The computation of the

TABLE 6-1

Diffusion Coefficients for Binary Systems*

SOLUTE (A)	SOLVENT (B)	$D_{AB}(cm^2/hr)$	$T(°C)$
Ag	Al	$7.47(10^{-7})$	466
		$1.26(10^{-5})$	573
Ag	Pb	$5.40(10^{-5})$	220
		$3.29(10^{-4})$	285
Al	Cu	$6.12(10^{-9})$	500
		$7.92(10^{-6})$	850
C	Fe(BCC)	$2.27(10^{-6})$	500
		$2.27(10^{-3})$	1000
Cu	Al	$1.80(10^{-7})$	440
		$5.04(10^{-6})$	540
Cu	Co	0.101 ± 0.01	700–900
Mg	Al	$3.96(10^{-8})$	365
		$6.84(10^{-6})$	450
Ni	Cu	$2.56(10^{-9})$	550
		$7.56(10^{-7})$	950
Mn	Cu	$7.2 \ (10^{-10})$	400
		$4.68(10^{-7})$	850
Pd	Ag	$4.68(10^{-9})$	444
		$4.32(10^{-6})$	917
Si	Al	$1.22(10^{-6})$	465
		$3.13(10^{-1})$	697
Sn	Cu	$1.69(10^{-9})$	400
		$1.40(10^{-5})$	850
Zn	Al	$9 \ \ (10^{-7})$	415
		$1.8 \ \ (10^{-5})$	555

*Data primarily taken from *Handbook of Chemistry and Physics, 46th ed.*, Chemical Rubber Company, Cleveland, Ohio, 1965–66.

self-diffusion coefficient, usually designated as D_{ii}, might seem to be of only academic significance. However, in many real isotope solutions, the binary pair diffusion coefficient D_{ij} approximates the self-diffusion coefficient so closely that the latter is used. A significant example of the separation of isotopic systems which are very close in mass is the gaseous diffusion separation of $U^{235}F_6$ from $U^{238}F_6$ in the atomic energy industry.

Liquid and solid diffusion coefficients are temperature dependent in a manner similar to the Arrhenius equation:

$$D = D_0 e^{-Q/RT} \qquad (6.16)$$

where D_0 is a constant, and Q is the activation energy for the diffusion process explained above. The temperature dependence of gaseous diffusion coefficients is markedly different [6] from that of liquids and solids because of the differences in the transport mechanisms of the two types of phenomena. In the gaseous state transport is due to the momentum interchange which occurs when dissimilar molecules collide with one another; in the liquid and solid states transport results when bound molecules become sufficiently activated to overcome the energy barrier created by the packing of their neighbors and move into an adjacent hole or vacancy in the structure. This latter process is governed by a probability rate factor of the Boltzmann type.

Because Eq. (6.16) is exponential, a plot of ln (D) vs $1/T$ results in a straight line, with ln (D_0) as the intercept and slope of Q/R (or $Q/2.303R$, if common logarithms are used). Table 6-1 contains diffusion coefficients for several binary solid systems.

Crystal Growth Mechanisms

If a crystal nucleus exceeds the critical size, the addition of more molecules will cause a net decrease in its total free energy, resulting in crystal growth. Growth mechanisms are one source of dislocations in crystalline materials.

VAPOR PHASE   A certain degree of supersaturation is necessary for vapor phase nucleation to occur and crystal growth to continue. Atoms striking a solid surface are more easily attached at sites where there are many nearest neighbors (see site A in Fig. 6-8), than at planar sites (see site B in the same figure). Continued condensation tends to move the "step" in the close-packed face layer toward the edge where it disappears, necessitating nucleation of a new surface layer if further growth is to occur.

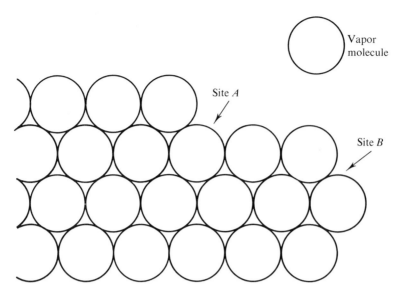

FIG. 6-8 Schematic illustration of the steps growing from the vapor phase in a close-packed crystal face. Site A is a step site, while B is an edge site.

The nucleation of new clusters of atoms on the surface occurs at a rate [2] given by

$$r = \frac{pN_0}{(2\pi kmT)^{1/2}} \exp\left(\frac{-\Delta G_c}{kT}\right) \qquad (6.17)$$

where p is the pressure, m is the molecular mass, N_0 is Avogadro's number, and ΔG_c is given by Eq. (6.4). Using Eq. (6.17), Volmer and Webber calculated [2] the critical nuclear supersaturation for vapor phase nucleation to be 1.25–1.50, a much larger value than is observed experimentally. The theoretical explanation of this phenomenon by Frank and his colleagues [7] led to the postulation that spiral dislocations terminate in the crystal growth face, previously discussed in Chapter 4.

The nucleation process discussed thus far occurs between free molecules in the vapor phase. Such a process is known as *homogeneous nucleation* because it involves only the molecules themselves. When foreign particles or impurity atoms serve as nucleation sites, this process then becomes *heterogeneous*. In the presence of heterogeneous nucleation sites, homogeneous nucleation frequently does not occur because $\Delta G^{\ddagger}$ (homogeneous) $> \Delta G^{\ddagger}$ (heterogeneous).

The Volmer theory of condensation from the vapor to the liquid phase is reasonably good if allowance is made for the fact that [8] some large nuclei may decompose and revert to the vapor state. As predicted by this theory, surface energies required for the fogging of several vapors undergoing adiabatic expansion agree well [8] with experimentally observed values.

LIQUID PHASE   The formation of nuclei from liquids cannot be as readily calculated as nucleation in the vapor state because no correspondingly developed kinetic theory of the liquid state exists. It seems reasonable that the rate of nucleation should involve a free energy change for the formation of the interface and the volume change to occur. However, ΔG_{max} is determined differently for liquids than for vapors. In liquids the surface energy of the nuclei is essentially independent of temperature, while the volume change of free energy is a negative function of temperature. Therefore, as the temperature decreases below the equilibrium value, ΔG_{max} also decreases and the rate of nucleation increases. With decreasing temperature, however, the free energy of activation increases for liquid phase diffusion. This competing process slows down the rate of nucleation and the maximum rate is determined by a balance between the two phenomena. This diffusion controlling phenomenon is responsible for the formation of the glassy state in glass-forming substances [9].

It is difficult to supercool massive samples of a liquid more than a few degrees before the impurities which are always present initiate heterogeneous nucleation. When liquids are dispersed into fine droplets, however, it is possible to obtain many, impurity-free droplets which can be supercooled hundreds of degrees before homogeneous nucleation occurs and solidification results [8].

Because the phenomenon of heterogeneous nucleation from the liquid phase affects the crystallization process so markedly, the resulting solid phase is almost always polycrystalline. This occurs because the liquid nucleates simultaneously at a large number of sites which correspond to the impurity distribution. As each of these nuclei grows larger in turn, crystals form with a random distribution of crystallographic planes relative to one another. As these crystals come together to compete for the last remaining drops of liquid at the end of the solidification process, they are not similarly oriented. The atoms which solidify last must bridge the gap between the nonoriented planes of adjacent small crystals and thus assume a strained position intermediate to the neighboring crystals. The nonoriented neighboring crystals are called *grains*, and the strained, bridging regions intermediate to them are *grain boundaries.* The energy of the grain boundary atoms is higher by the amount of this strain energy than the energy of those atoms in the fully oriented crystalline state. Consequently, the grain boundaries are important in determining many of the physical properties of materials. Their influence on material properties will be considered in subsequent chapters.

The size and distribution of grains in the solid state is greatly affected by both the rate of heat removal from the melt and the degree of supersaturation achieved in the nucleation process. High rates of cooling result in the formation of many small grains and a large amount of grain boundary surface area. Slow cooling with low supersaturation forms a few large grains with relatively little grain boundary area. Therefore, the rate of cooling from the melt significantly affects physical properties through its effect on grain boundary structure.

The growth of spherulitic structures in polymeric materials discussed in Chapter 4 is a process analogous to grain formation in metallic solids and ceramics, and likewise occurs when a melt is rapidly supercooled. The tent-shaped, pyramidal, single crystals previously described are grown from solutions by special techniques, using slow cooling and small supersaturations.

It has already been observed that in growth from the vapor phase the presence of spiral dislocations in the solid phase make low degrees of supersaturation possible. In crystallizing from the melt, the presence of imperfections in the surface of the crystalline phase also assists the nucleation and growth processes greatly, but growth in the crystal surfaces tends to form large, linear steps, rather than a spiral as in the case of vapor.

Figure 6-9 is a schematic representation of the edge of a crystal, showing the steps present. In Fig. 6-9(a), steps A and C are single steps, while B and D are multiple. Steps B and D grow at a slower rate than A and C because multiple steps provide sites where larger numbers of atoms can attach themselves; thus, an attendant, larger energy release locally heats the surroundings, decreasing the degree of supercooling and retarding the growth process. Although this same effect is present in A and C, it is not as pronounced as in the multiple-step regions; single steps, therefore, grow faster than multiple ones and soon overrun them to form even larger multiple steps (see Fig. 6-9(b)). In highly supercooled melts with widespread heterogeneous nucleation, the above mechanism often results in the growth of dendritic crystals previously described.

SOLID PHASE   In many metallurgical and ceramic systems of engineering importance, solubility limitations require that phase transformations occur in the solid phase as the temperature is changed. These processes are governed by considerations similar to those just described above, with various stages of the rate phenomenon controlling under different conditions.

In Eq. (6.5), both the surface energy and the strain energy terms are positive, and $\Delta G_v = 0$ at the equilibrium phase transformation temperature. This means that the rate of nucleation and, therefore, the rate

KNOW GRAINS AND
GRAIN BOUNDARIES (197)

FROM 195

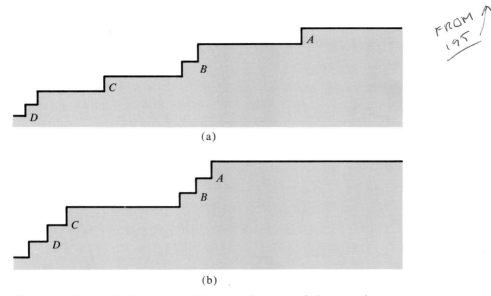

(a)

(b)

FIG. 6-9 Schematic illustration of the development of the growth steps formed from melts in crystal faces.

of transformation are both zero at this temperature, and some degree of supercooling is required. If the supercooling is too great, however, occurring so rapidly that it markedly retards diffusion, metastable solid phases can result. (This is much like the vitrification of a liquid into a glass, which is a metastable state in many cases.)

In a previous discussion of dendritic crystallization from the melt, it was shown that rapid supercooling results in the formation of many small crystals or grains in the solid polycrystalline material; that is, the cooling conditions govern the size of these grains. Because of their increased degree of grain boundary curvature, small grains have higher strain energies and are more unstable than larger grains. Therefore, if the temperature is increased in a solid, atoms diffuse across the boundaries from the smaller to the larger grains. The net result of this diffusion is the growth of the larger grains at the expense of the smaller ones, a phenomenon called *grain growth*. On the average, an increase in the size of the grains results in a decrease in the total amount of grain boundary area and, therefore, in the overall free energy of the solid.   KNOW   WHAT   RECRYSTALLIZATION   IS.

Recrystallization is a phenomenon associated with grain growth. If the temperature is raised on a material strained by plastic deformation, the thermalization of the atoms results in the formation of small, strain-free regions or nuclei [10]; these continue to grow in size at the expense of their strained neighbors, until the entire material recrystal-

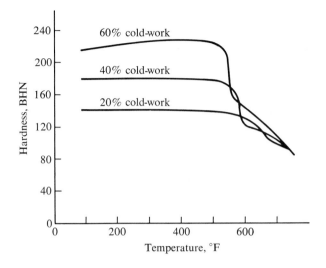

FIG. 6-10   Data for a 65% Cu-35% Zn brass, showing softening due to re-crystallization. The per cent cold work is the total plastic strain to which the sample was subjected prior to heating. (From *Elements of Materials Science, 2nd ed.*, by L. H. Van Vlack, copyright 1964, Addison-Wesley. Used by permission.)

lizes and the strain is removed. Recrystallization usually occurs over the relatively narrow temperature range which is roughly characteristic of the given material, but somewhat dependent upon the prior degree of strain. As a convenient rule of thumb, the recrystallization temperature is given *roughly* by

$$T_{\mathrm{recr}} = 0.4\,T_m \qquad\qquad (6.18)$$

where T_m is the absolute melting temperature of the solid. Figure 6-10 contains curves showing the recrystallization temperatures for a plastically deformed brass. It indicates, in clear terms, the effect recrystallization has in reducing the hardness of a material to its normal, unstrained condition, called *annealing* of the material.

Transformation Rates: Applications

Vapor-Liquid Systems

The applications of rate processes form the basis for numerous industrial operations that are of considerable, practical significance to the engineer. In Chapter 5, for example, we indicated that equilibrium exists between liquids and vapors of differing composition at a given temperature; many separational processes in the chemical industry are based on this equilibrium relation. However, actual operating equipment must be designed to take into account the diffusional transport of matter across the liquid-vapor interface under highly nonequilibrium conditions, and this involves a knowledge of the various factors that affect transport rate processes.

The amount of interfacial area between the phases is one very important factor in such separative processes, and equipment is usually designed to maximize it. This is often accomplished by using counter-current flow of the phases through a column packed with irregularly shaped, ceramic materials with high surface-to-volume ratios; these will greatly divide and disperse the two phases among one another. The volume of the tower determines the total *residence time* of contact which, in turn, governs the total amount of mass to be transferred across the interfaces [11].

The various drying operations are also interesting applications of rate processes involving vapor-liquid equilibria. In these, the diffusion of a liquid (frequently water) from the interstices of a solid to the surface, where it evaporates into an unsaturated stream of air, is dependent upon many complex variables [11]. Again, surface area of contact is important and greatly influences the design of drying equipment. One particularly interesting process is spray drying, where the liquid which is to be divested of undesired water is sprayed through an atomizing nozzle and divided into tiny droplets, thus greatly increasing the surface area for contact.

Equation (6-11) clearly indicates the form of the diffusional transport relation and, therefore, the variables which affect the rate of mass transport. Since the flux J is the rate N divided by the area normal to the transport direction, the mass transport rate is clearly dependent upon the area, the transport coefficient, and the concentration gradient. Therefore, any combination of factors which increases the product of these three variables will also increase the total transport rate. Since the temperature of operation, and thus the value of the transport coefficient, are usually fixed by the equilibrium relations involved in the particular separation, the interfacial area and the concentration gradient are the two variables most frequently manipulated in industrial systems.

Before continuing, it should be made clear to the student that an understanding of both equilibrium and nonequilibrium relations between the phases of multicomponent systems is of importance in industrial applications.

Liquid-Liquid Systems

Like vapor-liquid equilibria, liquid-liquid, multiphase equilibria give rise to many separative processes of especial interest to the nuclear and chemical industries. The process most frequently employed in such cases is liquid-liquid extraction [12], where the difference in the relative solubility of a desired component in two immiscible liquids is used to separate it from an undesirable liquid solvent. This frequently involves contacting an aqueous solution of the material with an organic solvent immiscible in water; from this, it is subsequently easier to remove the desired solute. The latter, organic solvent then *extracts* the solute from the aqueous phase, hence the term *extraction*.

As in the case of the vapor-liquid systems, Eq. (6.11) indicates that the important operational variables here are the interfacial area and the concentration gradient. Therefore, counter-current packed columns are frequently used in liquid-liquid extraction systems [12].

The emulsification of two liquids is an interesting example which utilizes some of the rate process knowledge involving immiscible liquid phases that has already been discussed. This process is frequently used in the pharmaceutical industry to prepare liquid-liquid medicinal suspensions in a nontoxic, inert media where the desired material is insoluble. If two insoluble liquids are violently agitated, the one present in a lesser amount can be divided into myriads of tiny droplets suspended in the other liquid. If they are smaller than the critical nucleation size, these droplets can acquire a stable form. Droplet stabilization is enhanced by the acquisition of electrostatic charges by the tiny droplets, keeping them separate and precluding the formation of supercritical nuclei. Such minute, electrostatically stabilized droplets are called *colloidal droplets*. Thus, an immiscible phase can be suspended indefinitely in another liquid phase (such as water or alcohol) as a *colloidal dispersion* or *emulsion*. Ordinary homogenized milk is a common example of an emulsion.

Except in the above instance of emulsion or colloidal dispersion, all of the vapor-liquid or liquid-liquid processes previously discussed are primarily separative systems, designed to concentrate a desired substance in one or the other of the two phases. This is due to the ease with which liquids and vapors may be handled and transported. In principle, separative processes could also be devised for solid systems, but the physical problems associated with handling and separating the phases are prohibitive.

In addition to the separative uses of vapor and liquid equilibria, the actual physical properties of these phases can be altered by the phase transformations which occur. These property changes, although usually of a secondary importance in vapors and liquids, can often be all-important in the solid state.

Liquid-Solid Systems

Two significant applications of rate process information to the liquid-solid transformation can be found in the electronics industry, where it is important to grow single, ultrapure crystals of materials such as silicon or germanium.

The effect of supercooling on nucleation rate is utilized in the growing of single crystals. The melt is maintained at the melting point, or very slightly below it, so that the nucleation rate is extremely small. The growth permitted to occur is initiated by introducing a small seed crystal which is oriented in the desired crystallographic direction for growth. This crystal is slowly removed from the melt as the material grows onto it, resulting in the formation of a large single crystal of proper crystallographic orientation.

The process for obtaining ultrapure materials such as silicon is one of the few solid-phase, separative operations; it is based on the differential solubility of impurities in the liquid and solid phases. When silicon is nearly pure, the concentration of impurities in the liquid phase exceeds that in the solid phase. A cylindrical piece of silicon is placed in a cylindrical crucible surrounded by an electric induction heater; the heater produces a moving zone of local melting in the silicon, as illustrated schematically in Fig. 6-11. This process is repeated several times until the lower end of the solid eventually becomes highly purified. The upper end can then be sawed off and discarded. When the lower end is remelted and subjected to the single crystal growing technique described above, single crystals of high purity can be obtained, which are necessary for the manufacture of many electronic components.

Solid-Solid Systems

ISOTHERMAL TRANSFORMATION CURVES   Since the rate of growth is inversely proportional to the transformation time which has elapsed, the Arrhenius equation suggests that a plot of $\ln(1/\text{time})$ vs $1/T$ should be nearly linear once nucleation has occurred. During the nucleation process, however, this plot is very nonlinear and, in fact, possesses a maximum which corresponds to the critical nucleation temperature. These results are commonly displayed in terms of a plot of reaction temperature as a function of reaction time (see Fig. 6-12, which is an *isothermal transformation curve* for a steel of eutectoid composition). The transformation which is of greatest interest in the steelmaking industry is the eutectoid pearlite reaction, also displayed in Fig. 6-12. Such curves are also called *C-curves* (because of their shape), or *T-T-T-curves* (time-temperature-transformation curves).

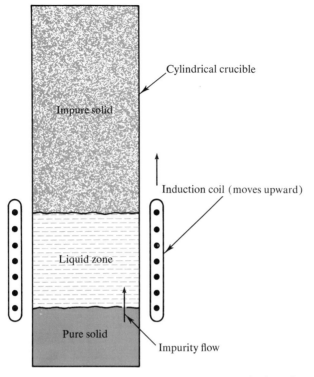

Cylindrical crucible

Impure solid

Induction coil (moves upward)

Liquid zone

Pure solid

Impurity flow

DON'T NEED TO KNOW ZONE REFINING

FIG. 6-11 Schematic illustration of the zone refining method used to obtain ultrapure solid materials. As the induction coil moves upward, the zone of lique-faction follows it. The differential solubility of impurities drives them from the solid phase into the moving liquid zone, and thus purifies the solid.

The region of the graph to the left of the first solid curve and below the eutectoid temperature represents thermodynamically unstable austenite which has not had sufficient time to nucleate. At any given temperature, the time corresponding to the solid curve labeled "start" is the required nucleation time. As the transformation proceeds iso-thermally, more and more of the thermodynamically unstable austenite phase transforms into pearlite, until, at the time given by the inter-section of an isotherm with the curve labeled "finish," the transforma-tion is essentially complete.

The information recorded in the isothermal transformation curves is obtained experimentally, using the *interrupted quench-technique*. In this method, a small sample of the metal is heated above the pearlite transformation temperature and held until all of it is in the stable austenite phase. The sample is then plunged into an isothermal bath, maintained at some lower temperature, and held for a predetermined time; after removal, it is immediately placed in a second isothermal

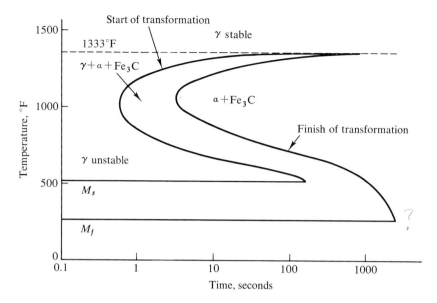

FIG. 6-12 The isothermal transformation curve for a steel of eutectoid composition. (Courtesy of The United States Steel Corporation. Used by permission.)

bath at room temperature to *freeze* the reaction. The cold sample is examined microscopically, and the fraction of the austenite grains which have transformed into pearlite is estimated visually and recorded for the temperature of the intermediate bath and the immersion time therein. These data constitute a single point on the isothermal transformation curve. The above process is repeated at the given, intermediate temperature for several immersion times, until a complete traverse of the transformation region is made. Similar experiments are performed at different, intermediate bath temperatures until sufficient information has been obtained to define the start and finish curves shown in Fig. 6-12. From this accumulated information, curves representing different, intermediate fractional conversions could also be shown, although this is not normally done.

If a noneutectoid sample of steel had been used, a set of curves would have resulted similar to those illustrated in Fig. 6-13 for 1045 steel.[4] The nucleation time for such a material is so short at the *knee* of the curve that it cannot be measured accurately by the interrupted-

---

[4]The American Iron and Steel Institute (AISI) has defined a four-digit numerical system to designate the carbon content in low (less than 1.0 $w/0$) carbon steels. The first two digits indicate the alloy type (10 for simple carbon alloys); the latter two represent the carbon content in hundredths of a per cent. (Thus, 1045 is a 0.45 $w/0$ carbon steel.)

KNOW

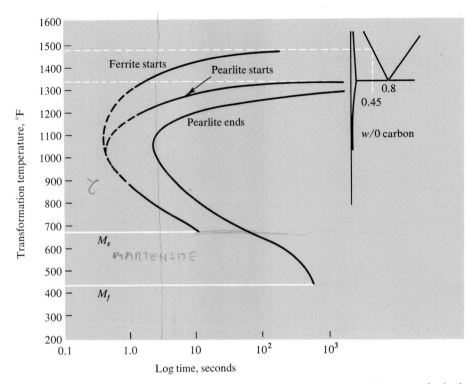

FIG. 6-13 The isothermal transformation diagram for a 1045 steel, showing curves for both primary α-phase and pearlite formations. (Courtesy of The United States Steel Corporation. Used by permission.)

quench technique; this uncertain portion of the curve is dashed in Fig. 6-13. The upper curve in the figure represents the transformation of some of the austenite into the stable primary grains of ferrite; this occurs before the remaining austenite begins the pearlite transformation indicated by the second "start" curve.

The knee of the isothermal transformation curve results from the operation of these competing mechanisms. As the transformation temperature is decreased below the pearlite equilibrium transformation temperature, the degree of supercooling, and thus the free energy difference driving force for nucleation, are increased. As a result, the nucleation rate increases, as evidenced by a decrease in the amount of time necessary for nucleation to occur. Since the grain boundaries are in a higher energy state than the bulk of the grains because of their strained conditions, their atoms are more readily activated, causing nucleation to occur along the austenite grain boundaries. Once the transformation nucleates, it is then sustained by the diffusion of carbon atoms from the austenite crystals to the growing cementite crystals.

As a result of this process, the cementite crystals grow *from* the grain boundaries *into* the body of the grain in dendritic fashion. As the carbon atoms diffuse from the austenite crystals to feed the growing cementite dendrites, the remaining carbon-depleted FCC-iron phase undergoes the FCC-BCC crystallographic shift, converting to ferrite and completing the austenite to pearlite transformation.

As the transformation temperature decreases, the nucleation rate continues to increase and the nucleation time continues to lessen. However, the diffusion of carbon atoms from the austenite crystals to the cementite dendrites is essential to the transformation. This diffusion rate also decreases as the temperature decreases. Consequently, a temperature is eventually reached where the decreasing diffusion rate overbalances the increasing nucleation rate and becomes the controlling factor in the transformation. This change in controlling mechanisms results in an increase in the "start" time for the reaction; the knee in the curve is then observable. Further decreases in reaction temperature result in continuous increases in reaction time.

METASTABLE PHASES: MARTENSITE   The isothermal transformation curves in Fig. 6-12 and Fig. 6-13 indicate that the unstable austenite region is bounded below by a curve marked M_s, where the increasing reaction times caused by diffusion retardation terminated abruptly. Below this temperature M_s, illustrated as a function of carbon content in Fig. 6-14, essentially instantaneous nucleation to a new metastable phase called *martensite* occurs. At the temperature M_s, the free energy of the FCC-austenite crystal is so much higher than that of the intermediate martensite phase that nucleation occurs almost instantaneously at many scattered sites throughout the crystal, not simply at the grain boundaries as in the pearlite transformation.

Crystallographic examination of the martensite grain structure reveals that it possesses a single-phase, body-centered tetragonal (BCT), platelike crystal structure with the *same carbon content* as the austenite from which it formed. The austenite-martensite reaction only requires an expansion in one of the crystal directions to change the FCC into the BCT arrangement without allowing the carbon atoms to diffuse out of the structure. The extreme diffusion retardation exhibited is due to the decreased temperature.

Thus, the martensite crystal is a single-phase structure, as compared with the two-phase, pearlite equilibrium mixture of ferrite and cementite. Martensite is extremely hard and brittle, partially because of the strain introduced into the crystal structure by the lack of carbon diffusion. The presence of the carbon in this distorted crystal greatly restricts atomic movement and likewise ductility.

Because of the short nucleation time periods required for the pearlite

[handwritten margin notes: CARBON CAN'T DIFFUSE OUT AS QUICKLY AND IS TRAPPED. INCREASES HARDNESS.

MUST DELAY PEARLITE FORMATION TO FORM MARTENSITE.]

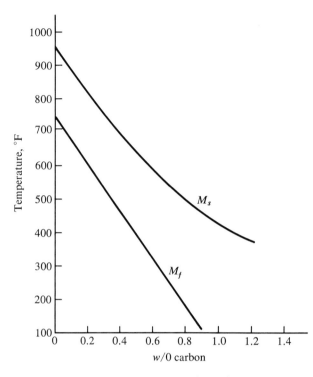

FIG. 6-14 Dependence of martensite transformation temperatures on the carbon content of plain carbon steels. M_s is the start temperature and M_f the finish temperature.

transformation, it is extremely difficult to encourage the martensite transformation to occur in the center of a large sample of steel. This is purely a heat transfer rate phenomenon determined by the thermal conductivity and geometry of a material. Since the hardness of pearlite is considerably less than that of martensite, it is very possible that the profile of hardness as a function of position in a steel bar might appear as shown in Fig. 6-15 for a 1080 steel. In order to achieve uniform hardness throughout the sample, it is necessary to introduce various alloying elements into the steel to interfere with the nucleation process at the grain boundaries. This will delay the pearlite nucleation time and permit the rapid removal of enough thermal energy from the inner regions of the sample for the martensite transformation to occur there also. For example, the addition of significant amounts of molybdenum to the steel will cause this delay, as illustrated in Fig. 6-16 for several concentrations of molybdenum in eutectoid steels [13].

That martensite is an unstable phase is evidenced by the transformation which occurs when it is heated sufficiently to permit carbon diffusion. Because this diffusional transformation occurs very slowly at room temperature, the martensite appears to be stable and is, therefore, a *metastable phase.* Many other systems exhibit martensitic transformations, but the Fe-C transformation is probably the most commonly encountered.

HEAT TREATMENT OF STEELS   Because of the effect of the pearlite microstructure on the hardness of low carbon alloys, it is possible to modify the properties of steels by the correct choice of both the alloying agents and the method of heat treating or cooling the material from austenitic to room temperatures. It has been found experimentally that various grades of pearlitic structures with varying values of hardness are formed as the austenite transformation temperature is decreased. As the degree of supercooling becomes great enough to pass beneath the knee of the isothermal transformation curve, the ferrite-cementite mixture which forms no longer possesses the lamellar pearlite structure.

Below about 1000°F, the rate of carbon diffusion from austenite is

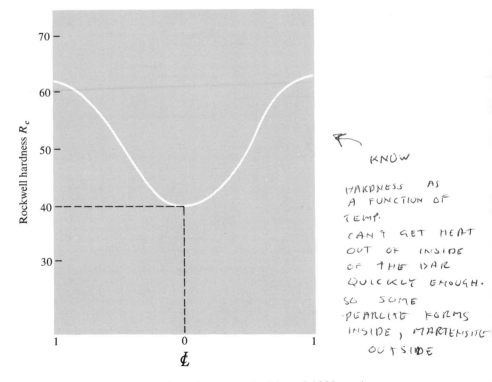

KNOW HARDNESS AS A FUNCTION OF TEMP. CAN'T GET HEAT OUT OF INSIDE OF THE BAR QUICKLY ENOUGH. SO SOME PEARLITE FORMS INSIDE, MARTENSITE OUTSIDE

FIG. 6-15 Schematic profile of hardness in a quenched bar of 1080 steel. The surface is martensite ($R_c = 62$), and the center is pearlite ($R_c = 40$). The abscissa is the fraction of the bar radius.

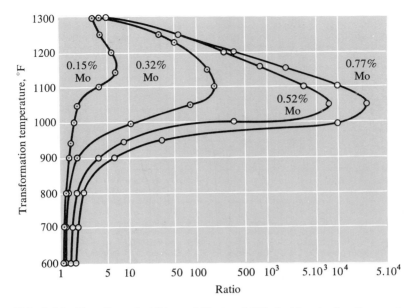

FIG. 6-16   The effect of additions of Mo to a 1080 steel in retarding the onset of pearlite transformation. The abscissae are the ratios of the transition time for the Mo steels to that for plain 1080 steel. (From *Molybdenum Steels, Irons, Alloys*, by R. S. Archer, J. Z. Briggs, and C. M. Loeb, Jr., copyright 1965, Climax Molybdenum Company. Used by permission.)

sufficiently slow that nucleation begins at numerous points in the interior of the grains and along shear planes in the crystal, in addition to occurring at the grain boundaries, thus forming many locations for cementite growth. Unlike pearlite, therefore, the dendrites of cementite extending from the grain boundaries into the crystal body are not the predominant growth form here. The resultant material, called *bainite*, has a much different microstructure than pearlite, as shown in Fig. 6-17 and Fig. 6-18. In bainite, the finely-divided cementite is dispersed more or less uniformly throughout the ferrite matrix, resulting in a much harder substance than pearlite.

In reality, the isothermal transformation diagram is composed of two sets of curves—one for the pearlite transformation and one for the bainite transformation. In the case of an alloy with a low carbon content and no diffusion-retarding alloying agent, such as molybdenum, present, the two sets of curves overlap and merge into the single smooth curve previously illustrated. However, if sufficient molybdenum is added to retard the diffusion-controlled, pearlite transformation significantly, the two sets of curves are distinctly revealed (see Fig. 6-19).

In many applications, bainite is a more useful material than martensite, approaching it in hardness, but lacking much of its brittleness.

KNOW

BELOW 1000° IF PEARLITE IS PREVENTED THEN YOU GET BAINITE. BAINITE NOT AS HARD AS MARTENSITE, BUT ALSO NOT AS BRITTLE.

FIG. 6-17   Photomicrograph of "upper" bainite formed from 1080 steel at 850°F (×2,500). (Courtesy of W. F. Kindle, United States Steel Corporation. Used by permission.)

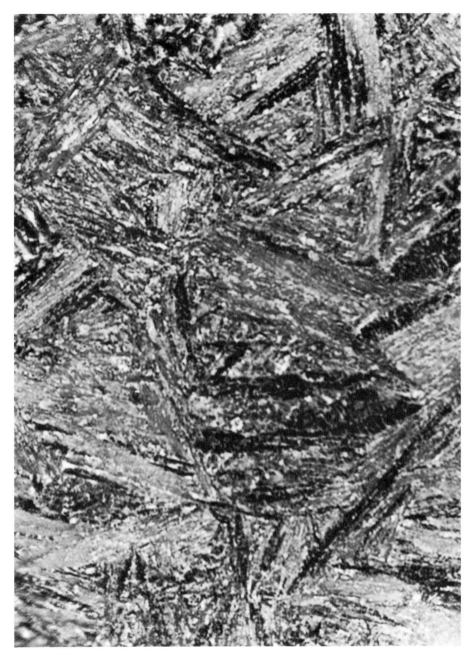

FIG. 6-18 Photomicrograph of "lower" bainite formed from 1080 steel at 550°F (×2,500). Note the difference between the appearance of this microstructure and the one in Fig. 6-17. (Courtesy of W. F. Kindle, United States Steel Corporation. Used by permission.)

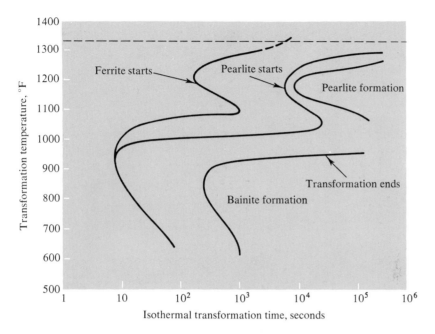

FIG. 6-19 The isothermal transformation diagram for a steel containing 0.43% C, 1.65% Mn, 0.29% Si, and 0.36% Mo, showing the separation of pearlite and bainite curves. (From *Molybdenum Steels, Irons, Alloys*, by R. S. Archer, J. Z. Briggs, and C. M. Loeb, Jr., copyright 1965. Climax Molybdenum Company. Used by permission.)

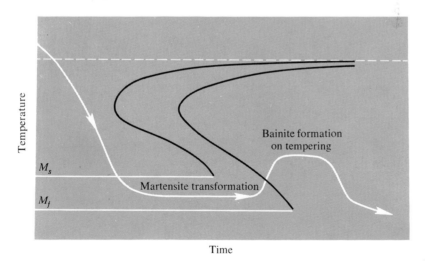

FIG. 6-20 Schematic representation of the cooling path for the martempering process.

This is the result of one common, heat treatment process, called *martempering*, in which the steel is quenched to form martensite and then reheated to a higher, "tempering" temperature where carbon diffusion transforms the metastable martensite to bainite; this relieves most of the strain-caused brittleness from the steel, while retaining most of its hardness. The cooling path for this process is schematically illustrated in Fig. 6-20, and the microstructures of *as-quenched* and *tempered* martensite are shown in Fig. 6-21 and Fig. 6-22, respectively.

If the pearlite transformation occurs very slowly at a temperature

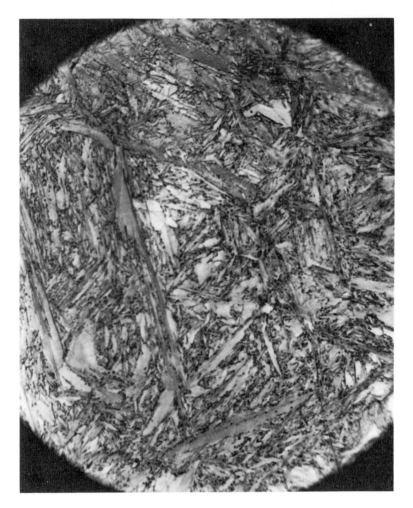

FIG. 6-21 Photomicrograph of an as-quenched martensite microstructure (×1,700). (Courtesy of W. F. Kindle, United States Steel Corporation. Used by permission.)

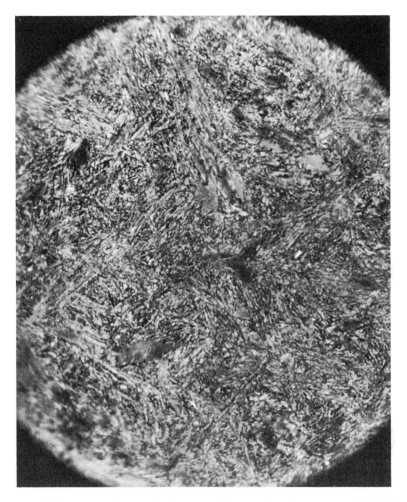

FIG. 6-22 Photomicrograph of a tempered martensite microstructure ($\times 1,700$). Tempering temperature is 450°F; note the similarity to Fig. 6-18. (Courtesy of W. F. Kindle, United States Steel Corporation. Used by permission.)

1300°

just below the equilibrium transformation temperature, still another phase distribution is observable, as illustrated in Fig. 6-23. The long transformation time accompanying this process permits the cementite phase to agglomerate diffusionally into isolated, roughly spherulitic nodules that are dispersed throughout the ferrite matrix. The resulting material, called _spheroidite_, is very soft in comparison with bainite or martensite; the process by which it is formed is called spheroidizing.

KNOW

The hardness obtainable by varying the cooling rates in steel is determined experimentally by means of the Jominy end-quench test

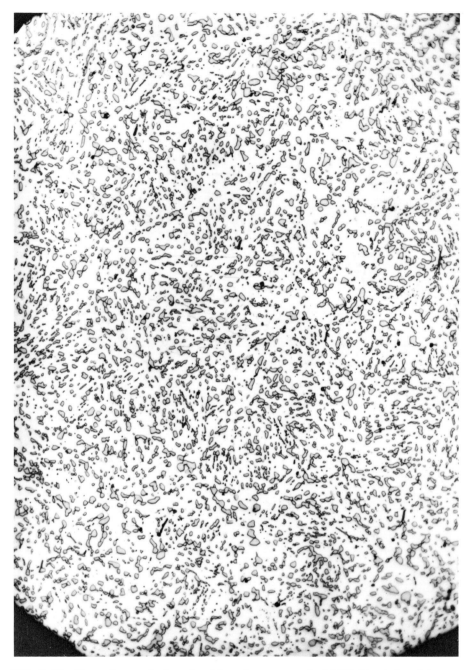

FIG. 6-23 Photomicrograph of a spheroidite microstructure (×1,000). The dark nodules are cementite; white background is the ferrite matrix. (Courtesy of W. F. Kindle, United States Steel Corporation. Used by permission.)

[14], illustrated schematically in Fig. 6-24. A controlled jet of water is directed against the end of a specifically sized bar of the test material that has been previously austenitized by heating. After cooling, the test piece is subjected to a series of hardness measurements at specific points along the length of the bar. The hardness profile thus obtained is directly related to the rates of cooling which occur at each particular location; it is useful in predicting *hardenability*, or the degree to which a steel will harden under given cooling conditions [14]. A set of Jominy curves for various steels is shown in Fig. 6-25. The results of such tests, when obtained for a variety of steels, permit the engineer to select one with the proper hardenability.

In actual practice, many different combinations of conditions are used to obtain the desired properties of a final product. The detailed mechanics of the heat treatment of steels lie outside the scope of this book.

AGE-HARDENING   In many alloys, the solubility boundary of a two-phase region of the equilibrium phase diagram is curved so that cooling the single-phase alloy of a given composition results in the precipitation of a new phase at lower temperatures. This is illustrated for the case of high Al-content, Al-Cu alloys in Fig. 6-26. An alloy containing 96 $w/0$

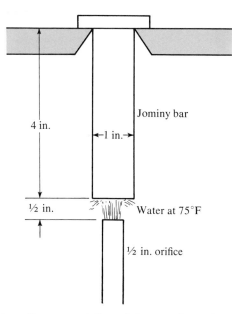

FIG. 6-24   Schematic representation of the experimental arrangement of the apparatus used in the Jominy end-quench test.

Al at 1050°F exists in the single-phase κ-region. If this material is cooled slowly from 1050°F to room temperature, the curved solubility boundary of the two-phase region indicates that some of the θ-phase must precipitate, a process controlled by nucleation and diffusion rates. By properly controlling the temperature cycle, considerable utility can be derived from this precipitation reaction through the process of *precipitation-* or *age-hardening.*

The mechanism responsible for the age-hardening phenomenon involves the degree to which the dispersed or precipitated phase particles resist the movement of a dislocation through the crystal matrix of the host phase. (The mechanisms of dislocation movement will be discussed in detail in Chapter 7; however, the age-hardening phenomenon will be explained here by a result that is developed in more detail in the next chapter.)

When a dislocation line moves through a crystal under the influence of an externally applied stress field, it cannot move through the dispersed solute atoms; it moves around them, encapsulating each such

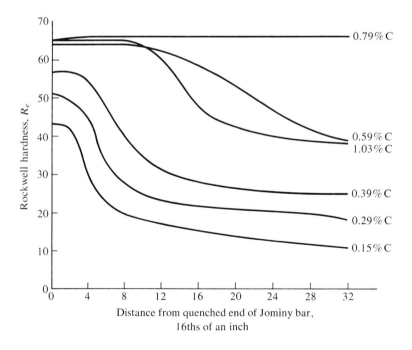

FIG. 6-25 Jominy end-quench curves for a series of Mo-Ni alloy steels (AISI number 4600) of base composition 0.5% Mn, 0.2% Si, 0.2% Cr, 0.25% Mo, and 1.7% Ni. (From *Molybdenum Steels, Irons, Alloys,* by R. S. Archer, J. Z. Briggs, and C. M. Loeb, Jr., copyright 1965, Climax Molybdenum Company. Used by permission.)

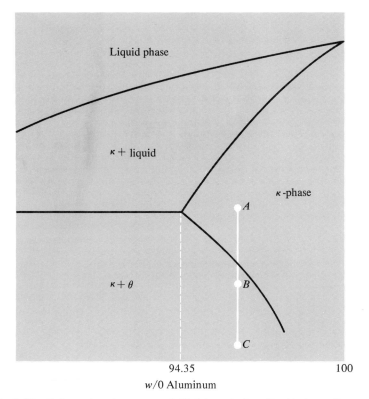

Liquid phase

κ + liquid

κ-phase

κ + θ

A

B

C

94.35   100

w/0 Aluminum

FIG. 6-26 Schematic enlargement of Al-rich end of an Cu-Al phase diagram, illustrating the cooling path for the age-hardening of a κ-phase alloy due to the precipitation of the θ-phase.

atom or group of atoms with a dislocation loop, as illustrated schematically in Fig. 6-27. The resistance to the passage of the dislocation line is proportional to the mean separation distance d between the solute atoms. When the dislocation loops encircle the atoms, the distance is effectively decreased, thus increasing the resistance. This observation now makes an understanding of age-hardening possible.

The thermal treatment path followed in precipitation-hardening involves the steps shown in Fig. 6-26. The material is first heated into the single-phase (or, in this figure, the κ-phase) region (point A) and annealed until all soluble precipitates dissolve. This is followed by a rapid quench (point C), forming an unstable, supersaturated κ-phase which is then reheated to the desired aging temperature (point B).

Because this single-phase mixture is unstable at these lower temperatures, the atoms begin diffusing to the second phase. Initially, tiny platelets on the order of one atom thick are formed. These atoms line

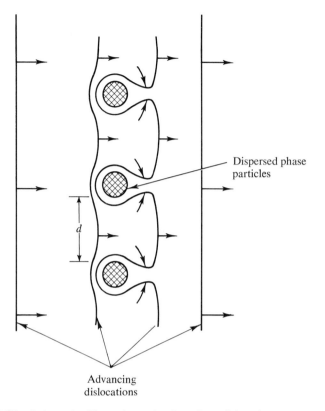

Dispersed phase
particles

Advancing
dislocations

FIG. 6-27 Schematic illustration of advancing dislocations encapsulating a row of dispersed phase particles in the age-hardening process.

up in a one–one correspondence, coherent with the host matrix. However, the disparity in atom sizes, valences, and other factors results in a strong distorted and strained effect, producing the influence on dislocation movement already mentioned. The resulting coherent particles are called Guinier–Preston zones after their discoverers. As is apparent from these tiny monolayer zones, nucleation and agglomeration of the final precipitate of the new phase take place with the passage of time. However, this process, called *overaging*, effectively increases the interparticle distance, relieves the strain introduced by the Guinier–Preston zones, and results in a decrease in the hardness and strength of the alloy.

Since the growth rate of the Guinier–Preston zones, the nucleation of the phase which precipitates from them, and the growth of this nucleated phase are all phenomena dependent upon diffusion, the age-hardening process relies strongly on the aging temperature, as illustrated in Fig. 6-28.

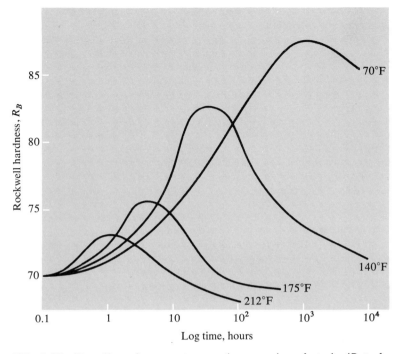

FIG. 6-28   The effect of temperature on the overaging of steels. (Data from E. S. Davenport and E. C. Bain, *Trans. A.S.M., 23,* p. 1061 (1935). Used by permission.)

References Cited

[1]   Keating, K. B., *Chemical Engineering Progress Symposium Series,* **60,** No. 48, p. 15 (1964).

[2]   Volmer, M., and A. Webber, *Zeit. Phys. Chem.,* **119,** p. 277 (1925).

[3]   Weinberg, F., and B. Chalmers, *Can. Jrl. Phys.,* **29,** p. 382 (1951); **30,** p. 488 (1952).

[4]   Glasstone, S., K. J. Laidler, and H. Eyring, *Theory of Rate Processes,* McGraw-Hill Book Co., Inc., New York, 1941.

[5]   Hirschfelder, J. O., C. F. Curtiss, and R. B. Bird, *Molecular Theory of Gases and Liquids,* John Wiley and Sons, Inc., New York, 1954.

[6]   Present, R. D., *Kinetic Theory of Gases,* McGraw-Hill Book Co., Inc., New York, 1958.

[7]   Frank, F. C., N. Cabrera, and W. K. Burton, *Nature,* **163,** p. 398 (1949).

[8] Sinnott, M. J., *The Solid State for Engineers*, John Wiley and Sons, Inc., New Ycrk, 1961, Chapter 5.

[9] Azaroff, L. V., *Introduction to Solids*, McGraw-Hill Book Co., Inc., New York, 1960, p. 197.

[10] *Ibid.*, p. 161.

[11] Treybal, R. E., *Mass Transfer Operations*, McGraw-Hill Book Co., Inc., New York, 1955.

[12] Treybal, R. E., *Liquid-Liquid Extraction*, McGraw-Hill Book Co., Inc., New York, 1960.

[13] Archer, R. S., J. Z. Briggs, and C. M. Loeb, Jr., *Molybdenum Steels, Irons, Alloys*, Climax Molybdenum Company, New York, 1965.

[14] Guy, A. G., *Physical Metallurgy for Engineers*, Addison-Wesley, Reading, Mass., 1962, p. 308*ff.*

PROBLEMS

6.1   Using the theory in Chapter 3 for computing the average internal energy of gas molecules, compute the probability that a gas molecule has an internal energy equal to 3, 5, and 10 times the mean.

6.2   In a certain vapor phase reaction, the interfacial energy is 600 ergs/cm^2 and the volumetric free energy change is -125 cal/cm^3. Using the Volmer–Webber theory, compute the critical nucleus size and the critical nucleation free energy change.

6.3   In a certain polymorphic transformation, the interfacial energy is 500 ergs/cm^2, and the values of ΔG_v are -100 cal/cm^3 at 1000°C and -500 cal/cm^3 at 900°C. Assuming that the Volmer–Webber theory applies for this solid state transformation, compute the critical radius of nucleation and the free energy change of the reaction at each temperature.

6.4   The nucleation of the CuAl$_2$ alloy from a supersaturated Cu solution is observed at a temperature of 212°F after 3 min and at 70°F after 3 hrs. At what temperatures will this reaction require 3 days, 4 days, and 5 days for initiation?

6.5   In a certain reaction, the following nucleation times are observed: 13.8 sec at 427°C, 0.316 sec at 527°C, and 0.001 sec at 727°C. Determine the activation energy for this reaction.

6.6   Draw a schematic, free energy diagram for a transformation reaction; indicate thereon the influence which can be expected from a catalyst. How does this catalyst influence the forward and reverse rates of the reactions?

6.7   The diffusion coefficients of carbon in alpha titanium at different temperatures are: $D = 2.0(10^{-9})$cm^2/sec at 736°C, $5.0(10^{-9})$cm^2/sec at 782°C, and $1.3(10^{-8})$cm^2/sec at 835°C. From these data, compute D_0 and Q/R. Compute D at 750°C and 800°C.

6.8   Compute the change in D_0 occassioned by an error of 0.5% in the determination of Q in Problem 6.7.

6.9   The diffusion coefficients of aluminum in silicon at various temperatures are: $3.11(10^{-10})$ at 1380°C, $7.1(10^{-11})$ at 1300°C, $4.1(10^{-11})$ at 1250°C, and $1.74(10^{-11})$ at 1200°C, with all values measured in cm^2/sec. From these data, determine D_0, Q/R, and the values of D at 800°C, 900°C, 1000°C, and 1100°C.

6.10   Using the data of Problem 6.9, determine the error which would result if the value of the diffusion coefficient at 1275°C was determined by simple linear interpolation, rather than by the logarithmic relation.

6.11   A block of zinc and a block of copper are placed in close contact in a constant temperature oven at 600°C and left until they interdiffuse somewhat. When examined, the copper block shows a zinc concentration profile, measured from the original interface and given by $c/c_0 = 1 - \sin(\pi x/2L)$, where x is the distance from the interface into the Cu block, $L = 1$ cm is the thickness of the Cu block, c is the concentration of Zn in g/cm^3 at any position, and c_0 is the concentration of Zn in the Cu block at $x = 0$. If $c_0 = 38$ w/0, compute the flux J_x in g-atoms Zn/cm^2 sec at $x = 0.25L$ and $x = 0.75L$. For this calculation, assume that the structure is the FCC α-phase, with Cu as the solvent and Zn as a substitutional solute.

6.12   A small piece of 1045 steel is heated to 1600°F, quenched to 1100°F, held at this temperature for 6 sec, and then quenched to room temperature. What phases are present?

6.13   What phases would be present if the sample in Problem 6.12 had been held for 10 sec at an intermediate bath temperature of 700°F?

6.14   What would be the nucleation time for the pearlite formation of a eutectoid composition steel containing 0.15 w/0 Mo? 0.30 w/0 Mo?

6.15   The steel for which the isothermal transformation diagram is given in Fig. 6-19 is quenched from austenitic temperatures to 1000°F and held for 10, 50, 100, 500, and 1000 sec. What phases are present at each of these times?

6.16   The quenched end of a Jominy bar is 50 R_c. What is its carbon content?

6.17   Draw a hardness traverse for a bar of 1040 steel 1.5 in. in diameter.

6.18   Estimate the hardness expected at the centerline if a 2.0 in. diameter bar of the steel in Problem 6.17 were quenched rapidly.

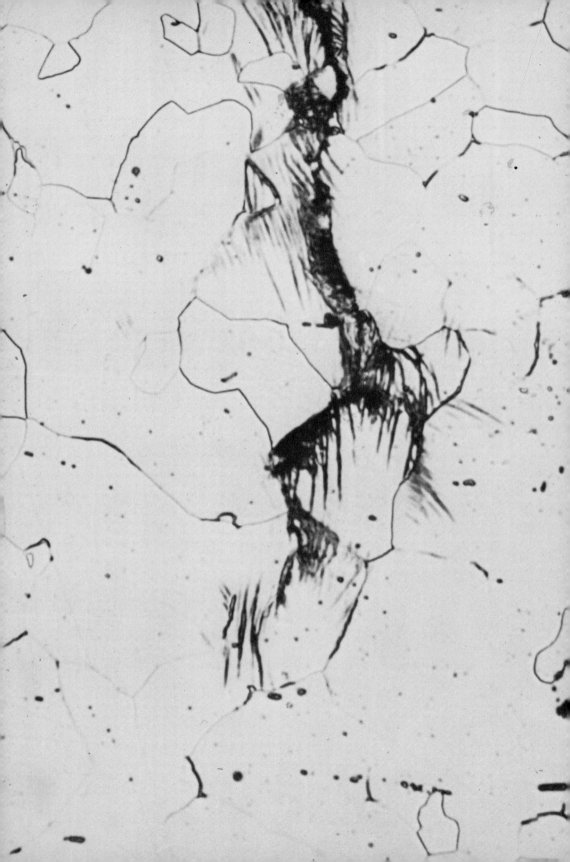

PART  II

MACROSCOPIC PROPERTIES OF MATERIALS: RESPONSE TO ENVIRONMENTAL CONDITIONS

THE RESPONSE OF MATERIALS TO STATIC MECHANICAL FORCES

In the preceding chapters, some of the molecular properties of matter and the relationships governing phase formation and distribution were discussed. In the next several chapters, the various ways in which materials respond to different environmental variables, such as mechanical forces, temperature fields, electric and magnetic fields, and corrosive chemical environments, will be considered in some detail. The practical materials engineer must be able to prudently apply the fundamental physical chemical principles in Part I to the areas which will now be considered in Part II.

Elastic Deformation

Elastic Moduli

In Chapter 3, a theoretical derivation of Young's modulus was presented for a perfect ceramic crystal of the Na^+Cl^- type, and a computation of E for NaCl was made. The E of 1.08×10^7 psi was then observed to be somewhat higher than the experimental value of 7.89×10^6 psi. This discrepancy arose because the requirement that

227

the interatomic distances must be shortened in the transverse direction was ignored when Eq. (3.14) was derived. Such a shortening must occur for atomic equilibrium to be maintained in the plane transverse to the direction of the stress. The negative ratio of transverse to axial strains,

$$\nu = \frac{-\epsilon_t}{\epsilon_a} \tag{7.1}$$

called *Poisson's ratio*, is a characteristic material parameter which measures this effect.

Generalization of Hooke's Law

The above results indicate that more than one parameter is required to describe the mechanical response of even simple ionic crystals. Any general stress can always be resolved into three components, acting on surfaces oriented normal to some conveniently chosen set of reference axes. Each of these three component stresses, in turn, will have components which act normal or parallel to the faces of a unit cell. The nine components thus obtained for a stress field can be related [1] to nine strain components by a generalized form of Hooke's law

$$\sigma_{ij} = \sum_{k,l=1}^{3} C_{ijkl}\epsilon_{kl} \tag{7.2}$$

where the 81 constants C_{ijkl} are elastic moduli. In this equation, $i = 1, 2, 3$ corresponds to $x_1 = x$, $x_2 = y$, $x_3 = z$ of an ordinary, rectangular, Cartesian coordinate system. Fortunately for the engineer, however, all 81 of these moduli are not required to describe the mechanical responses of engineering materials. In fact, considerations of symmetry usually reduce this number to two or three moduli.

For cubic symmetry the number of moduli is reduced to three, and for hexagonal symmetry to five. The most general case of a triclinic crystal requires a maximum of 21 of the possible 81 constants for complete specification, and the engineer most often deals with materials of either cubic or hexagonal symmetry.

In addition, an engineer usually assumes that his material is *iso-tropic*, that is, that both the moduli and the physical properties of a material will be the same, regardless of the spatial direction in which each is examined. However, *single* crystal isotropy is the exception rather than the rule, and pure single crystals are rarely isotropic. The engineer usually deals with polycrystalline materials in which the random grain orientation disperses the anisotropy inherent in each single crystal.

Under the assumption of isotropy, the number of cubic moduli is reduced to two, Young's modulus and Poisson's ratio being the most

TABLE **7-1**

Elastic Moduli for Selected Materials

MATERIAL	$10^{-6} E$, psi	ν
Al	9.9	0.34
Be	43.0	0.01
Cu	18.0	0.35
Au	11.4	0.42
Fe	28.5	0.28
Pb	2.3	0.45
Mg	6.5	0.33
Ni	30.0	0.31
Pt	22.0	0.39
Ag	11.0	0.38
Ti	16.0	0.34
W	52.0	0.27

commonly used. Typical values for these moduli are tabulated in Table 7-1 for several metals.

Interrelation of Elastic Moduli

Although only two moduli are required to characterize an isotropic cubic material, four moduli are commonly defined; the engineer must be familiar with these and with their interrelations.

Because Young's modulus involves changes in both shape and volume of the material, the notion of a transverse strain, as indicated by Poisson's ratio, had to be introduced. (It is possible to derive a modulus which involves only changes in volume.)

Consider the unit cube[1] in Fig. 7-1. The force F causes a change in volume, the new volume V' being

$$V' = 1 + \epsilon'_v = (1 + \epsilon)(1 - \nu\epsilon)^2 \qquad (7.3)$$

where ϵ'_v is the volumetric strain $\Delta V/V_0$, ϵ is the axial linear strain, and Poisson's ratio has been used to relate the strains in the two transverse directions to the strain in the axial direction. Thus, the volumetric strain is[2]

$$\epsilon'_v = \epsilon(1 - 2\nu + O(\epsilon^2)) \qquad (7.4)$$

$$\text{or } \epsilon'_v = \epsilon(1 - 2\nu) + O(\epsilon^2)$$

---

[1]The unit cube is used only for convenience. The shape of the volume is completely arbitrary, and choosing the unit cube does not reduce the generality of the result. The student should convince himself of this by repeating the derivation for some other shape.

[2]The notation $O(\epsilon^2)$ indicates a sum of terms involving ϵ raised to the 2nd or higher powers. Since these terms are negligible, the complete form is unimportant.

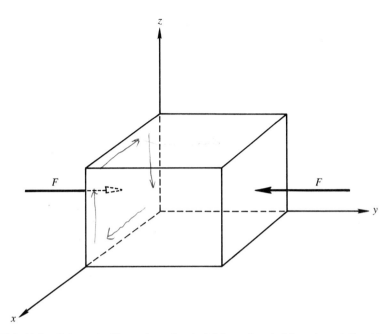

FIG. 7-1 Schematic illustration of uniaxial force for deriving the relationship between Young's modulus and the bulk modulus.

If it is assumed that the compressive stress σ is proportional to ϵ, then Eq. (7.4) leads to the linear approximation

$$\sigma = E\epsilon = \frac{E}{1 - 2\nu}\,\epsilon'_v \tag{7.5}$$

where E is Young's modulus. For the state of equal stresses in all three axial directions, called hydrostatic stress, the total volumetric strain ϵ_v is the sum of the strains due to stresses in each direction given by Eq. (7.5). Thus,

$$\sigma = \frac{E}{1 - 2\nu}\left(\frac{\epsilon_v}{3}\right) \tag{7.6}$$

When dealing with compressive stresses, the stress is conventionally related to the volumetric strain ϵ_v through the definition of the *bulk modulus K*. This modulus, defined by the equation

$$\sigma = K\epsilon_v \tag{7.7}$$

leads to the relation

$$K = \frac{E}{3(1 - 2\nu)} \tag{7.8}$$

A fourth elastic modulus, the *shear modulus*, is also definable. Consider the shear deformation illustrated schematically in Fig. 7-2.

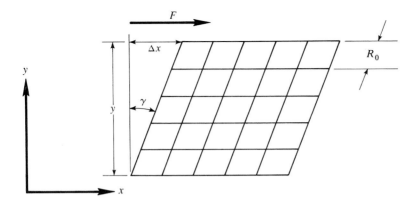

FIG. 7-2 Schematic illustration of the quantities used in defining the shear modulus G.

If displacements Δx are small compared with crystal dimensions y, the relationship between the shear stress $\tau = F/S$ (S being the tangential area of force application) and the shear strain $\gamma = \Delta x/y$ is[3]

$$\tau = G\gamma \qquad (7.9)$$

where G is the shear modulus. Since Δx is small, the change in R_0 is essentially zero and the deformation is nearly constant in volume. Therefore, linear shear deformation only involves changes in shape. The existing relationship between G, E, and ν is [2]

$$G = \frac{E}{2(1 + \nu)} \qquad (7.10)$$

In view of Eq. (7.5), the expression derived for E in Eq. (3.14) should have included the factor $(1 - 2\nu)$ to account for the change in shape caused by the transverse strain. For the NaCl crystal considered in Example 3.6, the value of ν is 0.16 [3]. If the E computed in this example is multiplied by $1 - 0.32 = 0.68$, $E = 7.35(10^6)$ psi is obtained; this is much nearer the experimental value.

Influence of Molecular Radii

A precise molecular analysis of deformation and strain is practically impossible due to the partial ionic character of bonding and the lack of a satisfactory mathematical representation of the metallic bond. However, Eq. (3.15) can serve as a guide in qualitatively evaluating the

---

[3] If the angle γ is sufficiently small, $\sin \gamma \doteq \gamma$. We have made this approximation here.

influences which the positions of elements in the periodic table can have on elastic moduli. By analogy, it may be written that

$$K = \frac{Ce^2}{R_0^4} \tag{7.11}$$

for any material, where C is an undetermined constant which may be a complex function of the electronegativity[4] of the element and the geometry of its crystal habit.

Taking logarithms of both sides of Eq. (7.11) gives

$$\ln(K) = \ln(Ce^2) - 4\ln(R_0) \tag{7.12}$$

which indicates that a plot of $\log(K)$ as ordinate against $\log(R_0)$ as abscissa should be linear with a slope of -4, if C can be assumed to be constant within a periodic group. Because of the approximate nature of this relation, it is not universally true for all elements. Although the inverse fourth power dependence of K on R_0 is observed for some of the groups in the periodic table, it fails for the heavier transition metals due to their increasing degree of partial ionic bonding character. Values for C which allow Eq. (7.11) to fit the data for these periodic groups are $C = 0.5$ (Group I), $C = 1.6$ (II), $C = 2.0$ (III), and $C = 1.3$ (IV).[5]

If the values of C followed the trend indicated by the factor $(n - 1)z^2$ in Eq. (3.15), a fourfold increase in C from Group I to Group II and a ninefold increase from Group I to Group III would be expected. That these do not occur is indicative of the strong influence electronegativity can exert. The fact that a single value of C represents all the elements of Group I is at variance with the indicated factor of $(n - 1)$ in the simple ionic formula. The value of the Born exponent is known to increase with the increasing number of closed subshells in the ion. Thus, such simple generalizations are not particularly useful.

Effect of Crystal Anisotropy

Additional elastic moduli are required to describe the mechanical responses of anisotropic materials. Because almost all materials in single crystal form are anisotropic, the complete description of a single crystal is more difficult than dealing with polycrystalline aggregates. Table 7-2 contains experimental [2] values of the three moduli required

---

[4]Electronegativity, as shown in Chapter 2, has considerable influence on the partial ionic nature of covalent bonds. It is also very influential in determining the properties of the metallic bond.

[5]Data taken from reference [3]. Used by permission.

TABLE **7-2**

Elastic Moduli for Selected Cubic Crystals, 10^{12} dynes/cm^2*

CRYSTAL	C_{1111}	C_{1222}	C_{4412}
Al	1.08	0.622	0.284
Fe (BCC)	2.37	1.41	1.16
C (diamond)	9.2	3.9	4.3
NaCl	0.49	0.124	0.126
KBr	0.35	0.058	0.050
KCl	0.40	0.062	0.062

*From *The Solid State for Engineers*, by M. J. Sinnott, copyright 1961, John Wiley and Sons, Inc. Used by permission.

for several anisotropic cubic crystals. If the condition of isotropy could prevail, the coefficients C_{1222} and C_{4412} would become identical [2]. This is nearly true for the ionic materials listed, but certainly does not apply to the other crystals. Clearly, nonionic materials in the cubic system are anisotropic rather than isotropic. (Similar results would be observed for hexagonal crystals.)

A further illustration of the basic anisotropy of crystal properties is found in the values of Young's modulus for BCC-iron, which vary from $28.3(10^{11})$ dynes/cm^2 in the [111] direction to $13.2(10^{11})$ dynes/cm^2 in the [100] direction. The corresponding values for several metals are tabulated in Table 7-3. (The isotropy of tungsten crystal has not been satisfactorily explained [3].)

Tables 7-2 and 7-3 clearly show that the actual stress distribution within a polycrystalline material varies markedly from point to point according to grain orientation. Thus, the value of Young's modulus,

TABLE **7-3**

Comparative Values of Young's Modulus in Various Crystallographic Directions for Selected Materials 10^{11} dynes/cm^2*

MATERIAL	[111]	[100]
Fe (BCC)	28.3	13.2
Cu	19.2	6.7
Al	7.6	6.3
W	39	39

*From *Introduction to Properties of Materials*, by D. Rosenthal, copyright 1964, D. Van Nostrand Company. Used by permission.

normally determined from experiments performed with polycrystalline samples, represents a mean value intermediate to the two extremes listed. This mean also varies somewhat, according to past heat treatment or any other operation capable of altering the grain size and/or grain distribution.

Plastic Deformation

Many important engineering applications involve large deformations for which the linearizations that lead to Eq. (3.13) are invalid. Such plastic deformations are frequently discussed as if they were *static* or independent of time. *However, this is only an approximation.* Fortunately, the approximation of static conditions is reasonably good for most materials of common interest to the engineer, provided that loading during the test is neither of a very short time duration (impul-

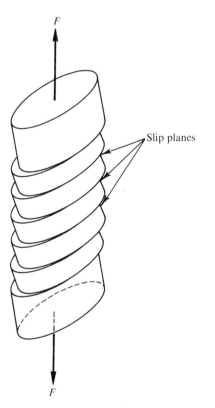

F

Slip planes

F

FIG. 7-3 Grossly exaggerated schematic illustration of slip planes in plastic deformation.

sive), or of an excessively long duration (creep testing). The influence
of the rate of deformation will be considered in some detail in Chapter 8.

Slip and Slip Lines

Upon microscopic examination, it can be seen that plastic deformations are the result of the slipping or gliding of portions of the crystal
structure over one another (see Fig. 7-3). Figures 7-4, 7-5, and 7-6 are
photomicrographs of these actual slip lines in crystals. Electron microscopy reveals that certain planes of the crystal slip in bands or groups of
planes which are separated by regions of unslipped planes several
thousand atomic distances thick. These *slip bands* appear under the
optical microscope as the slip lines shown in Figs. 7-4, 7-5, and 7-6.

FIG. 7-4 Photomicrograph of broad slip bands in the right side of an Fe (3% Si)
bicrystal which initiated slip in the left side ($\times$250). (Courtesy of R. E. Hook. From
R. E. Hook and J. P. Hirth, *Acta Metallurgica*, **15**, p. 535 (1967). Used by
permission.)

Usually these slip bands consist of the planes of the crystal lattice with the densest atomic packing. In HCP crystals, for example, the slip planes are usually the basal (001) planes, while in FCC crystals the slip planes are usually the octahedral (111) planes. Also, the direction in which the slip occurs on the slip planes usually corresponds to the linear direction of densest atomic packing. In HCP crystals this is [100]; in the FCC it is [110].

The reason for this method of selecting slip planes and slip directions is easily understood in terms of atomic packing. For a given configuration of lattice points, the number of points per unit volume is constant. Therefore, the more points in a given plane, the fewer and farther apart such planes can be and still traverse the total volume, including all points. Consequently, the smallest forces will be required to slip these densest packed planes past one another, and the atoms will move the shortest distances if these planes slip. (This corresponds to the shortest Burgers vector discussed in Chapter 4.)

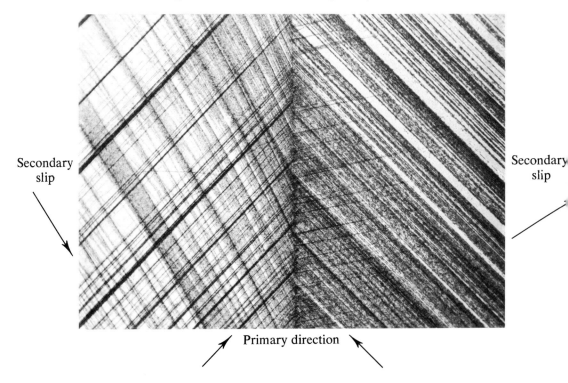

Secondary slip

Secondary slip

Primary direction

FIG. 7-5 Photomicrograph of slip bands in an Fe bicrystal. Compare the heavy, secondary slip lines in the left half of the crystal with the light, secondary slip lines in the right half (×150). (Courtesy of R. E. Hook. From R. E. Hook and J. P. Hirth, *Acta Metallurgica*, **15**, p. 535 (1967). Used by permission.)

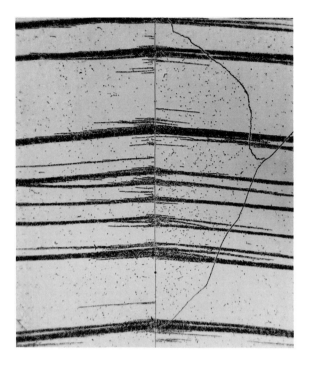

FIG. 7-6   Photomicrograph of slip bands due to screw dislocations ($\times 200$). (Courtesy of R. E. Hook. From R. E. Hook and J. P. Hirth, *Acta Metallurgica*, **15**, p. 535 (1967). Used by permission.)

Two very different mechanisms are involved in a crystal undergoing plastic deformation. The *slip mechanism* occurs when two planes discontinuously glide past one another. The other mechanism is more of a pure shearing action called *twinning*. These two mechanisms will be considered separately.

The Critical Shear Stress

The plastic deformation of solids always occurs either by slip or twinning along certain planes and in certain crystallographic directions. The *critical shear* which initiates this slip is the true measure of the strength of the material undergoing plastic deformation. Usually the externally applied force F will not coincide in direction with the crystallographic orientation of the slip plane, as illustrated in Fig. 7-7. The normal direction to the slip plane makes an angle ϕ with the force F, which, in turn, makes an angle λ with the slip direction. (In the most general situation, these three directions are *not* coplanar.)

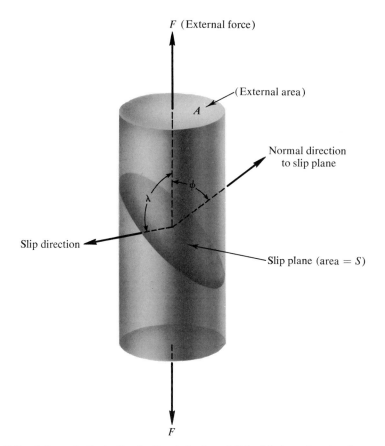

FIG. 7-7 Schematic illustration for the derivation of Schmid's law, showing the resolution of forces on the slip planes.

The area of the slip plane is $S = A/\cos \phi$. The shear stress acting on the slip plane is $F/S = (F/A) \cos \phi$, which must be diminished by the factor $\cos \lambda$ in the slip direction. Therefore, the critical shear stress is given by *Schmid's law:*

$$\tau_{cr} = \frac{F}{A} \cos \phi \cos \lambda \tag{7.13}$$

where F is the external force which initiates slip. If the slip direction, the normal direction to the slip plane, and the direction of the external force F happen to be coplanar, then $\cos \phi = \sin \lambda$. From Eq. (7.13), it follows that the *minimum critical shear stress* for this special case occurs when $d\tau_{cr}/d\phi = 0$, or when $\phi = 45°$. This is the minimum shear stress capable of initiating slip in a crystal. Table 7-4 contains values of both

TABLE **7-4**

Critical Stress for Slip (20°C) in Various Crystals*

METAL	IMPURITY CONTENT, %	SLIP PLANE	DIRECTION	CRITICAL STRESS, kg/mm²
Cu	0.1	(111)	[10$\bar{1}$]	0.10
Ag	0.01	(111)	[10$\bar{1}$]	0.060
Au	0.01	(111)	[10$\bar{1}$]	0.092
Ni	0.2	(111)	[10$\bar{1}$]	0.58
Mg	0.05	(0001)	[11$\bar{2}$0]	0.083
Zn	0.04	(0001)	[11$\bar{2}$0]	0.094
Cd	0.004	(0001)	[11$\bar{2}$0]	0.058
Cd	0.004	(11$\bar{2}$0)	—	0.03
β-Sn	0.01	(100)	[001]	0.19
β-Sn	0.01	(110)	[001]	0.13
β-Sn	0.01	(101)	[10$\bar{1}$]	0.16
β-Sn	0.01	(121)	[10$\bar{1}$]	0.17
Bi	0.1	(111)	[10$\bar{1}$]	0.221
Hg	10^{-6}	(100)	—	0.007
NaCl	—	(110)	[1$\bar{1}$0]	0.2
AgCl	—	(110)	[1$\bar{1}$0]	0.1

*From *The Solid State for Engineers*, by M. J. Sinnott, copyright 1961, John Wiley and Sons, Inc. Used by permission.

the critical shear stress and the Miller indices for the slip planes and directions in several materials [4].

If the critical shear stress for Cu in Table 7-4 is compared with 0.015 times the minimum value of E in Table 7-3, it is found that the latter is roughly 1000 times larger than the critical shear stress. In fact, this disparity is true for all materials.

It is evident from the above observations that *real crystalline materials fall far short of their ideal theoretical strength* in actual tests. (The reasons for their remarkable behavior will be discussed later in this chapter.)

The critical shear stress is influenced by numerous variables. It decreases with increasing temperature, for example, abruptly descending toward zero as the melting point of a solid is approached (see Fig. 7-8). The critical shear stress is also extremely sensitive to impurity levels, as illustrated by the data in Fig. 7-9 for single crystals of mercury.

The amount of prior plastic deformation is a considerably important variable in determining the value of the critical shear stress. If a material is subjected to plastic deformation, a certain amount of the work done on it is stored internally as *strain energy*. This storage of strain energy

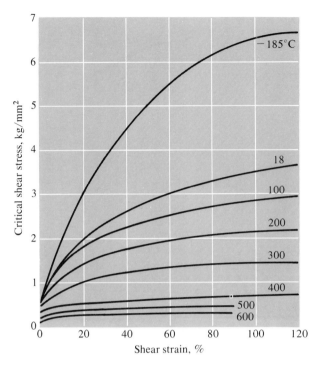

FIG. 7-8 Variation of the critical shear stress of a single aluminum crystal with temperature and per cent strain. (From W. Boas and E. Schmid, *Zeit. Physik*, **71**, p. 712 (1931). Used by permission.)

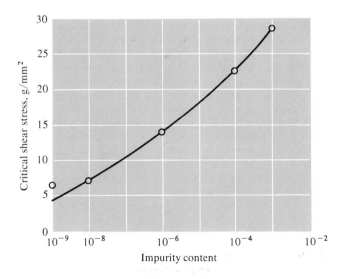

FIG. 7-9 The variation of critical shear stresses in single mercury crystals when related to impurity content. (From K. M. Greenland, *Proc. Roy. Soc.* (*London*), **A163**, p. 52 (1937). Used by permission.)

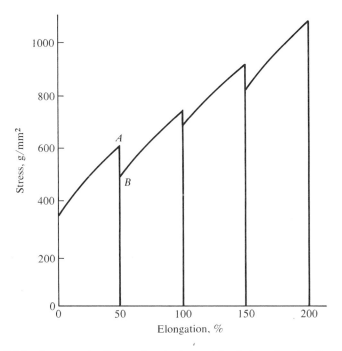

FIG. 7-10   Stress-strain diagram for single crystals of zinc, showing the effect of loading and unloading on sample response. At *A*, the load was removed for 30 seconds and then reapplied; yield occurred at *B* after reloading. Successive steps repeat this cycle. The distance *AB* is due to elastic recovery. (From O. Haase and E. Schmid, *Zeit. Physik,* **33,** p. 413 (1925). Used by permission.)

in a crystal results in the strengthening or *work-hardening* of the solid.[6] When the stress is relieved, an elastic recovery occurs which leaves a permanent plastic deformation. Because the modulus of elasticity is not appreciably affected by the straining, the elastic portion of the stress-strain curve is retraced when the material is restressed. However, the crystal will not yield until a stress is reached which corresponds to the intersection of the elastic curve with the normal portion of the engineering stress-strain curve for plastic strain, as illustrated in Fig. 7-10 for single crystals of zinc.

From Fig. 7-10, it can be noted that a slight offset (*AB*) occurs in the curve between unloading and reloading; this is caused by the process of *recovery* [2], during which some of the strain energy diffuses out of the crystal. Because diffusion is a rate process, the degree of recovery depends on both temperature and time allowed. This is clearly illustrated in Fig. 7-11, where tests are performed on the same zinc

---

[6]The mechanism of this phenomenon will be discussed later in the chapter.

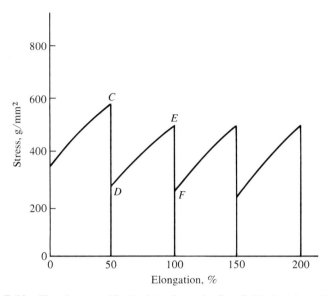

FIG. 7-11   The data are identical to those in Fig. 7-10, but here the time interval between *C* and *D* was one day. (From O. Haase and E. Schmid, *Zeit. Physik,* **33**, p. 413 (1925). Used by permission.)

single crystals with time lapses of one day apiece between reloadings. Clearly, an engineering stress-strain diagram such as the one illustrated in Fig. 1-1 represents the summation of plastic strain minus the recovery which occurred during the test; it is only an average result obtained over a reasonable period of time.

Normally the rate of stress applied to a crystal does not significantly affect the critical shear stress. However, if the application time of the load is very short (as in impulsive loading), or very long (as in creep testing), the results obtained may vary substantially from the moderately timed or *static* tests.

Twinning

The atomic motions involved in twinning are quite different from those observed in the slip or glide mechanisms. In the twinning process, each successive, atomic plane moves a successive fractional amount, and the resulting twinned crystal appears to be two crystals separated by a reflection plane. Visually, these parts are the mirror images of one another across this boundary or *twinning plane*, as illustrated schematically in Fig. 7-12.

Because the twinned crystals bear different crystallographic orienta-

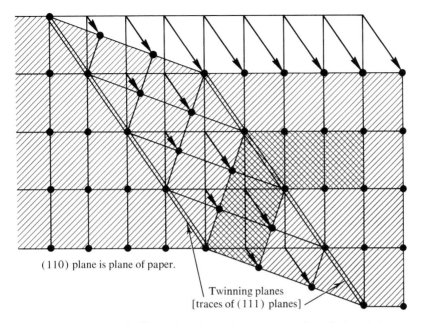

(110) plane is plane of paper.

Twinning planes
[traces of (111) planes]

FIG. 7-12   Schematic illustration of atomic movements in twinning deformation. (From *Metallurgy for Engineers,* by J. Wulff, *et al.,* copyright 1954, John Wiley and Sons, Inc. Used by permission.)

tions, each reflects light at a different angle when illuminated, thus making the twins easily visible under the microscope as distinct bands within a suitably etched grain (see Fig. 7-13).

Twinning does not normally occur in FCC crystals, but it is often observed in BCC structures and hexagonal metal crystals. Just as a critical shear stress for slip is observable, a *critical twinning stress* also exists. The factors which affect twinning are believed to be similar to those which affect slip, but twinning data are scarcer and more difficult to interpret [2] than slip data.

Crystal Imperfections

It has been observed that the critical shear stresses for real crystalline materials are several orders of magnitude less than the theoretical perfect crystal strengths, suggesting that perfect crystals either do not exist or only occur rarely. Indeed, perfect crystals can only be found in rare instances where very special care has been taken to produce tiny, essentially perfect, single crystals.

Dislocations were among the various types of crystal imperfections discussed in Chapter 4. They not only account for growth behaviors,

FIG. 7-13 Deformation twins in brass (×145). (Courtesy of T. T. Moran, The International Nickel Company. Used by permission.)

but for the reduction in critical shear stresses, as well. In describing the geometry of a dislocation, the term slip plane was introduced because the dislocation appears to be the boundary between the region of perfect crystalline material and an area of crystal that has slipped relative to this perfect region. The dislocation boundary then represents a region of strained atomic configuration. The use of the term *slip* in describing the geometry of a dislocation is not fortuitous, as will be seen shortly.

Distinctions between different types of dislocations are accomplished by using the Burgers vector. The crystallographic specifications of this vector are most conveniently shown in terms of a slip direction and the number of fractional lattice distances moved in that direction by the slip plane.

EXAMPLE 7.1

Specify the Burgers vector if slip occurs on the (001) plane of an FCC crystal from a corner to an FCC location.

Solution:

The components would be $a/2$, $a/2$, and 0. The Burgers vector would be $\frac{1}{2}a$ [110], or a vector of magnitude $a/\sqrt{2}$ pointing in the [110] direction.

For the cubic system in general, the Burgers vector would be designated by (a/n) [uvw]. The *strength* of the dislocation is given by the magnitude $(a/n)(u^2 + v^2 + w^2)^{1/2}$, which is also the magnitude of the Burgers vector. A dislocation of unit strength causes atomic displacement from one equilibrium position to another; its Burgers vector has a magnitude of one lattice spacing [5].

Slip Mechanisms

The low experimental values for critical yield stresses suggest that the dislocation motion in crystal structures causes slip to occur at reduced stress levels. Figure 7-14 illustrates an edge dislocation moving transversely under the action of a shear stress. Careful examination of Fig. 7-14(a) shows that, in traversing the A_i atomic plane from A_1 to A_7, the mismatching due to dislocation amounts to one full lattice spacing relative to the B_i atoms directly below the slip plane. Because of the distortion caused when interatomic forces attempt to correct this dislocation, atoms A_2 and A_3 are in successively more slipped orientations, atom A_4 is actually at the top of the slip position, and

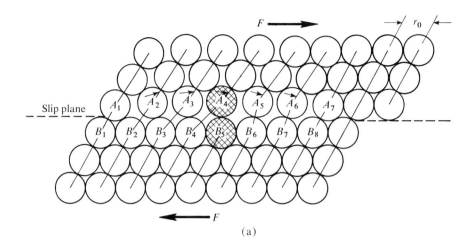

(a)

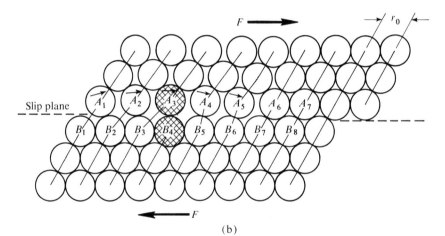

(b)

(c)

atoms A_5 and A_6 are over this peak and are moving down the other side into the depressions between the B_i atoms. Thus, atom A_4 seems to require zero force to move it in either direction, while the atoms immediately to either side need less force to initiate dislocation movement than those further away require. Although zero force appears to be required in order to move atom A_4, a detailed analysis of the forces involved will show [6] that this dislocation is actually *locked* in position until a finite, external force is applied.

The presence of the dislocation configuration in a crystal lattice corresponds to the partial slipping of some of the atoms in its immediate neighborhood. In Fig. 7-14, parts (b) and (c) show the dislocation moving to the left as the slip plane moves *segmentally* to the right under the action of the shear force couple. A slipping of the upper atoms A_i past the lower atoms B_i is the net result of this dislocation motion through a crystal. Thus, rather than the entire row of atoms slipping simultaneously, *only a small part of the plane must be slipped at a given time*. Therefore, a much smaller magnitude stress is necessary to move the dislocation than to slip an entire plane of atoms through the crystal.

Figure 7-14(c) shows that, once the dislocation passes through the crystal and reaches a boundary, it is annihilated and a perfect crystal in a unit slipped position remains. For further slip to occur, the theoretical shear stress must be applied to the crystal or a source of additional dislocations must be found. In real materials, both the existence of multiple slip bands involving many planes and the fact that repeated plastic deformations are possible at less than theoretical shear stresses imply that such a dislocation source must exist.

MULTIPLICATION OF DISLOCATIONS   The multiplication of dislocations in a plane can be explained by several mechanisms; one of the simplest of these is the *Frank–Read generator* illustrated in Fig. 7-15. In this, ABC is a mixed dislocation with a Burgers vector lying in the CDE plane. If it is assumed that only planes parallel to CDE are active slip planes, the AB portion of the dislocation line will be immobile, while the BC part will be free to rotate about pivot point B. As this dislocation line moves through the crystal to position BD, for instance, it causes the CBD area to undergo unit slip. Thus, when the dislocation

FIG. 7-14   Schematic representation of the slip process resulting from the movement of an edge dislocation through a close-packed crystal. (a) Initial position of the dislocation before slip; dislocation terminates at B_5. (b) Dislocation after unit slip. (c) Dislocation ready to leave the crystal. (From *Introduction to Properties of Materials*, by D. Rosenthal, copyright 1964, D. Van Nostrand Company. Used by permission.)

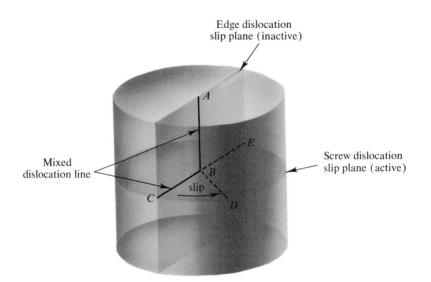

FIG. 7-15 Schematic illustration of the Frank–Read generator. The mixed dislocation lies in two planes at right angles to one another. The axial plane is inactive, anchoring that part of the dislocation. The transverse plane is active and its dislocation rotates; using the anchored axial portion as a pivot point, it continuously slips transverse planes and regenerates itself.

has completed a revolution about point B, the entire plane will have undergone unit slip. With each subsequent revolution, the BC dislocation will slip an entire plane, effectively generating an indefinitely large amount of slip in time.

In order to understand the other mechanisms that generate dislocations, it is necessary first to define the concept of the force which acts upon these dislocations. Mott and Nabarro [7] developed a simple expression for such a force. If the potential energy of a particle assemblage, such as the atoms in a dislocation configuration, is increased by an amount dU when one of the particles moves a distance dx, a force $F_x = -dU/dx$ acting on the array can be defined.

The force acting on a dislocation can also be determined in this manner. Let the shear stress acting in the slip direction on the slip plane of area S be τ. Then, the applied force which causes an element ds of a dislocation of strength b to move forward by an amount b is

$$dF_x = \tau(b\,ds) \qquad (7.14)$$

and the amount of work[7] dW is

$$dW = b\,dF_x = \tau b^2\,ds \qquad (7.15)$$

---

[7]and, therefore, the negative change in internal energy

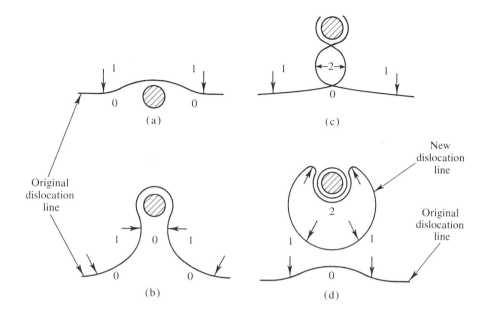

FIG. 7-16 Schematic illustration of a second mechanism of dislocation generation. The shaded circle represents an impurity atom. Arrows indicate the direction of motion of the dislocation. (From *The Solid State for Engineers*, by M. J. Sinnott, copyright 1961, John Wiley and Sons, Inc. Used by permission.)

Since b is constant along the dislocation line, we can integrate, from Eq. (7.14) obtaining

$$F_x = \tau b s \qquad (7.16)$$

or the force per unit length of dislocation

$$\frac{F_x}{s} \equiv F = b\tau \qquad (7.17)$$

Since work is defined in terms of force acting in the *same* direction as displacement, it follows that the direction of the force F is *normal* to the dislocation line everywhere, *regardless of its orientation*. Also, because the magnitude of b is constant along the entirety of the dislocation line,[8] it follows from Eq. (7.17) that this normal force acting on the dislocation is of *constant magnitude*. This means that, regardless of the orientation of the dislocation line, it will continue to move forward under the action of the constant force F.

With the above concepts in mind, we can now discuss the other mechanisms of dislocation generation. Consider the dislocation line shown schematically in Fig. 7-16(a). The shaded circle represents some

---

[8]See the discussion in Chapter 4.

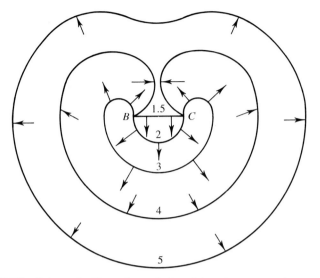

FIG. 7-17 Schematic illustration of the dislocation generation mechanism with two anchor points. Note that this mechanism produces a series of concentric circles, rather than spirals. (From F. C. Frank and W. T. Read, *Phys. Rev.,* **79,** p. 722 (1950). Used by permission.)

local disturbance in the crystal structure[9] which will impede the passage of the dislocation. As the dislocation line reaches this *anchor point,* its central portion is retarded or stopped; the outer portions of the line, however, continue to advance under the action of the force *F.* Since the dislocation line must remain continuous and part of it has anchored, the other portions swing around the anchor point, as illustrated in Fig. 7-16(b). In Fig. 7-16(c), these two fronts have passed through one another, and the points of intersection now form a new set of dislocation lines: one replaces the original advancing front and the other—a new, closed loop—encapsulates the anchor point. This separation into new lines, shown in Fig. 7-16(d), can obviously proceed indefinitely as the new encapsulating loops continue to grow under the action of the force *F.*

Another multiplication mechanism is shown in Fig. 7-17, where the dislocation line *BC* is anchored at two points. Here, the progression of the dislocation loop until it subsequently closes on itself is illustrated in five different stages.

The latter two types of multiplication mechanisms result in dislocation rings rather than in the spirals produced by the Frank–Read generator discussed earlier. The photomicrographs in Fig. 7-18 and

---

[9]This could be any number of local imperfections—for example, a precipitated second phase, as in the case of age-hardening discussed in Chapter 6.

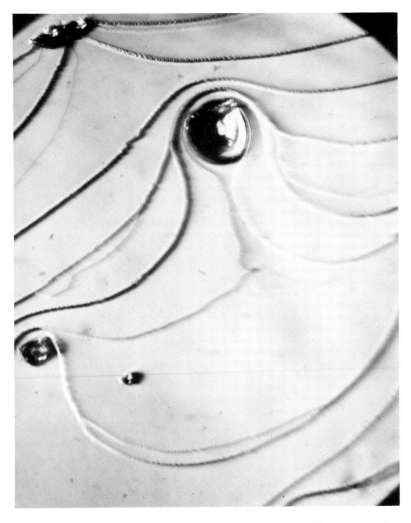

FIG. 7-18   Photomicrograph of dislocation loops generated from two anchor points. (Courtesy of A. A. Burr, Rensselear Polytechnic Institute. From M. J. Freser, *et al., Acta Metallurgica,* **4,** p. 194 (1956). Used by permission.)

Fig. 7-19 illustrate actual dislocation loops generated by each of the mechanisms discussed here.

STRAIN-HARDENING   As the total strain increases, greater stresses are required to initiate slip, implying that dislocation movement has been restricted. This phenomenon, known as work or *strain-hardening,* is a characteristic behavior in nearly all materials. This increased resistance to dislocation movement results from the interaction of

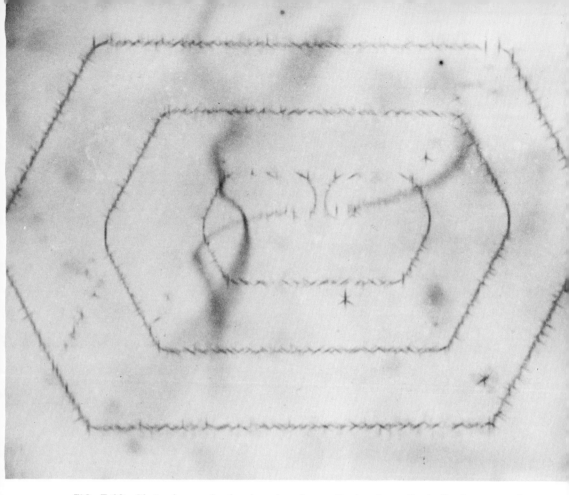

FIG. 7-19 Photomicrograph showing ringed growth due to a Frank–Read source of dislocation in a Si crystal. (Courtesy of The General Electric Research and Development Center. From the cover of *J. Appl. Phys.*, by Dr. William Carter Dash (November 1956). Used by permission.)

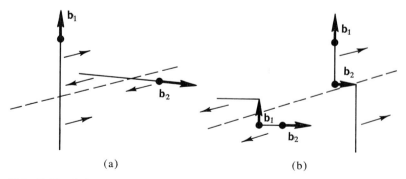

(a)   (b)

FIG. 7-20 Schematic representation of the interaction of two dislocations at right angles to one another. Each imparts a jog to the other; both jogs will have the Burgers vector of its respective parent dislocation.

existing dislocations with one another or with other crystal imperfections. When dislocations interact with each other, anchor points
develop which tend to immobilize the dislocations, making further
motion difficult. To illustrate this, consider the schematic representa-
tion of two screw dislocations with Burgers vectors $\mathbf{b}_1$ and $\mathbf{b}_2$ oriented
at right angles to one another and moving along the dashed line in
Fig. 7-20(a). When these dislocations intersect, each imparts a *jog* or
kink with its respective Burgers vector to the other, as illustrated in
Fig. 7-20(b). Each new jog dislocation is edge-oriented with respect to
the original screw dislocation. Further motion of such a now mixed
dislocation would require the edge dislocation to move out of its slip
plane, a process not only necessitating an external stress, but the
diffusion of atoms from the dislocation into interstitial positions nearby
as well. This phenomenon, called *dislocation climb* [6] or *jog*, is illustrated schematically in Fig. 7-21.

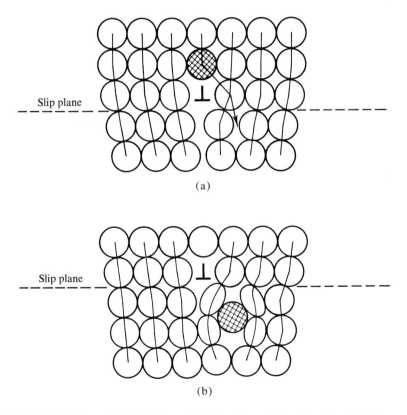

FIG. 7-21 Schematic illustration of dislocation climb. The terminal atom
(shaded in both parts of the figure) must move to the interstitial position shown
in (b) for dislocation ⊥ to move up one row.

In order for the edge dislocation, designated by the symbol ⊥ in Fig. 7-21(a), to climb up to the next plane and, thereby, move the slip plane upward one row, the terminal atom of the extra plane (shaded in the figure) must move into the interstitial position shown in Fig. 7-21(b), thus producing an unstable condition unless interstitialcy can be eliminated by diffusion. This entire process requires additional energy, particularly at lower temperatures; as a result, dislocation jog or climb is a higher energy process than ordinary slip, and the application of larger stresses is required. Thus, the material is said to have strain-hardened because it appears harder here than in its unstrained condition. The existence of the dislocation generating mechanisms discussed above insures that increasing numbers of dislocations will interact and contribute to the strain-hardening process as plastic deformation continues.

Another process contributing to strain-hardening is the phenomenon of *dislocation pileup* which often occurs near the crystal boundaries (see Fig. 7-22). As a dislocation traversing the crystal approaches a boundary, the slip plane tends to move away from the boundary in the opposite direction, creating a slip band on the surface. As the surface structure of the crystal begins to collapse under this action, the normal tendency of this dislocation will be to annihilate itself. Very often, however, the surfaces of crystals, particularly in polycrystalline materials, will be covered with films of various types such as oxides, which can impede or completely block dislocation escape and cause the pileup seen in the figure. As a result of this phenomenon, polycrystalline materials for a given amount of plastic deformation strain-harden much more readily than do single crystals.

The phenomenon of strain-hardening is temperature dependent due to the temperature dependency of diffusion. As the temperature level

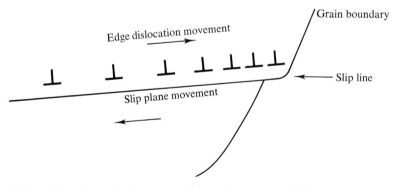

FIG. 7-22   Schematic illustration of the pileup of edge dislocations at a grain boundary.

is raised, the process of recrystallization discussed in Chapter 6 will occur with increasing ease. This nucleation of new, unstrained crystal regions counteracts the effect of dislocation interaction and relieves strain-hardening. Above a certain more or less well-defined temperature (see Fig. 6-10), the recrystallization process occurs more rapidly than the strain-hardening process, with the result that any increases in the yield stress are immediately relieved. Consequently, it is impossible to strain-harden a material above the recrystallization temperature defined by Eq. (6.18). Any mechanical working of a material performed at a temperature in excess of T_{recr} is termed *hot-working;* if performed at a temperature below this limit it is known as *cold-working.* Many common industrial operations employ combinations of hot- and cold-working cycles to achieve the desired properties in a finished product.

Grain Boundaries

One of the few portions of dislocation theory amenable to direct experimental investigation is the prediction [6] of the energies of small angle grain boundaries. A small angle of misfit grain boundary is illustrated schematically in Fig. 7-23(a), where θ is the angle of misfit.

The only physically reasonable way in which grains can be joined together is by the array of evenly spaced dislocations in Fig. 7-23(b). For small angles of misfit θ, the spacing D of the dislocations is related to the magnitude of the Burgers vector b by the relation

$$\frac{b}{D} = 2 \sin \frac{\theta}{2} \doteq \theta \qquad (7.18)$$

This is an easily measurable prediction. The presence of the dislocations corresponds to a series of sites along the grain boundary, having higher energies than the bulk of the grains, which should be preferentially attacked when the material is subjected to an etching process. Such a process results in a row of uniformly spaced etch pits with calculable spacings, such as those illustrated by the photomicrograph in Fig. 7-24 for a germanium crystal. The measurements made from such photographs are in reasonable agreement with the theory expressed in Eq. (7.18), lending weight to its argument.

The energy of the small angle of misfit grain boundary has been shown to be [6] of the form

$$E = E_0\theta[A - \ln (\theta)] \qquad (7.19)$$

where E is the energy per unit area of the boundary, E_0 is a theoreti-

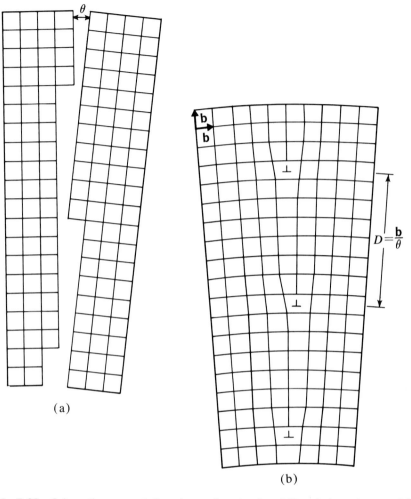

(a)

(b)

FIG. 7-23   Schematic representation of a small angle of a misfit grain boundary, modeled by edge dislocations. (From *Dislocations in Crystals*, by W. T. Read, Jr., copyright 1953, McGraw-Hill Book Co., Inc. Used by permission.)

cally calculable and measurable quantity [6] dependent on the elastic distortion of the boundary, and A is an unknown integration constant which must be determined from individual measurements of E and θ.

The mathematical form of Eq. (7.19) indicates that a plot of E/θ vs ln (θ) should be linear. On linear coordinates, the plot of E vs θ possesses a maximum, as illustrated in Fig. 7-25. If the coordinates of this maximum are designated as E_m and θ_m, they are related by

$$E_m = E_0 \theta_m \qquad \qquad \textbf{(7.20)}$$

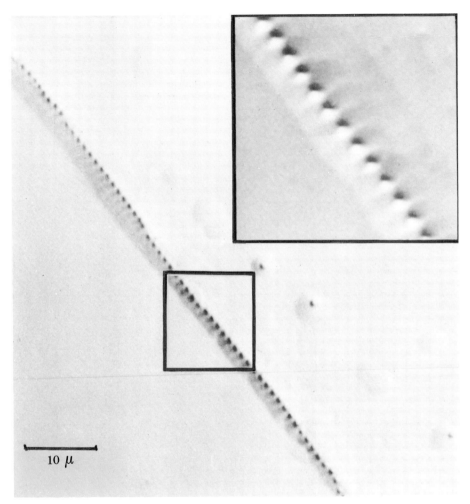

FIG. 7-24 Photomicrograph of etch pits in germanium, showing the small angle of misfit grain boundary. (Courtesy of F. L. Vogel. From F. L. Vogel, *et al.*, *Phys. Rev.*, **90**, p. 489 (1953). Used by permission.)

if $A = 1 + \ln (\theta_m)$. In terms of the parameters E_m and θ_m, Eq. (7.19) can be rewritten

$$\frac{E}{E_m} = \frac{\theta}{\theta_m}\left[1 - \ln\left(\frac{\theta}{\theta_m}\right)\right] \qquad (7.21)$$

The data for energies of small angle of misfit grain boundaries should superimpose on a single plot if the terms of Eq. (7.21) are correct; data obtained on a number of different systems are compared with Eq. (7.21) in Fig. 7-26. Since the derivation of this equation is valid

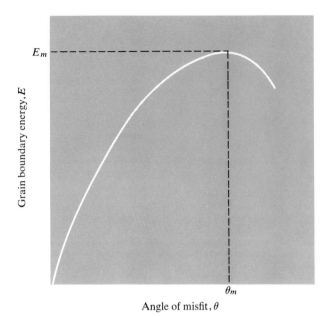

FIG. 7-25 Schematic representation of a small angle of misfit grain boundary energy as a function of the misfit angle.

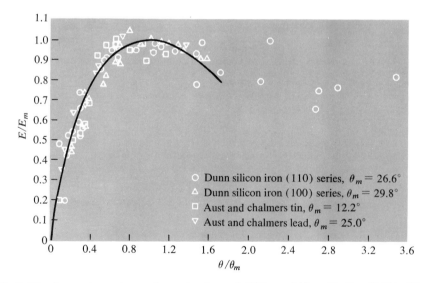

FIG. 7-26 Comparison of experimental values of E/E_m vs θ/θ_m with Eq. (7.21). (From *Dislocations in Crystals,* by W. T. Read, Jr., copyright 1953, McGraw-Hill Book Co., Inc. Used by permission.)

only for small angles of misfit, a breakdown in the correlation for large angles is predictable. The theoretical computation of similar quantities for large angles of misfit grain boundaries is virtually impossible [6].

Fracture

When the stress applied to a crystal exceeds the strength of the interatomic forces holding its atoms together, the material will separate or fracture into two or more parts. Such *ductile fracture* ultimately terminates the slip process. On the other hand, if the stress normal to the slip plane can exceed the bonding strength before the resolved critical shear stress is reached for a material, the crystal planes cleave in the process of *brittle fracture.*

Referring to the derivation of Schmid's law previously discussed, the normal component of the applied stress can be seen as $(F/A)\cos^2\phi$. The value of this stress component sufficient to cause crystal failure is the *critical normal fracture stress.*

Like the critical shear stress, the critical normal fracture stress has been found to be a more or less constant material property. (Table 7-5 contains critical normal stress data for an exemplary series of materials.) As in the case of the critical shear stress, cleavage usually occurs on the planes of higher atomic density. The critical fracture strength also appears to be relatively insensitive to temperature, as indicated by the curve in Fig. 7-27 which was computed on the assumption [2] that the critical normal fracture stress is independent of temperature.

Many materials, ceramics in particular, behave in a brittle fashion, failing by brittle fracture before undergoing appreciable plastic deformation. The fact that such materials possess a critical normal fracture stress several orders of magnitude less than the theoretical, cohesive material strength, however, suggests that the existence of dislocations is even responsible for the failure of these materials under load.

Dislocation movement is greatly hampered in brittle materials by their types of interatomic bonding. The metallic bonds permit the nuclear cores to arrange themselves in various close-packed arrays, but as their structures change, the partially covalent natures of these bonds increase. They become relatively restricted in space, requiring much larger forces to disrupt them than the nonoriented bonds do. The dislocations which are present, even in completely covalent materials such as diamond [8], are, therefore, severely constrained in their mobility. The application of external stress fields to such covalently bonded materials tends to produce *microcracks*, rather than induce

TABLE **7-5**

Critical Normal Fracture Stresses for Various Crystals*

METAL	CLEAVAGE PLANE	TEMPERATURE (°C)	CRITICAL NORMAL STRESS, kg/mm²
Zinc + 0.03% cadmium	(0001)	−80	0.19
	(0001)	−185	0.19
	(10$\bar{1}$0)	−185	1.80
Zinc + 0.13% cadmium	(0001)	−185	0.30
Zinc + 0.53% cadmium	(0001)	−185	1.20
Bismuth	(111)	20	0.32
	(111)	−80	0.32
	(11$\bar{1}$)	20	0.69
Antimony	(11$\bar{1}$)	20	0.66
Tellurium	(10$\bar{1}$0)	20	0.43
Magnesium	(10$\bar{1}$2)	—	—
	(10$\bar{1}$1)	—	—
	(10$\bar{1}$0)	—	—
α-iron	(100)	−100	26
α-iron	(100)	−185	27.5
Rock salt (dry)	(100)	—	0.44

*From *The Solid State for Engineers*, by M. J. Sinnott, copyright 1961, John Wiley and Sons, Inc. Used by permission.

slip in their structures. When a sufficient number of these microcracks develop, or when they exceed a certain critical size, they will propogate through the crystal, causing brittle cleavage along the cleavage plane.

If an external hydrostatic pressure of sufficient magnitude to offset microcrack formation is applied to brittle materials, large plastic deformations by the slip process can be induced, as in metallic substances. An example of such plastic deformation in a brittle material under large hydrostatic pressure is seen in the folding and warping of whole strata of rocks in the earth. Laboratory experiments [9] of this nature have been successfully performed on such normally brittle ceramic materials as rock salt, quartz, marble, and MgO.

If the dislocation structure of a ductile material can be locked in place, it will behave in a brittle fashion. Thus, if a material is subjected to repeated cold-working, the strain-hardening process will eventually raise the critical shear stress to such a level that brittle failure will ensue before further slip occurs. A common example of this phenomenon is found in the breaking of a normally ductile iron wire by repeated flexural bending. If a cross-section of broken wire is examined carefully, it will appear coarse and granular, as is typical in brittle fracture.

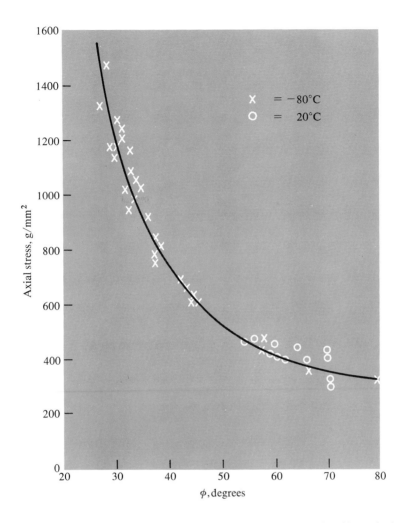

FIG. 7-27 Variation in the critical fracture stresses of Bi crystals with angle ϕ at two different temperatures. (From M. Georgieff and E. Schmid, *Zeit. Physik,* **36,** p. 759 (1926). Used by permission.)

From this discussion, the distinction between ductile and brittle materials seems artificial. All materials, even diamond [10], will undergo plastic deformation, given the proper conditions. Similarly, all materials which normally behave in a ductile manner, can be induced to react in a brittle fashion under the precise conditions. Of course, the latter observation only applies to crystalline materials. Amorphous, non-crystalline materials may never exhibit such a phenomenon as brittleness because they lack well-defined crystal planes where cleavage may occur.

Because of ever-increasing technological demands, it is probable that the application of the facts discussed here will become increasingly common. The use of hydrostatic pressures to facilitate low temperature, cold-forming operations such as extrusion and drawing, for example, will probably expand the application range of many heretofore brittle materials such as high strength, high hardness alloy steels.

References Cited

[1] Prager, W., *Introduction to Mechanics of Continua*, Ginn & Company, Boston, 1961, Chapter 8.

[2] Sinnott, M. J., *The Solid State for Engineers*, John Wiley and Sons, Inc., New York, 1961, Chapter 10.

[3] Rosenthal, D., *Introduction to Properties of Materials*, D. Van Nostrand Company, Princeton, N. J., 1964, Chapter 6.

[4] Sinnott, *op. cit.*, p. 215. (*See also*, C. S. Barrett, *Structure of Metals*, McGraw-Hill Book Co., Inc., New York, 1952.)

[5] Cottrell, A. H., *Dislocations and Plastic Flow in Crystals*, Oxford University Press, New York, 1956, Chapter 1.

[6] Read, W. T., Jr., *Dislocations in Crystals*, McGraw-Hill Book Co., Inc., New York, 1953.

[7] Mott, N. F., and F. R. N. Nabarro, *Report on Strength of Solids*, London Physical Society, p. 1 (1948).

[8] Rosenthal, D., *op. cit.*, Chapter 9.

[9] Bridgman, P. W., *Studies in Large Plastic Flow and Fracture*, McGraw-Hill Book Co., Inc., New York, 1952.

[10] Van Bueren, H. G., *Imperfections in Crystals*, 2nd ed., North-Holland Publishing Company, Amsterdam, 1961, Chapter 30.

PROBLEMS

7.1 A copper bar 1 in. square in cross-section and 10 in. long has a 1000 lb tensile load in the direction of its longest dimension. If $E = 16(10^6)$psi and $\nu = 0.36$, compute:

(a) The stress developed. $10^3 \, psi$

(b) The elongation in the direction of the load. $\Delta \ell = 0.625 \times 10^{-3}$ in.

(c) The elongation at right angles to the direction of the load. $\delta y = -2.25 \times 10^{-5}$ in.

7.2 The cube of a mild steel 1 in. on an edge is subjected to a uniform hydrostatic pressure of 10,000 psi. Compute the per cent change in volume experienced by the steel. Use the elastic moduli for Fe listed in Table 7-1.

$\Delta V = 0.0463 \, \%$

7.3   A copper cube 4 in. on each side is subjected to a shearing force of $8(10^4)lb_f$. The resultant shear angle is 2.68 min of arc. Calculate Young's modulus and the bulk modulus for Cu, using only these data and the value of Poisson's ratio listed in Table 7-1.

$$E = 17.32 \times 10^6 \, psi \qquad K \doteq 19.25 \times 10^6 \, psi$$

7.4   A sample of steel undergoes a series of tests. In *one pair* of tests it is subjected to a tensile stress and then to a shear stress. For a ratio of the tensile to shear stresses of 2.82, the tensile strain is found to exceed the shear strain by 5.5%. In a *separate* test, a com-pressive load producing a stress of 50,000 psi results in a volumetric strain of 0.001545. Compute the three elastic moduli and Poisson's ratio from these data.

$-3.25 \times 10^7 psi$

$E = 1.217 \times 10^7 psi \qquad \nu = 0.3375 \qquad K = 3.34 \times 10^7 psi \qquad G = 1.217 \times 10^7 psi$

7.5   A stress of 815 psi, applied in the [100] direction, causes slip in the [$\bar{1}10$] direction on the (111) plane of a certain FCC material. Compute the magnitude of the critical shear stress of this material.

7.6   Using the data of Table 7-5, compute the magnitude of the stress in the [111] direction needed to produce cleavage in a NaCl crystal on the (100) plane.

7.7   Using the data of Table 7-4, compute the magnitude of the stress in the [112] direction which results in slip on the (111) plane of Cu in the [110] direction.

7.8   Using the data of Table 7-4, compute the magnitude of the stress in the [112] direction which results in slip on the (111) plane of Bi. What fraction of the critical normal stress is the normal component of this stress?

7.9   In a series of single crystal tension tests, slip is in the direction of the steepest angle between the slip plane and the direction of the loading force. In one particular test, the angle between the normal to the slip plane and the direction of application of the load changes from 45° to 55°. If the distance d between slip planes does not change, compute the amount of elongation which will occur during such a test.

7.10   An aluminum sample shears on the (110) plane in the [$1\bar{1}0$] direction when a stress of 500 psi is applied in the [$1\bar{1}1$] direction. Compute the critical shear stress of aluminum.

7.11   For the data of Problem 7.5, compute the magnitude of the force per unit length which acts on the dislocations causing slip. The atomic radii of this metal are 1.35 Å.

7.12   Compute the magnitude of the force per unit length acting on the dislocations which cause slip in the NaCl crystal in Problem 7.6.

7.13   Compute the magnitude of the force per unit length acting on the dislocations which cause slip in the aluminum crystal in Problem 7.10.

7.14   Using the data of Table 7-1, compute the volumetric strain a cube of nickel 1 cm on an edge will experience if it is subjected to the following stress distribution: 10,000 psi tensile in the z direction, 20,000 psi tensile in the y direction, and 20,000 psi compressive in the x direction. (The cube is oriented with its faces perpendicular to the coordinate axes.)

$$\varepsilon_v = 0.000126$$

7.15   A cube of steel 1 cm on an edge is subjected to 10,000 psi tensile stress in the x direction, 12,000 psi tensile stress in the y direction, and 8000 psi compressive stress in the z direction. Compute the strain in the direction of each stress for this material, using the data of Table 7-1.

7.16   For lead, the value of θ_m is 25.0° and $E_m = 550$ ergs/cm^2. Compute the energy of a grain boundary which has a mismatch angle of 10°.

7.17   Using the parameters of Problem 7.16, compute a reduced curve of grain boundary energy as a function of the angle of mismatch. From this curve and from the fact that $\theta_m = 12.2°$ for tin, compute the ratio of grain boundary energies for boundaries having mismatch angles of 6.0° and 7.5°.

THE RESPONSE OF
MATERIALS TO DYNAMIC
MECHANICAL FORCES

Because the modern engineer often deals with a large variety of materials under far more complex conditions than those revealed in a static loading situation, he must be familiar with the response of such materials to dynamic loading forces. In the following discussion, three main classes of materials will be considered under dynamic loading conditions: crystalline solids, liquids and gases, and amorphous non-crystalline materials such as polymers.

Crystalline Solids

Creep

In many experiments, constant loading of the sample for a considerable period of time causes plastic deformation to slowly increase in magnitude with the passage of time. That is, the material *creeps* or flows very slowly under constant load. Upon closer examination, it is found that this phenomenon consists of four major divisions:

1. The initial or *instantaneous strain* (the ordinary static load response discussed in Chapter 7).
2. A period of decreasing rate or *primary creep.*

3. A period of constant minimum rate or *secondary creep*.
4. A period of accelerating rate or *tertiary creep* which terminates in the fracture of the material.

These characteristics are illustrated schematically by the plot of observed strain as a function of time of loading in Fig. 8-1. The slope of the curve is the rate of creep $d\epsilon/dt$.

PRIMARY CREEP During the period of time shortly after the loading of the sample, the material undergoes creep at a decreasing rate; that is, the slope of the strain-time curve decreases with increasing time. During the period of *primary creep*, this strain-time curve is frequently found empirically to be represented by a power law expression of the type

$$\frac{d\epsilon}{dt} \equiv \dot{\epsilon} = At^{-n} \qquad (8.1)$$

where A and n are empirical constants. n is usually found to be less than unity, although some materials (a few metals, glass, and rubber) are represented by the extreme case $n = 1$. This leads to the so-called

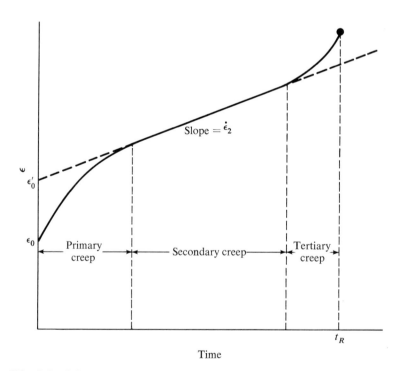

FIG. 8-1 Schematic representation of a creep curve, showing primary, secondary, and tertiary periods. t_R is the time required for the material to rupture.

logarithmic creep law [1]

$$\epsilon = \alpha \ln (t) \qquad (8.2)$$

where α is a constant. For fast creep rates and large strains, this logarithmic variation is *not* the normal relation. In such cases, n is usually less than unity and often has the value $n = \frac{2}{3}$, which leads to Andrade's creep law [2]

$$\epsilon = \beta t^{1/3} \qquad (8.3)$$

In these relationships, the constants A, α, and β are dependent on temperature and stress level [1].

Like the static plastic deformations in Chapter 7, the instantaneous plastic deformation which initiates creep begins when an avalanche of dislocations is set into motion throughout the crystal. The interactions of these mobile dislocations, with one another and with various other crystal imperfections, retard and anchor an increasing number of them resulting in strain-hardening. Eventually, this process of dislocation pileup, at the grain boundaries and at other anchoring points, builds up a backlog of pressure caused by the immobile dislocations; such a pressure restricts further dislocation movement and, therefore, plastic deformation. This dislocation pileup, illustrated in Fig. 8-2, results in a continuous reduction of the creep rate. If this were the only operative mechanism, the creep rate would approach zero and further plastic deformation would cease. Such behavior is actually observable at low temperatures. However, as the temperature increases at constant stress levels, or as the stress levels increase at constant temperature, other mechanisms become operative to counteract this dislocation blocking.

Thermally activated recovery and dislocation climb are two such mechanisms for counteracting strain-hardening and dislocation blocking. Both release dislocations from their anchor points so that they are free to move again. As the total plastic strain increases, a condition is eventually reached where the rates of the two competing dislocation and dislocation blocking mechanisms will balance if the temperature or stress levels are sufficiently high.

SECONDARY CREEP When these rates of strain-hardening and recovery or dislocation climb balance, the overall creep rate becomes constant at the *minimum* value which corresponds to the roughly linear portion of the strain-time curve. The mathematical representation of this part of the curve is the simple linear function

$$\epsilon = \epsilon_0' + \dot{\epsilon}_2 t \qquad (8.4)$$

where ϵ_0' is the intercept at $t = 0$ of the straight dashed line drawn through the linear portion of the creep curve in Fig. 8-1, and $\dot{\epsilon}_2$ is the constant slope of the line or the *secondary creep rate*.

The secondary creep rate is found empirically to be dependent on

FIG. 8-2 Photomicrograph of dislocations piled up at an inclusion. (Courtesy of A. A. Burr, Rensselear Polytechnic Institute. From M. J. Fraser, *et al.*, *Acta Metallurgica*, **4**, p. 194 (1956). Used by permission.)

both temperature and stress level. Its temperature dependency is expressed by a rate expression of the Arrhenius type

$$\dot{\epsilon}_2 = A_2 \exp\left(\frac{-Q}{RT}\right) \qquad (8.5)$$

This empirical relationship is convenient for interpolation between experimental values of secondary creep rates obtained at different temperatures.

The stress level dependency of the secondary creep rate is found empirically to be of the power law form

$$\dot{\epsilon}_2 = k\sigma^n \qquad (8.6)$$

Here k and n are both constants, with n usually having values between 3 and 4 depending upon the crystal structure involved. These empirical relations are useful for *interpolation* between existing data, but should only be used *very cautiously* to *extrapolate* data beyond the experimental conditions.

TERTIARY CREEP   As the total plastic strain continues to increase, sample necking (described in Chapter 1) becomes significant. The true stress begins to increase rapidly as this necking occurs under constant load, resulting in a similar rapid increase in the generation rate of new dislocations and, hence, in the creep rate. This period of *tertiary creep* is of a relatively short duration and terminates with the failure of the sample, soon after its inception. The time at which the sample ruptures, indicated by t_R in Fig. 8-1, is found empirically to be related to the stress level by a power law

$$t_R = k' \sigma^{m'} \tag{8.7}$$

where k' and m' are constants which differ from k and n in Eq. (8.6). (In general, $m' < 0$, while $n > 0$.)

Since the mechanisms of primary and secondary creep are strongly influenced by grain boundaries due to dislocation pileup, polycrystalline materials with small grains and large grain boundary areas might be expected to show a greater resistance to creep than coarse-grained materials. This is true at low temperatures, but thermally activated dislocation climb at the boundaries becomes more active as temperature increases. The grain boundaries represent dislocation sources and diffusional sinks for interstitial atoms. Eventually an *equicohesive temperature* is reached where the influence of the grain boundaries reverses, aiding rather than hindering dislocation motion. Above this temperature, coarse-grained materials are more creep resistant than fine-grained polycrystallines. Below this temperature, the reverse is true.

FIG. 8-3 Freshly fatigue-fractured shaft, showing smooth and rough fracture areas. Note how the rough fracture fans out into the shaft from a point on the surface. (Courtesy of T. J. Dolan. From *Metal Fatigue*, by G. Sines and J. L. Waisman, copyright 1959, McGraw-Hill, Inc. Used with permission of McGraw-Hill Book Company.)

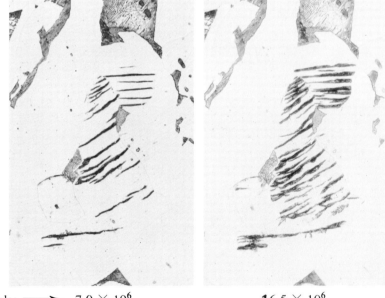

No. of cycles ⟶ 0.025×10^6 No. of cycles ⟶ 7.0×10^6

0.05×10^6 16.5×10^6

FIG. 8-4 Photographs of slip band concentrations in isolated grains, with numbers of stress cycles indicated. (Courtesy of M. R. Hempel, Max Planck Institut für Eisenforschung, Dusseldorf, Germany. From "Fracture," by B. L. Averbach, *et al.* (eds.), *Proc. Intern. Conf. Technology,* MIT, Cambridge, Mass., copyright 1959, John Wiley and Sons, Inc. Used by permission.)

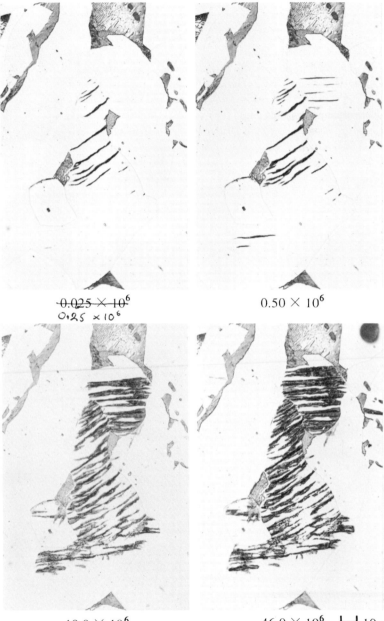

0.025 × 10⁶
O.25 × 10⁶

0.50 × 10⁶

19.0 × 10⁶

46.0 × 10⁶   ⊢⊣ 10μ

Fatigue

Another phenomenon of dynamic material response, *fatigue failure*, is also of great interest to the engineer. This is characterized by the apparently sudden, brittle failure of a piece after a period of time, under the force of oscillating stresses significantly smaller than the normal fracture stress of the material. As examples, either an alternating stress developed in a steel shaft through rotation at a high speed or vibrations set up in an otherwise rigid member can result in failure of the part under otherwise acceptable stress conditions.

When fatigue was first observed, it was thought to be a sudden occurrence due to unknown causes. It is now known that fatigue is a neither sudden nor mysteriously hidden result of the action of dislocations. Fatigue failure occurs when cracks form on the *surface* of a sample during cyclic loading. Following the initiation of these surface cracks, the fracture spreads inward over the cross section. Eventually the unfractured portion of the cross section is so reduced that the true stress level in the piece exceeds the critical normal stress of the material and cleavage results. Careful examination of a freshly broken cross section of a fatigued part reveals two characteristic surfaces: one which is smoothly polished or rippled, and one which is rough and granular—characteristic of brittle cleavage.

The smooth regions partially result from the frictional abrasion of the cleaved surfaces against one another as each stress cycle closes the crack temporarily; this smooths off the roughness caused by the failure. The last portion cleaves at the last stress cycle, however, and the rough surface is never rubbed smooth. The photograph of a freshly broken section (Fig. 8-3) clearly shows both types of surfaces.

The appearance of slip lines which multiply with the increasing number of stress cycles always precedes surface cracking. In the case of fatigue, however, the slip lines preferentially concentrate in certain grains, causing severe local strain-hardening. If the stress level is sufficiently high, these slip bands will open into very fine cracks. The development of slip bands in preferential grains and the opening of cracks on the band surfaces are illustrated in Fig. 8-4 and Fig. 8-5, respectively. These microcracks will continue to propagate, joining together and developing into a major crack which ultimately leads to failure, as shown in Fig. 8-6.

Another surface phenomenon believed to contribute to fatigue failure is the development of *surface intrusions* and *extrusions* under the action of relatively low cyclic stresses. The operation of this mechanism is indicated by the sequence of models in Fig. 8-7. At location *A* in the first model, *t* and *c* represent slip planes which would be active under tension and under compression, respectively. At location *B*, the reversal of these planes indicates that the dislocation orientation is inverted. The remaining models illustrate the sequence of events which occurs as a cyclic stress is applied, as indicated by the horizontal arrows. The motion of the slip planes at the surface results from dislocation annihi-

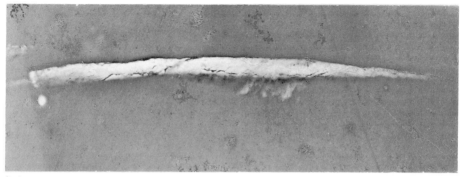

⊢ 1μ

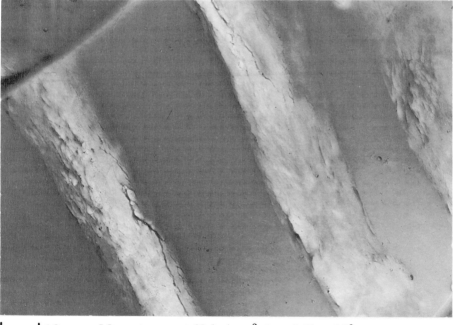

⊢——⊣ 10μ Mean stress $= \pm 19 \; kg/mm^2$, $N = 1.56 \times 10^6$

FIG. 8-5 Cracks on the surface of slip bands. (Courtesy of M. R. Hempel, Max Planck Institut für Eisenforschung, Dusseldorf, Germany. From "Fracture," by B. L. Averbach, *et al.* (eds.), *Proc. Intern. Conf. Technology,* MIT, Cambridge, Mass., copyright 1959, John Wiley and Sons, Inc. Used by permission.)

lation as they pass out through alternate sets of planes under an alternating stress field. It is clear that the net result of the movement is to create an intrusion or fissure at *A* and an extrusion or protuberance at *B*.

FIG. 8-6 Development of a major crack leading to failure. (Courtesy of M. R. Hempel, Max Planck, Institut für Eisenforschung, Dusseldorf, Germany. From "Fracture," by B. L. Averbach, *et al.* (eds.), *Proc. Intern. Conf. Technology,* MIT, Cambridge, Mass., copyright 1959, John Wiley and Sons, Inc. Used by permission.)

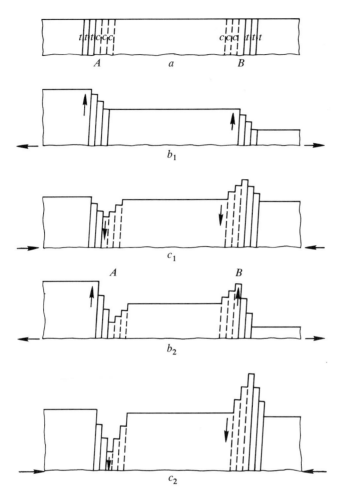

FIG. 8-7 Schematic model showing the formation of surface intrusions and extrusions. (From *Introduction to Properties of Materials* by D. Rosenthal, copyright 1964, D. Van Nostrand Company. Used by permission.)

Figure 8-8 is a photomicrograph of several of these intrusions and an extrusion on the surface of a fatigued material.

Neither of the above mechanisms can fully account for fatigue; each is probably the manifestation of a more complex mechanism. One of the most characteristic features of fatigue failure is its statistical nature. Seemingly identical samples of a material subjected to fatigue tests fracture at a widely differing number of cycles, some an order of magnitude or more [3] apart.

Two basic types of behavior can be observed when fatigue-breaking stress amplitude is plotted against the number of cycles at the break on logarithmic coordinates, as illustrated in Fig. 8-9. Aluminum and its alloys exhibit curves similar to the smooth curve in the figure; others, such as ferrous alloys, show a linear relation followed by a definite stress cut-off below which failure cannot be induced by fatigue, illustrated by the two-piece, linear curve in the figure. In the case of ferrous materials, the horizontal curve defines a lower limit called the *endurance* or *fatigue limit*. In the case of aluminum alloys where no well-defined limit exists, a stress level at which the material can withstand a very large number of cycles without failure (10^7, for example) is arbitrarily defined as the *fatigue strength*. (This is analogous to the arbitrary definition of the 0.2 per cent offset yield strength discussed in Chapter 1.)

Although no completely satisfactory correlation between fatigue limits and other structurally sensitive, mechanical properties has been developed, a rough rule of thumb for *ferrous* materials is that *the reverse fatigue limit is equal to half the tensile strength,* but is *not* true of non-

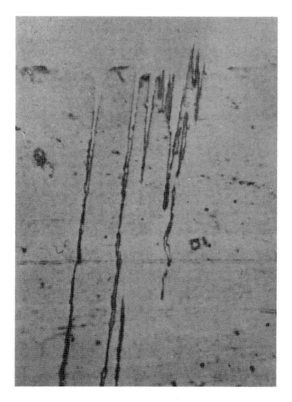

FIG. 8-8 Photomicrograph of material surface, showing intrusions and extrusion on fatigued portion. (From "Fracture," by B. L. Averbach, *et al.* (eds.), *Proc. Intern. Conf. Technology,* MIT, Cambridge, Mass., copyright 1959, John Wiley and Sons, Inc. Used by permission.)

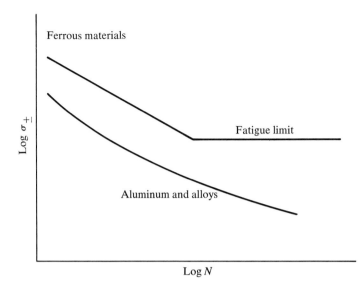

FIG. 8-9 Schematic illustration of fatigue curves for Fe and its alloys and Al and its alloys.

ferrous materials such as aluminum. The reverse fatigue limit is the fatigue strength defined in terms of an alternating stress whose magnitude is σ and whose average value is zero.

The fatigue strength of a material increases when it is simultaneously subjected to a net compressive stress on its surface, leading to one of the more commonly used protective measures called *peening*. The surface of a metal is subjected to an impingement of steel or sand shot which covers the surface with many tiny indentations, effectively compressing the surface layers and increasing their resistance to fatigue. Other protective techniques include surface hardening by quenching, nitriding (the diffusion of nitrogen into the surface layers of a steel at high enough temperatures to produce hardened alloys on its exterior), and the application of protective coatings to reduce corrosion.

Liquids and Gases

Viscous Flow

The considerations above primarily apply to the responses of crystalline solid materials to dynamic loads. However, a great many operations of engineering importance involve the transporting and handling of

liquids, gases, and amorphous semisolid materials. Since the responses of these substances to mechanical stresses vary considerably
from those of crystalline solids, the engineer must be familiar with such
differences and know how to deal with them. The full treatment of the
response of fluids to mechanical forces is far outside the scope of this
text. However, certain basic concepts of fluid mechanics essential to the
materials engineer will be discussed here.

THE CONCEPT OF VISCOSITY   In considering the effect of stress on
solids, we are concerned with the magnitude of the resultant displacement or strain. If the stress level is less than the yield strength, a predictable elastic strain results and remains in the solid until the stress is
removed. This is not true in liquids, however.

Consider the following mental experiment. A liquid is contained in
a region bounded above and below by two flat parallel planes separated
by a distance L, which is small in comparison with their transverse
dimensions. Initially, the liquid is quiescent. At some time t_0, a force F
moves the upper plane parallel to itself by an amount dx, as illustrated
schematically in Fig. 8-10. If the liquid between the planes wets their
surfaces, it will adhere to them; therefore, as the upper plane is displaced, so is the adjacent layer of liquid. This displacement propagates
downward through the fluid in amounts proportional to its distance
from the lower plane, so that, at time $t_0 + dt$, the system appears as
shown in the figure.

The liquid has experienced a shear strain $d\gamma = dx/L$. As discussed
in Chapter 3, however, the liquid structure cannot sustain a permanent
deformation because it is full of holes and massive dislocations. Consequently, if the upper plane is allowed to rest after the deformation dx,

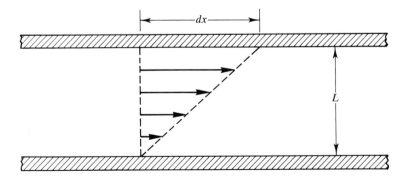

FIG. 8-10   Schematic representation of the geometry for the flow which leads
to the definition for viscosity.

molecular realignment of the holes by diffusion will quickly restore the liquid to its original quiescent state, and all evidence of the deformation will disappear.

If this experiment is repeated sequentially several times so that the upper plane moves to the right at a constant velocity, the fluid will be sheared by an amount dx/L in each time interval dt. Thus, the response of interest is not the shear strain $d\gamma = dx/L$, but the *rate of shearing strain*

$$\frac{d\gamma}{dt} = \frac{1}{L}\frac{dx}{dt} = \frac{v}{L} \qquad (8.8)$$

where v is the motion velocity of the upper plane.

Some quantitative relationship must exist between the shear stress $\tau = F/A$ and the rate of shearing strain $d\gamma/dt$. The simplest relation possible is a direct proportionality called *Newton's law of viscosity*

$$\tau = \mu\frac{d\gamma}{dt} = \mu\dot{\gamma} \qquad (8.9)$$

where the constant of proportionality is called the coefficient of viscosity, or simply the *viscosity*, and the shorthand notation $\dot{\gamma} = d\gamma/dt$ is used. If Eq. (8.9) is valid for a given fluid, with μ a temperature dependent constant independent of either τ or $d\gamma/dt$, it is called a *Newtonian fluid*.

If $\tau = 1$ dyne/cm^2, $v = 1$ cm/sec, and $L = 1$ cm, the value $\mu = 1$ dyne sec/cm$^2 = 1$ g/cm sec can be calculated from Eq. (8.9). This latter unit, defined as 1 *poise*, is the standard CGS unit of viscosity. Since water is often used as a reference material in determining physical properties, and since the viscosity of water at normal atmospheric conditions is roughly 0.01 poise, the more conventional unit of viscosity is the centipoise (cp), where 1 cp = 0.01 poise.

For a Newtonian fluid, Eq. (8.9) indicates that a plot of τ as ordinate vs $d\gamma/dt$ as abscissa on linear coordinates will be linear, will pass through the origin, and will have a slope numerically equal to the viscosity of the fluid. Such a plot, illustrated schematically in Fig. 8-11, is called a *rheogram*. From this figure, it is evident that the viscosity of fluid A is greater than that of fluid B; thus, A is said to be more viscous than B. To illustrate the viscousness of liquids, water has a viscosity of 1 cp, corn syrup one of about 60 cp, and glycerine one of about 300 cp; some road tars have viscosities of several million cp.

TEMPERATURE DEPENDENCY OF VISCOSITY   The mechanism of viscous flow in liquids is closely associated with the diffusion of holes through a liquid to relieve the strains caused by applied shear stresses. On a much larger scale and during a much shorter time, viscousness is similar

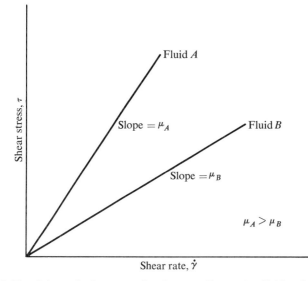

FIG. 8-11   Schematic rheogram showing two Newtonian fluids with differing viscosities.

to the motion of dislocations through a solid in slip. Therefore, it is not surprising that viscosity, like some of the coefficients in creep-stress relations, is strongly temperature dependent.

It has been found experimentally that many liquids possess exponential temperature dependencies of the form

$$\mu = \mu_\infty \exp\left(\frac{E_{vis}}{RT}\right) \tag{8.10}$$

where μ_∞ is a constant with viscosity units, and E_{vis} is the energy which activates viscous flow. It should be noted that the exponent here is positive, unlike the rate factors previously encountered. This means that the viscosity *decreases* with increasing temperature; therefore, *fluidity*, the reciprocal of viscosity, *increases* with increasing temperature. Viscosity measures resistance to motion; fluidity measures ease of motion and is analogous to a diffusion coefficient for holes.

Equation (8.10) suggests that a plot of ln (μ) as ordinate vs $1/T$ as abscissa should be linear, with a slope of E_{vis}/R (or $E_{vis}/2.303R$, if common logarithms are used) and ln (μ_∞) as an intercept at $1/T = 0$ (see Fig. 8-12).

The above set of data was purposefully chosen to illustrate that Eq. (8.10) does not always represent a temperature dependency over an entire temperature range. A decreasing value of E_{vis} caused by thermal effects results in the data curving away from the linear dependency at low values of $1/T$ (large temperatures which approach the boiling point). As the liquid approaches its boiling point, more and more of its interatomic bonds are weakened or broken by thermal activation, and large numbers of holes are generated. These, in turn, decrease the additional energy which must come from the stress field to generate flow. (This process is somewhat analogous to the thermal dislocation activation at grain boundaries which leads to the phenomenon of secondary creep in solids, as we have seen.) At the boiling point,

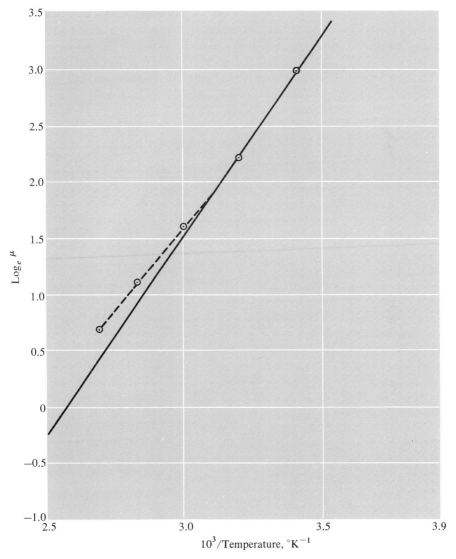

FIG. 8-12   The variation of viscosity with temperature for ethylene glycol. (From *Handbook of Chemistry & Physics, 46th ed.,* copyright 1966, Chemical Rubber Publishing Co.)

this effect is at a maximum, and the activation energy for viscous flow is small. Therefore, Eq. (8.10) should be used very cautiously for extrapolation, although it is satisfactory for interpolation over moderate ranges, as seen from Fig. 8-12.

The viscosity of a gas is the result of an entirely different phenomenon. The fundamental mechanism involved in the liquid state is the diffusion of vacancies or the rapid propagation of dislocations. In gases, however, the molecules are sufficiently dispersed to move essentially independently of one another, and viscous resistance to shear stress is associated with molecular collisions and intermolecular momentum transport. Momentum is proportional to molecular speed which was seen in Chapter 3 to be proportional to the square root of the gas temperature. Thus, it is not surprising to find that the viscosity of a gas as calculated from the simple kinetic theory [4] is similarly dependent upon temperature:

$$\mu_{\text{gas}} = \frac{2}{3\pi^{3/2}N_0}\left(\frac{MRT}{d^4}\right)^{1/2} \tag{8.11}$$

where N_0 is Avogadro's number, M is the molecular weight of the gas, R is the gas constant, T is the absolute temperature, and d is the molecular diameter. More detailed analyses of gas viscosity result in the introduction of a weak, temperature dependency function into the denominator of this expression, but the above result adequately illustrates the fundamental difference between the temperature dependency of liquid and gas viscosities. Clearly, from Eq. (8.11), the gas viscosity *increases* with increasing temperature.

Just as the interaction of dislocations with other crystal imperfections, such as foreign solute particles, retards dislocation motion and increases critical shear stress in a crystalline solid, so the introduction of insoluble particles into a liquid has a pronounced influence on its viscous flow resistance or viscosity.

A suspension of insoluble particles which are large by atomic dimensions, but still invisible to the unaided eye is called a *colloidal dispersion*. Colloidal particles frequently range in size from 10^{-3} to 1 micron (μ) (a micron is 10^{-6} m or 10^{-4} cm). Many colloidal dispersions behave as Newtonian fluids. These solid particles are solvated (surrounded by electrically polarized molecules of the suspending liquid), increasing the viscosity of the suspension until it is greater than that of the pure, suspending liquid. Einstein showed that the viscosity of a dilute suspension of spheres is related to their volume fraction by

$$\frac{\mu}{\mu_s} = 1 + 2.5\phi \tag{8.12}$$

where μ_s is the viscosity of the suspending medium and ϕ is the volume fraction of the spheres.

Equation (8.12) indicates that this increase of suspension viscosity over pure liquid viscosity is independent of the properties of the dispersed phase, the liquid phase, or the particle size or shape.[1] However, the assumptions under which Einstein derived this relation are rather stringent:

1. The particles are spheres.
2. They are large, compared with the solvent molecules.
3. They are rigid.
4. They are readily wetted by the liquid.
5. The suspension is very dilute, so that ϕ is very small.

As the concentration of particles enlarges, a theoretical calculation of the viscosity increase becomes impossible. However, many industrial processes are based on the fact that adding solids to liquids increases the viscosity of a mixture. For this reason, many plastics, rubbers, and asphalts are compounded with finely divided materials called *fillers*, which impart considerably more stiffness and rigidity to their mixtures than are characteristic of the pure materials. For example, fillers added to the semiliquid asphalt used in road surfacing form a stiff plastic mass called mastic, which, in turn, is mixed with sand and gravel or crushed rock to form a sufficiently solid substance. However, temperature even exerts a powerful influence on this filled asphalt, as evidenced by the softening of a road surface in the heat of the summer sun.

For almost all practical cases, filler particles are neither spherical nor uniform in size, making possible a greater degree of packing than the 74 per cent spherical limit. Moreover, the surface characteristics of filler particles may frequently play an important role. These factors indicate that increases in suspension viscosity far greater than those suggested by Eq. (8.12) are possible.

Non-Newtonian Materials

In Eq. (8.9), the proportionality coefficient between the shear stress and the rate of shearing strain is assumed to be a simple, constant property of the material which is independent of stress or rate of strain. This is analogous to assuming a linear stress-strain relationship such as Hooke's law in solids. However, as already seen in crystalline solids, the stress-strain relation becomes severely nonlinear when the stress levels exceed certain critical values. It is now obvious that the linear plot which passes through the origin in Fig. 8-11 is only a very special case in the infinitude of possible rheograms.

---

[1]Note that the maximum packing factor for uniform spheres is 0.74, while higher packing factors are possible for spheres of variable sizes.

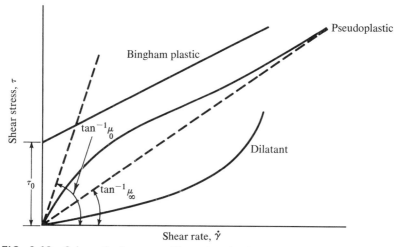

FIG. 8-13 Schematic rheograms illustrating the three general classes of non-Newtonian behavior.

If a Newtonian fluid has a linear rheogram passing through its origin, then in a rather negative fashion, a *non-Newtonian fluid* may be defined as any fluid without this special type of rheogram. A very large class of real materials of interest to the engineer fall into this all-inclusive category. Most of these substances can be grouped into one of the three general classes of rheological behavior illustrated schematically in Fig. 8-13.

THE BINGHAM BODY   The simplest generalization of the Newtonian proportionality is the completely linear equation proposed by Bingham [5].

$$\tau = \tau_0 + \eta\dot{\gamma}, \tau > \tau_0$$

$$\dot{\gamma} = 0, \tau \le \tau_0 \tag{8.13}$$

In this equation, τ_0 represents a small, but nonvanishing, *critical yield stress* which must be exceeded before the material can flow. Once this stress is reached, the material flows with very low viscosity η. Fluids to which Eq. (8.13) apply are called *Bingham plastic fluids* or *Bingham bodies*. Such substances as printer's ink, most paints, some greases, liquid suspensions of particulate solids larger than colloidal size, food purees, and sewage sludges are examples of Bingham bodies.

Since viscosity, as initially defined by Eq. (8.9), is the ratio of shear stress to rate of shearing strain, when dealing with non-Newtonian materials it is instructive to define an *apparent viscosity* similarly as

$$\mu_{\text{app}} = \frac{\tau}{\dot{\gamma}} \tag{8.14}$$

If Eq. (8.13) is introduced into Eq. (8.14), the following expression for the apparent viscosity of the Bingham plastic fluid results:

$$\mu_{app} = \frac{\tau_0}{\dot{\gamma}} + \eta, \tau > \tau_0$$

$$= \infty, \tau \leq \tau_0 \qquad (8.15)$$

From this equation, it is clear that the viscosity of a Bingham plastic material is very dependent upon the rate of shearing strain. In fact, as $\dot{\gamma}$ becomes very small, μ_{app} increases in value until it attains infinity at zero rate of strain, thus corresponding to the finite stress required to initiate flow. Additionally, as the rate of strain becomes very large, μ_{app} approaches an asymptotic limit of η. These two limiting behaviors characterize the Bingham plastic model.

THE PSEUDOPLASTIC FLUID   A very large class of fluids, including melts and aqueous solutions of high molecular weight polymers, exhibit rheograms similar to the middle curve in Fig. 8-13. At very high rates of shearing strain, the apparent viscosities of such fluids asymptotically approach a limiting Newtonian value μ_∞, but these apparent viscosities increase as this strain rate decreases toward zero. This behavior is very similar to the conduct of Bingham plastic fluids indicated in Eq. (8.15); however, these so-called pseudoplastic fluids differ in that their apparent viscosities again approach a much higher Newtonian limit μ_0 as zero rate of strain is approached. (These materials are called pseudoplastic because of the similarity between their behavior at high shear rates and that of Bingham plastic fluids.)

Many pseudoplastic fluids have been observed empirically to obey a power law relation

$$\tau = k\dot{\gamma}^n \qquad (8.16)$$

over limited ranges of strain rate. In this equation, k is a temperature dependent parameter called the consistency factor, and n is a constant ranging between 0 and 1 which is apparently independent of temperature. This phenomenon is illustrated in Fig. 8-14 for an aqueous solution of a high molecular weight polymer [6], where the data are plotted on logarithmic coordinates in accordance with Eq. (8.16).[2] The slope of such a plot is the parameter n, and the value of $\ln (\tau)$ corresponding to $\ln (\dot{\gamma}) = 0$ is $\ln (k)$. It is clear from the figure that this data at two different temperatures will fall on parallel lines, indicating the insensitivity of n to temperature.

Although the data in Fig. 8-14 are very well represented by the power law or Ostwald–DeWaele equation, it can easily be shown that

---

[2]It can easily be shown that this is the proper plot by taking logarithms of both sides of Eq. (8.16).

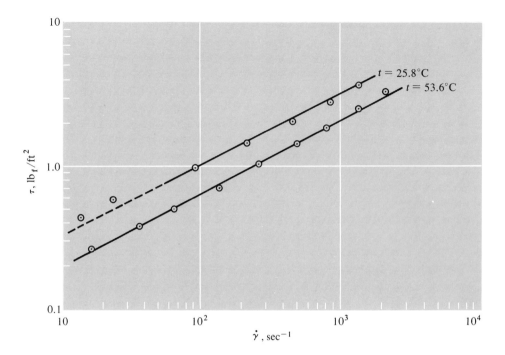

FIG. 8-14 Rheological data for an aqueous solution of a polymer, showing power law behavior. The parallel lines for different temperatures indicate the temperature independence of exponent *n*.

this will not continue to be the case for all shear rates. From Eq. (8.16), the apparent viscosity is

$$\mu_{app} = k\dot\gamma^{n-1}, \text{ power law} \tag{8.17}$$

which clearly shows that μ_{app} decreases for $n < 1$ with increasing $\dot\gamma$. As $\dot\gamma$ becomes very large, however, μ_{app} approaches zero rather than some limiting viscosity. Similarly, as $\dot\gamma$ approaches zero, μ_{app} becomes infinite instead of approaching a limiting value. Clearly, then, Eq. (8.17) is only a fortuitous engineering approximation that will work in some cases, but not in general. This is clearly indicated by the data in Fig. 8-15, again illustrated for an aqueous solution of a high molecular weight polymer [6]. Here, the power law fits a portion of the data, but the data drift away from the straight line in the expected manner as the rate of strain increases.

   This clearly points out the need for extreme caution when extrapolating beyond the data range for which the power law parameters were determined empirically. Although extremely useful for interpolation, this formulation is simply an empirical approximation, and suffers from the usual defects of such expressions.

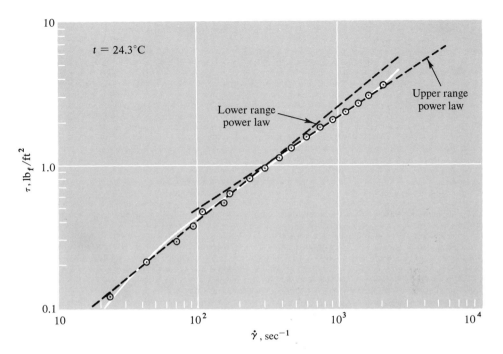

FIG. 8-15 Rheological data for an aqueous solution of a different polymer, showing power law failure.

Obviously, a more complex mathematical formulation is required to describe the rheological behavior of pseudoplastic fluids in general. The ingenuity shown by researchers in developing such complex formulae is remarkable.

THE DILATANT FLUID   In contrast to the two previous classes of non-Newtonian behavior, the third fluid shown in Fig. 8-13 is characterized by an apparent viscosity which increases with the increasing rate of shearing strain. Such a material, called *dilatant*, is exemplified by concentrated suspensions of sand in water (such as wet beach sand or quicksand), concentrated suspensions of cornstarch in water, and some organic solutions.

This material apparently dilates or expands when subjected to rapid shearing, thus, the derivation of dilatant. An example of such expansion is the swelling and apparent drying out of wet sand when it is stepped on. The mechanism of dilatancy is evidently associated with the particle-particle interference due to packing limitations which arises when a very concentrated suspension is sheared. The particles involved attempt to move past one another under the action of the shear stress. Due to close packing, however, these particles interfere with each other, causing the formation of a semirigid structure in which the liquid is

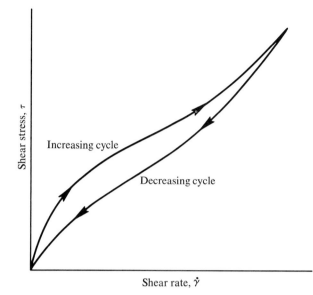

FIG. 8-16   Schematic rheogram illustrating thixotropic behavior.

interstitially trapped. As the stress is maintained, the structure will gradually deform and flow, allowing the liquid to percolate through the interstices. The more rapidly suspension shear is attempted, the greater the resistance of the structure will be. In fact, if sufficiently large rates of strain are applied, such materials can actually exhibit the characteristics of brittle fracture. This is easily observed by attempting to stir a very concentrated suspension of ordinary cornstarch in water rapidly.

If the value of n in Eq. (8.17) were greater than unity, the apparent viscosity would increase with increasing $\dot{\gamma}$. Some dilatant fluids can be approximated by a power law equation where $n > 1$. Thus, out of the spectrum of possible power law representations, Newtonian fluids only represent the special case $n = 1$.

THIXOTROPY   Many materials possess a definite structure which breaks down upon agitation or shearing, but can be restored upon standing quiescent. If subjected to a variable shear stress, such a material shows a hysteresis loop[3] such as the one illustrated schematically in Fig. 8-16. Here, variation of stress with time is indicated by the direction of the arrows. The stress is increasing with time on the upper curve and decreasing with time on the lower curve. Evidently, the apparent

---

[3]Any cyclic curve in which the reverse cycle differs from the forward cycle; derived from the Greek word *hysterein*, meaning to lag or be behind.

viscosity of this fluid is continuously decreasing as shear increases in duration. If this fluid is allowed to stand unsheared for a period of time, the upper curve will be retraced upon reshearing, indicating the restoration of the initial structure.

In some applications, thixotropy is highly desirable. Paint, for example, should exhibit several of the features of a non-Newtonian fluid. So that it will not run off a vertical wall or ceiling, paint should possess a yield stress similar to a Bingham plastic fluid. The magnitude of the yield stress governs the thickness of the film of paint which will adhere to a surface without *sagging* or running. During application, however, it is desirable for the paint to be fluid; that is, it should spread easily by brushing or rolling (shearing). Thus, paints must also be thixotropic, with low, high-shear limiting viscosities.

Noncrystalline Semisolids

Viscoelasticity

A large category of materials which includes many polymers or plastics cannot be classified either as liquids or solids. Such materials, called *semisolids*, exhibit solidlike creep properties sometimes and liquidlike viscous properties at other times.

Just as the application of a continued load leads to plastic flow in crystalline solids under the proper conditions, all liquids will exhibit elastic properties under given conditions. *The time scale for load application is the determining factor which separates dynamic material response into the modes of elastic or viscous flow.*

In the cases of both solid creep and liquid viscous flow, the mechanism for continuous material deformation is the motion of dislocations or imperfections (holes in the liquid) through the more or less ordered structure of the material. If a material experiences continuous permanent deformation under the application of forces moderate in magnitude for moderate time intervals, it will respond either in creep or viscous flow. In the case of solids, however, an elastic component to the deformation also exists which is recoverable upon removal of the stress; no such elastic recovery appears in the normal flow of viscous liquids.

Because of the high fluidity and low molecular weight of many liquids, the actual time scale necessary to relieve elastic strain by viscous flow is considerably shorter than the experimental time scale, and no elastic response is observed. However, if an experiment is performed involving very short stress application times, such as the oscillations caused by high frequency sound waves, pronounced elastic effects can be observed in common viscous liquids.

With an increase in viscosity, the response time characteristic of elastic strain becomes increasingly longer, until the elastic behavior predominates over the behavior of viscous flow in materials with extremely high viscosities, such as glass, pitch, and some polymers. For example, under prolonged stress glass exhibits the true viscous flow of a high viscosity liquid. Under a short stress application time, however, glass behaves as a brittle elastic solid (a well-known fact to any boy who has ever mistaken a window for a catcher's mit).

Thus, it appears that a distinction between the mechanical responses of solids, semisolids, and liquids relies primarily on the magnitude of the applied force and the time scale of its application in relation to certain characteristic material response times. In reality, all materials fall somewhere along a continuous scale of combined elastic and viscous responses, and the boundaries separating these three, somewhat arbitrary groups of response cannot be sharply defined.

VISCOELASTIC MATERIALS   Many materials of practical interest to the engineer exhibit a combination of elastic and viscous responses to applied stresses; both of these types of response are significant within the time scale of ordinary usage. Such *viscoelastic materials* cannot be treated either as solids or liquids; mucous, some types of glue, most polymers in the molten state, many "solid" rocket-propellant binder materials, and elastomeric or rubbery materials are a few examples.

In order to deal quantitatively with viscoelastic materials, mathematical models are needed to represent the observed strain-time behavior of real materials. The profusion and sophistication [7] of these models is great and, for the most part, outside the scope of this book. Two of the simplest linear models will be presented here to acquaint the student with certain basic properties of viscoelastic materials.

These simple, mechanical models are composed of *Hookean springs*[4] and *Newtonian dashpots* (defined below) whose mathematical descriptions result in strain-time curves analogous to the two common types of viscoelastic behavior. The student must realize that real viscoelastic materials are made up of molecules, not springs and dashpots. The description of actual materials from a molecular viewpoint can become extremely complex [8], however, making a resort to these simple mechanical analogs almost imperative.

A Newtonian dashpot consists of a cylinder filled with a Newtonian

---

[4]A Hookean or ideal linear spring represents purely elastic behavior because its deformation is related to its stress by Eq. (7.9).

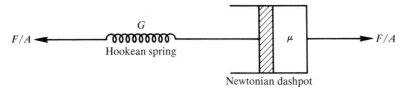

FIG. 8-17 Schematic representation of a Maxwell model of the strain-time response in viscoelastic fluid.

fluid of viscosity μ and fitted with a frictionless, leak-proof piston. The fluid chamber is connected through a small opening to a large reservoir of similar fluid. As the piston is withdrawn or inserted, fluid respectively flows into or out of the cylinder through the small hole. The rate of flow is restricted and proportional to μ. The force F moving the piston, divided by the piston area A, is proportional to the viscosity of the fluid and the rate of motion of the cylinder. Consequently, the mathematical description of the mechanical response of a Newtonian dashpot is the same as that of a viscous Newtonian fluid. Dashpots using air as the operative fluid, for example, are frequently attached to doors to retard their rate of closing.

THE MAXWELL MODEL   The *Maxwell model*, illustrated schematically in Fig. 8-17, consists of a Hookean spring in series with a Newtonian dashpot. At time $t = 0$, a load is applied to the model, causing it to experience an instantaneous elastic deformation γ_e initiated by the spring element, followed by a steady viscous flow due to the dashpot (see the AB portion of the curve in Fig. 8-18). If the load is removed at point B, the elastic deformation of the spring will be recovered as it contracts; again this occurs essentially instantaneously, as illustrated by the BC portion of the same curve. With removal of the force, the spring returns to its original relaxed state. The dashpot, however, remains extended and corresponds to a permanent viscous deformation which is irrecoverable.

The mathematical analysis of this model is simple and straightforward. The instantaneous deformation due to the spring is

$$\gamma_e = \frac{\tau}{G} \tag{8.18}$$

where G is the shear modulus of the spring. Meanwhile, the dashpot begins to deform at a steady rate

$$\frac{d\gamma_v}{dt} = \frac{\tau}{\mu} \tag{8.19}$$

The series arrangement within the model requires that each element experience the same stress. Consequently, the total strain of the model

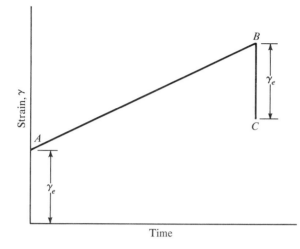

FIG. 8-18   Schematic strain-time response curve for a Maxwell model. γ_e is the initial elastic strain which is recovered when the load is removed at time *B*.

is the sum of each of the elemental strains

$$\gamma = \dot{\gamma}_e + \gamma_v \qquad (8.20)$$

Differentiating Eq. (8.20) with respect to time, we obtain the following first-order, linear differential equation which describes the strain-time behavior of this model

$$\frac{d\gamma}{dt} = \frac{1}{G}\frac{d\tau}{dt} + \frac{\tau}{\mu} \qquad (8.21)$$

Consider two special cases of this equation. If the stress is maintained at a constant value so that it corresponds to constant loading, Eq. (8.21) reduces simply to Eq. (8.9). That is, under a constant stress, the response of the Maxwell model of viscoelastic fluids is indistinguishable from the response of a viscous Newtonian liquid. This result has many important consequences in practical engineering [7], since viscoelastic materials of this type can be dealt with as if they were purely viscous liquids under many constant stress situations.

If, rather than the stress, the total strain were maintained at a constant value, Eq. (8.21) would reduce to

$$\frac{d\tau}{dt} + \frac{G}{\mu}\tau = 0 \qquad (8.22)$$

which has the solution

$$\tau = \tau_0 \exp\left(-\frac{G}{\mu}t\right) \qquad (8.23)$$

where τ_0 is the initial stress applied to affect the initial strain, which is then maintained at a constant value.

Equation (8.23) indicates that the stress gradually disappears or *relaxes* so that, if left to itself, the model will eventually lose all internal stresses due to spring contraction and further extension of the dashpot. When the time $\theta_r = \mu/G$ elapses, the stress will have decayed to a value τ_0/e, where $e = 2.73 \ldots$ is the base of natural logarithms. This particular time interval is defined as the *relaxation time* of the model. Thus, Eq. (8.23) can be rewritten as

$$\tau = \tau_0 e^{-t/\theta_r} \tag{8.24}$$

Many materials exhibit this phenomenon of *stress relaxation*.

THE KELVIN–VOIGT MODEL   In the Maxwell model, the mechanical elements are arranged in series. If these elements are parallel to one another, the *Kelvin–Voigt model* results (see Fig. 8-19).

In the latter model, both elements must experience the same total strain. The total stress $\tau = \tau_e + \tau_v$ is, therefore, the sum of the two

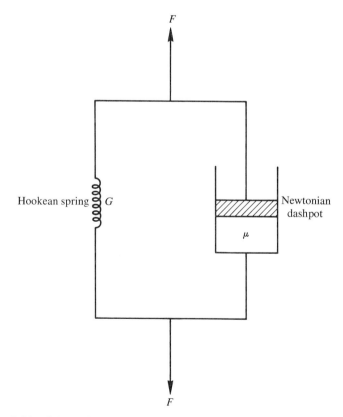

FIG. 8-19   Schematic representation of a Kelvin–Voigt model of the strain-time response in viscoelastic fluid.

component stresses. This can be expressed mathematically by the differential equation

$$\tau = \mu \frac{d\gamma}{dt} + G\gamma \qquad (8.25)$$

which has the solution

$$\gamma = \frac{\tau}{G}(1 - e^{-Gt/\mu}) \qquad (8.26)$$

for the special case of constant stress. If $T_r = \mu/G$, Eq. (8.26) can be written

$$\gamma = \frac{\tau}{G}(1 - e^{-t/T_r}) \qquad (8.27)$$

The reason for the time constant T_r, called the *elastic retardation time* in this case, is evident from the equation. When stress is applied in the Kelvin–Voigt model, the spring does not deform instantaneously as it does in the Maxwell model. Instead, the deformation is retarded by the action of the dashpot, only slowly approaching its asymptotic value τ/G at very large times.

If the stress is removed at time t_0 when the strain has the value γ_0, the material again recovers from its strain slowly. From Eq. (8.25), it is clear that this recovery can be described by the differential equation

$$0 = G\gamma + \mu \frac{d\gamma}{dt} \qquad (8.28)$$

which has a solution

$$\gamma = \gamma_0 \exp\left[-\frac{G}{\mu}(t - t_0) \right] = \gamma_0 \exp\left[-\frac{(t - t_0)}{T_r} \right] \qquad (8.29)$$

Although the time constant T_r in this equation is composed of the same two physical parameters that appear in the Maxwell model relaxation time, it is associated with the physically different phenomenon of *retarded elastic recovery*.

By considering the special case of constant strain in this model, it can be seen immediately that Eq. (8.25) reduces to Eq. (7.9). This observation signifies that under conditions of constant strain the response of the Kelvin–Voigt model is indistinguishable from that of an elastic solid.

The behavior of the Kelvin–Voigt model under constant stress conditions is illustrated schematically in Fig. 8-20. This model indicates the behavior to be expected from materials exhibiting *retarded elastic deformation* (also called *creep recovery*, *elastic after-effect*, or *elastic memory*). Many polymers, nylon and polyvinylchloride, for examples, exhibit elastic memory.

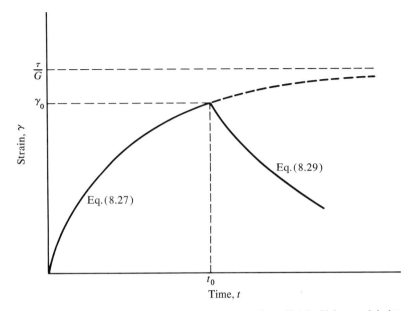

FIG. 8-20   Schematic strain-time response curve for a Kelvin–Voigt model. At time t_0, stress is removed and the model relaxes from γ_0, according to Eq. (8.29).

REAL VISCOELASTIC MATERIALS   As indicated above, real viscoelastic materials are composed of molecules, rather than springs and dashpots. Although the quantitative molecular analysis of viscoelastic behavior is extremely complex [8], it is possible to obtain some qualitative insight into the molecular origin of such phenomena as stress relaxation, retarded elastic strain, and elastic after-effect or memory.

In Chapter 12, the structures of several high molecular weight, polymeric materials and the relation of some of their properties, such as viscosity, to their molecular weights and weight distributions will be discussed in some detail. Here some of the results of that chapter will be borrowed without full exposition to explain the above viscoelastic phenomena.

Polymeric materials consist of extremely long molecules of varying individual molecular weights which are highly entangled, similar to the individual strands in a bowl of spaghetti. When such a mass of molecules is subjected to a shear stress, it is not free to flow immediately as a low molecular weight liquid filled with holes and dislocations would be. (For the sake of comparison here, think of the low molecular weight liquid as a partially filled box of marbles.) Initially, the polymer molecules simply uncoil along the long axis of their structure, much like springs. As the stress continues, however, the van der Waals forces which hold them together are overcome, and the molecules begin to untangle themselves and slide past one another. The viscous deformation (modeled by the dashpot) arises from this untangling and sliding. The elastic restoring force (modeled by the spring) is caused by the forced bending of the covalent bonds which hold the individual molecular atoms together along the chain.

The phenomenon of stress relaxation is now seen as the result of a delay in the sliding and untangling of molecules which occurs when a fixed strain is applied to the assemblage. The elastic stresses induced in the molecules cause them to recoil slowly and slide past each other, re-establishing new van der Waals forces in their newly deformed locations.

In addition to the van der Waals forces, many polymeric materials such as vulcanized rubber possess primary bonds between their molecular chains. These intermolecular primary bonds restrict both the axial uncoiling to slow occurrences and the sliding to certain fixed limits, causing the material involved to exhibit retarded elastic deformation. When the external stress on such a material is removed, these intermolecular primary bonds couple with the elastic axial stresses of the molecules to slowly restore the mass to its more or less original condition. Such a restoration is called an elastic recovery or memory phenomenon. (It should be obvious that the degree of intermolecular bonding will have a pronounced effect on material response and, thus, that all grades of response will exist between the two pure extremes discussed here.)

Each of the mathematical derivations presented above for these simple linear models has resulted in a characteristic time constant called the relaxation time or the retarded recovery time. In real materials, however, the existence of molecular weight distributions results in a spectrum of relaxation times, rather than a single one. Usually, though, one of these times will predominate over all the others, so that the simple mathematical results presented in this section can approximate the behavior of real viscoelastic materials.

In Chapter 12, the viscosity of polymeric liquids will be related to the distribution of individual molecular weights. Since the relaxation times of the above models are directly related to dashpot viscosity, it should not be surprising to find that the relaxation times of real materials are likewise related to molecular weight distributions [8]. When polymeric solutions are considered, these relaxation times are also strongly dependent upon polymer concentration. The analysis of these effects in relation to molecular parameters is a fascinating subject for advanced study. With a rapid increase in the use of polymeric materials in modern technology, many engineers will become increasingly involved with viscoelastic phenomena in several contexts.

References Cited

[1] Cottrell, A. H., *Dislocations and Plastic Flow in Crystals*, Oxford University Press, New York, 1956, Chapter 5.

[2] Andrade, E. N. da C., *Proc. Roy. Soc. (London)*, **A84**, p. 1 (1910); **A90**, p. 329 (1914).

[3] Rosenthal, D., *Introduction to Properties of Materials*, D. Van Nostrand Company, Princeton, N. J., 1964, Chapter 10.

[4] Present, R. D., *Kinetic Theory of Gases*, McGraw-Hill Book Co., Inc., New York, 1958.

[5] Buckingham, E., *Proc. A.S.T.M.*, **21**, p. 1154 (1921).

[6] Hanks, R. W., *PhD Thesis in Chemical Engineering*, University of Utah, 1960.

[7] Fredrickson, A. G., *Principles and Applications of Rheology*, Prentice-Hall, Inc., Englewood Cliffs, N. J., 1964.

[8] Williams, M. C., *AIChE Journal*, **12**, p. 1064 (1966).

PROBLEMS

$\dot{\varepsilon}_2 = .51 \times 10^{-4} \frac{\%}{hr}. \quad \sigma = 18600 \ psi \quad T_R = 4100 \ hr.$

8.1 A sample of 0.5% molybdenum steel subjected to creep testing yielded the following data:

σ (psi)	$\dot{\varepsilon}$ (%/hr)	t_{rupt} (hr)
15,400	$0.45(10^{-4})$	4738
17,600	$0.75(10^{-4})$	2712
19,800	$1.35(10^{-4})$	1550

Compute the minimum creep rate and rupture time at a stress of 16,000 psi. What stress level gives rise to a minimum creep rate of $1.0(10^{-4})$ %/hr?

8.2 A sample of Inconel "X" subjected to creep testing gave the following data:

Time (hr)	Elongation (%)
10	1.0
200	2.0
2000	4.0
4000	6.0
5000	(neckdown starts)
5500	(rupture)

$\dot{\varepsilon}_r = 1.050 \times 10^{-3} \frac{\%}{hr}.$

Determine the creep rate from these data.

8.3 In a creep test performed under tensile loading conditions, the flow of the material is related to the tensile stress by $\sigma = 3\mu\dot{\varepsilon}$; the factor of 3 arises because flow occurs in the direction of the stress, as well as in the other two transverse directions. For the data in Problem 8.2, estimate the viscosity which may be assigned to the Inconel for a stress of 2000 psi.

8.4   Using the data of Problem 8.1, compute the viscosities which may be assigned to the steel at various stress levels. Is this material Newtonian or non-Newtonian? If it is non-Newtonian, determine an approximate representation of its viscosity-stress relation.

8.5   A sample of aluminum subjected to a constant stress of 3000 psi exhibits the following creep behavior as a function of temperature:

$\dot{\epsilon}$ (in./in. hr)	T (°C)
0.04	151
0.122	205
2.41	258

Compute the creep rate at 175°C and 235°C. Also compute the activation energy for viscous flow associated with the assigned viscosities of aluminum over this temperature range.

8.6   Using the data of Table 7-1 and Problem 8.5, estimate the relaxation time of aluminum at 235°C. If a sample of aluminum were stressed to 2500 psi at this temperature, how much time would be required for the stress to relax to 75% of this value?

8.7   The following data are derivable from the creep curves of A. E. Johnson and N. F. Frost (*Creep and Fracture of Metals at High Temperatures*, Philosophical Library, New York, 1957, p. 370) for a 0.5% molybdenum steel at 550°C.

		Stress, psi	
	15,400	17,600	19,800
Time, hrs		Creep strain, in./in.	
0	0.00014	0.00027	0.00049
200	0.00045	0.00052	0.00103
400	0.00056	0.00067	0.00138
600	0.00063	0.00081	0.00167
800	0.00073	0.00093	0.00193
1000	0.00081	0.00107	0.00224
1200	0.00089	0.00121	0.00284
1500	0.00100	0.00149	0.00370
2000	0.00125	0.00207	(fractures
2500	0.00156	0.00300	at 1550 hr;
3000	0.00184	(fractures	elongation = 1.8%)
3500	0.00221	at 2712 hr;	
4000	0.00298	elongation = 4.0%)	
4738	(fractures;		
	elongation = 6.25%)		

From these data, determine the minimum creep rates at the three stress levels. Compute the viscosities assignable to the steel at these conditions. Is this material Newtonian or non-Newtonian? If non-Newtonian, determine the constants in a power law representation of the stress-viscosity relation. Also compute the creep rate and rupture time at 16,500 psi, 18,000 psi, and 22,000 psi.

8.8   A steel bolt joining two parts is initially stressed to 25,000 psi at a temperature similar to that at which the data in Problem 8.1 were obtained. Using the data for Fe in Table 7-1 as typical, estimate both the relaxation time of this steel and the length of time necessary for the stress to relax to 65% of its initial value. How reliable do you think these estimates are? Justify your answer.

8.9   The viscosities of castor oil at 10°C and 20°C are 2420 cp and 986 cp, respectively. Estimate its viscosity at 15°C and 30°C. Is your estimate at the higher temperature as reliable as your estimate at the intermediate temperature? Why or why not?

8.10   An asphalt specimen showed the following viscosity temperature dependence:

T (°C)	μ (poise)
0	$2(10^{11})$
20	$3(10^8)$
40	$3.5(10^6)$
60	$5(10^4)$
80	2000
100	200
140	9

From these data, determine the activation energy for viscous flow at 50°C and 90°C. Compute the viscosities of the asphalt at these same temperatures. Account for the change in E_{vis}.

8.11   A polyvinyl acetate resin has the following viscosities:

T (°C)	μ (poise)
200	$1.25(10^4)$
150	$1.25(10^5)$
100	$1.6 \ (10^7)$
80	$3.1 \ (10^8)$
60	$6.5 \ (10^9)$

Prepare a plot of E_{vis} as a function of temperature for this material. What are its viscosities at 125°C and 75°C?

8.12   A certain high molecular weight polymer is found to have a viscosity of $2.0(10^{12})$ cp and a Young's modulus of $0.75(10^6)$ psi at 80°C. It behaves like a Maxwell material and is to be used as a gasketing material in a device which will be subjected to a cyclic stress $\tau = \tau_0 \cos \omega t$. Derive an equation for the strain as a function of time for this material, and plot a curve of γ/γ_{e0} vs time for a few stress cycles, where $\gamma_{e0} = \tau_0/G$ is the initial elastic strain experienced by this material at $t = 0$ when the stress is initially applied. The frequency of oscillation in this device is to be 10 cycles per min. (HINT: In computing G from E, assume $K = \infty$.)

8.13 The following data were obtained with a rotational viscometer for a solution of sodiumcarboxymethylcellulose in water at 28.1°C.

τ $(\text{lb}_f/\text{ft}^2)$	$d\gamma/dt$ $(\sec^{-1})$
0.01194	7.05
0.0199	9.85
0.0298	18.61
0.0499	33.0
0.0696	41.5
0.0896	52.0
0.1135	70.0
0.1493	94.9
0.189	120
0.229	146
0.309	206
0.392	273
0.469	346
0.549	419
0.628	510
0.747	611
0.866	726
1.024	915
1.184	1173

Determine the rheological classification of this fluid. Prepare a plot of apparent viscosity as a function of shear rate using cp units. Determine both the type of rheological equation which best fits these data and the constants therefor. Using these constants, interpolate for the shear stresses which result in the shear rates of 250 $\sec^{-1}$ and 800 $\sec^{-1}$.

THERMAL PROPERTIES OF MATERIALS

We have previously discussed the influence of mechanical stresses on several "structure sensitive" properties of materials. Structure sensitive properties are those whose magnitudes are materially affected by molecular orientation, an arrangement which is very sensitive to the state of the mechanical stress of the material.

When compared to mechanical properties, thermal properties, such as specific heat, thermal conductivity, thermal expansion coefficients, thermodynamic properties, and so on, are relatively structure insensitive. Many of these thermal properties are of very great engineering importance because the processes involved in transferring thermal energy to, from, or through a given material are of vital practical importance.

Specific Heats

A great deal of work has been performed, both theoretically and experimentally, to obtain information concerning the specific heats of numerous materials. Engineers make extensive use of these data because it allows them to transfer various thermodynamic properties from one temperature range to another.

Because of its simplicity, a rigorous kinetic, theoretical structure that permits precise calculation of specific heats has been developed for the gaseous state. The liquid state, however, is much less well understood theoretically, and the solid state is the least well understood. Although much is known about the gaseous state, it is of considerable interest only to a small number of engineers and therefore will not be examined here. The interested reader may find excellent discussions in texts dealing with kinetic theory, physical chemistry, and thermodynamics.[1] The majority of information available concerning the specific heats of liquids is empirical in nature and is presented in the form of empirical correlations. Thus, much of this discussion will concern the solid state.

From thermodynamic considerations, the two separate specific heats C_p and C_v can be defined. The former is the amount of thermal energy that must be added as heat to a given mass of material at constant external pressure in order to raise its temperature by 1°K at any particular temperature. The latter is the quantity of thermal energy necessary to perform the same task under the condition of constant volume. In the solid state, these two thermal energies differ by only a small magnitude, but the constant pressure heat capacity is the one commonly tabulated because it can be more easily measured experimentally. All of the existing theories, however, present expressions for C_v. The thermodynamic relation between C_p and C_v is

$$C_p - C_v = TV \frac{\alpha_v^2}{\beta} \tag{9.1}$$

where

$$\alpha_v = \frac{1}{V}\left(\frac{\partial V}{\partial T}\right)_P \tag{9.2}$$

is called the *volume coefficient of thermal expansion* and

$$\beta = -\frac{1}{V}\left(\frac{\partial V}{\partial P}\right)_T \tag{9.3}$$

is the *volume coefficient of compressibility*. The most frequently encountered units for heat capacities are cal/g-atom degree.

Dulong and Petit observed that the gram atomic heat capacities of many elemental solids with higher atomic masses were very nearly the same—6 cal/g-atom degree. This empirical rule is equivalent to what one could compute from the theories of classical (as opposed to quan-

---

[1] A good introduction to this subject is found in O. A. Hougen, K. A. Watson, and R. A. Ragatz, *Chemical Process Principles, Part II, Thermodynamics*, John Wiley and Sons, Inc., New York, 1966.

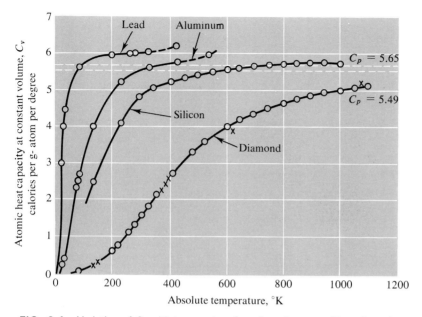

FIG. 9-1   Variation of C_v with temperature for a few elements. (From *Introduction to Modern Physics, 5th ed.,* by F. K. Richtmeyer, E. H. Kennard, and T. Lauritsen, copyright 1955, McGraw-Hill Book Co., Inc. Used by permission.)

tum) physics under the assumption that equipartition of energy[2] exists among the three degrees of freedom or possible modes of vibration available to a molecule or atom in a solid. According to this concept, the constant volume heat capacity should be $3R$, where R is the universal gas law constant. Since $R = 1.987$ cal/g-mole degree, this agrees well with the empirical rule of Dulong and Petit. However, this theoretical result is valid only for high temperatures and is seriously in error even at relatively high temperatures for highly covalent materials such as diamond and silicon (see Fig. 9-1). Using the quantum theory, numerous attempts have been made to account for the low temperature variation of the heat capacity data, the most successful being those of Debye, Born, and Von Karman.

Because the metallic bond consists primarily of a free electron gas permeating the ionic core structures (see Chapter 2), the contribution of the electrons to the specific heat should be an additional $(\tfrac{3}{2})R$, which would be far in excess of experimental observations. This puzzling fact was accounted for when Einstein realized that the quantum laws that electrons obey were quite different from the simple classical laws that

---

[2]i.e., all modes of vibration are equivalent: no mode tends to possess more energy than another, and the energy is equally divided, or partitioned, among the possible modes or ways of vibration.

ordinary gases obey. Einstein assumed that each atom in the metallic solid vibrated about its equilibrium position independently of its neighbors and with its own characteristic vibrational frequency ν. This frequency was then assumed to be proportional to the atomic mass. In terms of quantum theory, the energy of such an oscillating atom with three degrees of vibrational freedom is

$$E = \frac{3N h\nu}{e^{h\nu/kT} - 1} \tag{9.4}$$

where N is Avogadro's number, h is Planck's constant, k is Boltzmann's constant, and T is the absolute temperature. Differentiation of Eq. (9.4) with respect to temperature gives Einstein's expression for C_v.

Although this relationship is essentially correct, near the absolute temperature it predicts too rapid a decrease in C_v with temperature. Debye, Born, and Von Karman saw the cause of the difficulty almost simultaneously when they recognized that atoms in a crystal lattice could not vibrate independently of one another because coupling forces exist between the atoms as the result of bonding; thus, whole series of neighboring atoms can vibrate in sympathetic fashion. Such a model would be difficult to analyze mathematically if Debye and the others had not changed their viewpoint. Rather than an assemblage of discrete particles, their new model described the crystal as a continuous medium[3] in which sound waves of varying wavelength could be excited. Debye treated the different modes of vibration due to the sympathetic movements of groups of atoms as waves traveling in different directions with different energies, each wave being subject to Einstein's energy considerations. In this model, there are numerous vibrational frequencies associated with an atom, rather than the single characteristic one that Einstein assumed. To account for this, it was necessary to add them all up by integration to obtain the expression

$$E = \frac{9N}{\nu_m^3} \int_0^{\nu_m} \frac{h\nu}{[e^{h\nu/kT} - 1]} \nu^2 \, d\nu \tag{9.5}$$

where ν_m is a maximum upper-frequency limit which corresponds to a wavelength approximately equal to the interatomic distance. By expressing the variables in Eq. (9.5) as a dimensionless quantity $X = h\nu/kT$ and differentiating with respect to temperature, as before, the Debye expression for the specific heat can be obtained as

$$C_v = 9R \left\{ 4\left(\frac{T}{\theta}\right)^3 \int_0^{\theta/T} \frac{X}{e^X - 1} \, dX - \frac{\theta}{T}\left[\frac{1}{e^{\theta/T} - 1}\right] \right\} \tag{9.6}$$

---

[3]This is a mathematical model of the real material. It possesses certain mathematical properties useful in calculation, and it is expected that results calculated from this model will agree with the observed behavior of the real material.

where $\theta = h\nu_m/k$ is a quantity with the dimensions of temperature, known as the *Debye characteristic temperature* of the solid.

The Debye theory of solid-state specific heats is one of the early triumphs of the quantum theory. Although it is of considerable interest by itself, it will be discussed here in connection with the concept of the wave motions of sympathetically vibrating atoms. These waves, which can have an entire quantized spectrum of wavelengths or frequencies, are necessary to the explanation of the phenomenon of heat conduction in metallic solids.

Each vibration of the atoms of the crystal can be thought of as a traveling wave which carries energy through the crystal. By analogy with the quantum light theory, which depicts the electromagnetic radiation of light waves in some instances as particles called *photons*, the very high frequency sound waves traveling through the crystal as the result of atomic lattice vibrations can be considered quantized particles called *phonons*. Thus, the phonon is a packet, or bundle, of wave energy that travels through a crystal carrying thermal energy. Its interactions with other phonons and with the imperfections of the crystal structure are very important in determining the thermal behavior of metallic solids. Figure 9-2 is a schematic illustration of the origin of the lattice waves in a crystal. The solid circles represent a sympathetic vibration

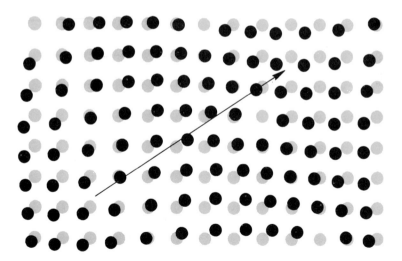

FIG. 9-2 Schematic representation of the origin of phonon waves in a solid crystal lattice. Solid circles represent the bunched atoms during a sympathetic vibration. Open circles are the mean undisturbed atom sites. (From "The Thermal Properties of Materials," by J. Ziman, in *Materials, A Scientific American Book*, copyright 1967 by Scientific American, Inc., published by W. H. Freeman & Co. Used by permission.)

cycle, while the open circles represent the mean, undisturbed positions of the atoms. The compression and rarefaction portions of the wave can be clearly seen.

Thermal Expansion of Solids

In Chapter 2 it was observed that the energy of a pair of atoms within a crystalline solid can be expressed in the form

$$E = -\frac{a}{R^m} + \frac{b}{R^n}, n > m \tag{9.7}$$

where a, b, m, and n are constants characteristic of the particular material, and R is the interatomic or interionic distance. Figure 9-3 is a schematic plot of this type of energy curve. It should be clear that the constants in Eq. (9.7) will govern the steepness of the trough at the minimum point, that is, the point which corresponds to the equilibrium state of the atoms. In addition, these constants will determine the degree

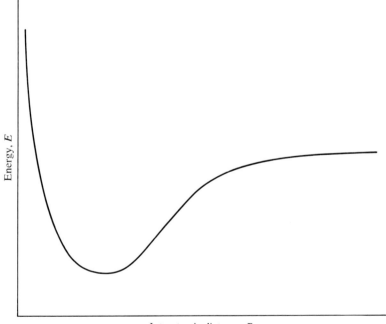

Interatomic distance, R

FIG. 9-3 Schematic representation of an interatomic potential energy curve.

TABLE 9-1

Coefficients of Linear Thermal Expansion*
for Selected Materials†

MATERIAL	$10^4 \alpha$	$10^6 \beta$	TEMPERATURE RANGE (°C)
Al	0.22	0.009	0–500
Cd	0.526	—	20–100
C (diamond)	0.214	—	20–100
Cr	0.086	—	20–500
Cu	0.161	0.0040	0–625
Fe (cast)	0.1158	0.0053	0–750
Fe (steel)	0.118	0.0053	0–750
Pb	0.269	0.011	100–240
Mg	0.248	0.0096	20–500
Mn	0.216	0.0121	20–300
Mo	0.0501	0.0014	19–305
Ni	0.1236	0.0066	20–300
Ni	0.1346	0.0033	300–1000
Ag	0.1939	0.00295	20–500
Sn	0.2033	0.0263	8–95
W	0.0428	0.00058	−105–(+502)
Zn	0.354	0.010	0–400

*The coefficients α and β are for use in the formula $L = L_0(1 + \alpha t + \beta t^2)$, where L_0 is the length at 0°C and t is the temperature in °C.
†Data from *Chemical Engineer's Handbook, 4th ed.*, by R. H. Perry, *et al.*, copyright 1963, McGraw-Hill Book Co., Inc. Used by permission.

of asymmetry exhibited by the material as its atoms move from their equilibrium positions under the action of heat addition and thermal expansion. If the bonds between the atoms are strong and highly directional, as in covalent and ionic solids, the degree of thermal expansion will be relatively small. If, however, the atoms are more loosely bound relative to one another, as in a metal, a greater degree of expansion is possible. In molecular solids, where the bonding least restricts the movement of the molecules, the thermal expansion will be the greatest. The relative values of the thermal expansion coefficients for a number of different materials are shown in Table 9-1.

Since thermal expansion arises from the storage of energy in the atoms and their subsequent movement away from their equilibrium positions (Fig. 9-3), it should not be surprising to find that the temperature dependency of the coefficient of linear thermal expansion and the heat capacity are very nearly the same. To demonstrate this, we begin with Eq. (9.1), which may be rewritten in the form

$$\frac{C_p}{C_v} = 1 + \frac{\alpha_v^2 VT}{\beta C_v} \qquad (9.8)$$

TABLE 9-2

Numerical Values of Gruneisen's Constant for Selected Materials*

MATERIAL	γ	MATERIAL	γ
Na	1.25	Ag	2.40
K	1.34	Pt	2.54
Al	2.17	NaCl	1.63
Mn	2.42	KF	1.45
Fe	1.60	KCl	1.60
Co	1.87	KBr	1.68
Ni	1.88	KI	1.63
Cu	1.96		

*Data from *Introduction to Solid State Physics*, 2nd ed., by C. Kittel, copyright 1956, John Wiley and Sons, Inc. Used by permission.

where the various terms have the same significance as before. In 1910, Gruneisen developed an empirical rule for the ratio

$$\frac{C_p}{C_v} = 1 + 3\gamma\alpha_L T$$

where α_L is the linear coefficient of thermal expansion, and γ, called the *Gruneisen constant*, has a numerical value of the order of 2 (see Table 9-2). Gruneisen assumed γ to be independent of temperature. Combining these relationships, it can be seen that

$$\gamma = \frac{\alpha_v^2 V}{3\alpha_L \beta C_v} \tag{9.9}$$

The relationship between α_v and α_L is easily deduced, as shown in the following calculation:

EXAMPLE 9.1

For small thermal strains, show to a first approximation that the coefficient of volume expansion is three times the coefficient of linear expansion.

Proof:

$$V = V_0(1 + \epsilon_v) \qquad \epsilon_v = \alpha_v \Delta T \qquad V_0 = L_0^3$$

$$L = L_0(1 + \epsilon_L) \qquad \epsilon_L = \alpha_L \Delta T$$

$$\therefore L_0^3(1 + \epsilon_v) = [L_0(1 + \epsilon_L)]^3 = L_0^3[1 + 3\epsilon_L + 0(\epsilon_L^2)]$$

$$\therefore \epsilon_v = 3\epsilon_L + 0(\epsilon_L^2) \doteq 3\epsilon_L$$

$$\therefore (\alpha_v - 3\alpha_L)\Delta T \doteq 0$$

$$\therefore \alpha_v \doteq 3\alpha_L \qquad\qquad \text{Q.E.D.} \qquad (9.10)$$

TABLE 9-3

**Thermal Expansion Coefficients
of Nonsymmetrical Crystals***

MATERIAL	$10^6 \alpha$ (PARALLEL)	$10^6 \alpha$ (PERPENDICULAR)
Mg	27.0	25.0
Cd	49.0	17.0
As	28.0	3.0 (parallel to (0001))
Ca(OH)$_2$	33.0	10.0 (parallel to (0001))
CaCO$_3$	25.0	−6.0 (parallel to (0001))
ZnO	5.9	6.9
SiO$_2$ (quartz)	9.0	14.0
KAlSi$_3$O$_8$	19.0	−1.0

*From *The Solid State for Engineers*, by M. J. Sinnott, copyright 1961, John Wiley and Sons, Inc. Used by permission.

Using this relationship in Eq. (9.9), we find that

$$\alpha_L = \frac{\beta\gamma}{3V} C_v \qquad (9.11)$$

which shows that the temperature dependence of C_v and α_L are essentially the same.

The simple development given above suggests that the thermal expansion of a solid is isotropic. However, the form of Eq. (9.7) and the shape of Fig. 9-3 should indicate that considerable anisotropy could be possible except for very symmetric structures. The data in Table 9-3 clearly indicate that marked anisotropy exists in many crystals. In fact, in some complex structures it is possible to observe expansion in two directions, coupled with contraction in the third.

Thermal Conductivity

Materials transport thermal energy from one location to another when subjected to a difference in temperature at the two locations—a fact that has long been of considerable practical importance to engineers. In terms of a simple one-dimensional system, the flow of thermal energy in an isotropic material is expressed by Fourier's law of conduction

$$q = -k \frac{dT}{dx} \qquad (9.12)$$

where $q = (1/A)(dQ/dt)$ is the heat flux, or rate of flow, of thermal energy per unit of area normal to the direction of the flow, k is the thermal conductivity, T is the temperature, and x is distance measured in the direction of energy flow.

If single crystals of pure materials are considered, it is evident that isotropy is the exception rather than the rule, and therefore the simple constant k must be replaced by a number of constants—perhaps as many as six in the completely asymmetric, triclinic crystal system. Because the majority of practical engineering materials are polycrystalline, most engineering tabulations of thermal conductivity data list only a single value. These values are polycrystalline averages, which mask the true anisotropy of the thermal conductivity.

The existence of a simple kinetic theory of gases has made theoretical computation of the thermal conductivity for gases relatively easy. Although the simple kinetic theory formula for thermal conductivity of gases is not especially of interest *per se*, it is included here because its formulation embodies a number of concepts which will be helpful in understanding certain crystalline, solid-state theories. The average distance (weighted with respect to the distribution of molecular speeds) a molecule travels after it has collided with another molecule, but before it collides with a second one or a wall of its container, is called the *mean free path* and is usually given the symbol λ. In terms of this mean free path notion, the kinetic theory expression for the gaseous thermal conductivity is

$$k = \tfrac{1}{3}\rho C_v \bar{v} \lambda \tag{9.13}$$

where C_v is the heat capacity of the gas, and $\bar{v}$ is the mean thermal speed of the gas molecules as defined by Eq. (3.1). In a gas, this mean thermal speed is also the velocity of sound wave transmission.

The theory of thermal conductivity in the liquid state is not nearly as well developed as the gaseous theory because of the difficulty in creating a model for a randomly ordered liquid. Formulation of the theory of the solid state might be expected to be even more complex than that of the liquid state. However, the concept of phonons introduced by Debye and others to account for the specific heat of crystalline solids provides a very simple picture for the calculation of the thermal conductivity.

As we have previously noted, the thermal energy transported by sympathetic lattice vibrations in a metal may be represented by a gas of phonons. When viewed as such, the formula for the thermal conductivity of a perfect crystalline metal would be given by Eq. (9.13), with $\bar{v}$ replaced by the velocity of sound in the solid. A rather simple formula for the mean free path has been devised [1] by combining several theories, and is given by

$$\lambda = \frac{20T_m d}{\gamma^2 T} \tag{9.14}$$

where T_m is the absolute melting point of the solid, d is the characteristic lattice parameter, γ is Gruneisen's constant, and T is the absolute temperature. For all its simplicity and lack of real theoretical rigor, this formula is nonetheless reasonably accurate [1] for a number of materials. If Eq. (9.13) and Eq. (9.14) are combined, the thermal conductivity should decrease with increasing temperatures, a fact in accord with experimental observations of a great many materials.

It should be obvious that the magnitude of λ will be very closely related to the degree of crystal imperfection, to crystal size, and to the presence of dissolved material or other phases. Thus, the more nearly perfect a crystal, the longer λ is. Similarly, a complex crystal structure tends to have lower, mean free phonon paths than a simple one. In fact, the lack of crystalline regularity in amorphous materials results in very low mean free phonon paths. In materials such as glasses, the effect is so pronounced that an increase in lattice perfection (apparently occurring upon heating) more than offsets the expected degree of decrease in λ due to the temperature increase, and the thermal conductivities of these materials will actually rise with increasing temperature.

The theoretical calculation of the influence of various types of crystal imperfections on the mean free phonon path, and hence on the thermal conductivity of metals, is a very formidable mathematical problem [1] due to the difficulty in analyzing the interaction of two colliding phonons. In an idealized *harmonic crystal* (that is, one in which the various modes of lattice vibrations are dynamically independent), colliding phonons do not interact. However, in real metals, the nonvanishing value of Gruneisen's constant insures anharmonicity, resulting in phonon interaction. The interesting mechanics of such phonon interaction illustrate some of the difficulties a theoretician faces.

Because phonons are wave bundles, they interact according to wave laws. When two waves of differing frequency and, therefore, energy content interact with one another, they produce a new wave whose frequency is the sum of the frequencies of the two original waves (see Fig. 9-4). This is somewhat analogous to the collision of gas molecules; however, gas molecules share their energy, thereby randomizing its distribution among all the molecules involved, while phonons conserve their energy so that the net current of heat is neither decreased nor deflected by their interactions. In principal, therefore, if all phonon-phonon interaction processes were of this normal or N-type, the

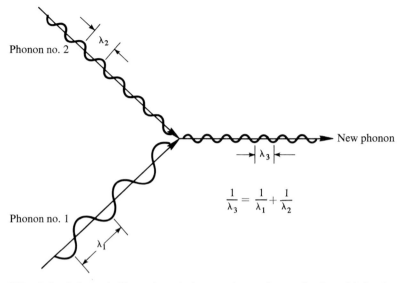

$$\frac{1}{\lambda_3} = \frac{1}{\lambda_1} + \frac{1}{\lambda_2}$$

FIG. 9-4 Schematic illustration of phonon-phonon interaction in solids by the normal or N-process in which momentum is conserved. Here the frequency of the new phonon is the sum of those of the original two that interacted. The inverse process, one phonon splitting into two new ones with conservation of momentum, is also possible.

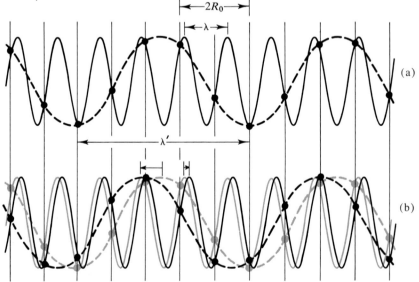

FIG. 9-5 Schematic illustration of the umklapp process of phonon-phonon interaction. (a) When the phonon wavelength is $\lambda < 2R_0$, a second wavelength (dashed curve) $\lambda' > 2R_0$ can be assigned. (b) The relative motions of these two waves. (From "The Thermal Properties of Materials," by J. Ziman, in *Materials, A Scientific American Book*, copyright 1967 by Scientific American, Inc., published by W. H. Freeman & Co. Used by permission.)

TABLE **9-4**

Selected Thermal Constants for Various Solids*

SOLID	ρ (g/cm^3)	$10^4\alpha_v(20°C)$ (°K^{-1})	$10^{12}\beta(20°C)$ (cm^2/dyne)	k (cal/sec cm°C)	C_v (cal/g°C)
Aluminum	2.70	0.672*	1.36	0.485	0.214
Antimony	6.68	0.407*	—	0.044	0.050
Bismuth	9.80	0.042*	—	0.020	0.029
Cadmium	8.65	1.62*	2.32	0.223	0.0551
Chromium	7.10	0.204*	0.618	—	0.110
Copper	8.92	0.50	0.72	0.927	0.092
α-Iron	7.86	0.355	0.618	0.148	0.107
Lead	11.34	0.840	2.43	0.084	0.030
α-Manganese	7.2	0.699*	0.806	—	0.0725
Magnesium	1.74	0.761*	—	0.270	0.245
Molybdenum	10.2	0.159*	—	0.349	0.065
α-Nickel	8.90	0.378	0.582	0.140	0.105
Palladium	12.0	0.353*	—	0.161	0.058
Platinum	21.45	0.265	0.379	0.166	0.032
Silicon	2.40	0.229	0.331	—	0.211
Silver	10.5	0.583	1.04	1.00	0.056
Sodium	0.97	2.13	15.9	—	0.225
Tin	7.31	0.642*	—	0.157	0.054
Tungsten	19.3	0.1332*	0.31	0.382	0.033
Zinc	7.14	0.893	1.79	0.270	0.092

* α_v is computed from $\alpha_v = 3\alpha_L$.

thermal conductivity of a perfect crystal would be infinite. Since this is not the case, there must be another type of interaction involved.

When the energy of the combined wave in Fig. 9-4 reaches a point where the wavelength of the phonon is less than twice the lattice spacing of the crystal, the direction of wave motion can no longer readily be distinguished. In fact, the motion of the atoms in the crystal can be interpreted equally well as the result of a wave of much longer wavelength, with a lower frequency and, hence, lower energy, moving in the *opposite* direction (see Fig. 9-5). In this phenomenon, called an *umklapp* or U-process,[4] phonons differ markedly from the gas molecule analogy, for here momentum is not conserved as it is when gas molecules collide. Phonon scattering lowers phonon energy, resulting in observed decreases in thermal conductivity. Sophisticated theories show [1] that the rate of scattering due to U-processes is proportional to the absolute temperature. In addition to temperature increases,

[4]From the German word meaning *flopover*.

interatomic force anharmonicity caused by dislocations, impurities, new phases, and phase boundaries, increases both the rate of phonon scattering and the complexity of the problem of computing this rate. Thus, the theoretical calculation of the thermal conductivity of real engineering materials is virtually impossible at the present time, and the engineer must rely almost exclusively upon reliable experimentally determined values of conductivity. Table 9-4 contains a short tabulation of several thermal properties of a number of solids.

Thermal Diffusivity

The physical significance of thermal diffusivity, originally introduced and defined by Eq. (1.10), can now be discussed.

As its name implies, thermal diffusivity is associated with the diffusion of thermal energy. Thus, if J_x in Eq. (6.11) is replaced by q_x (the heat flux in the x direction) and c is replaced by the energy concentration $C_p T \rho$, then the proper diffusion coefficient or diffusivity would be the thermal diffusivity α. The resulting equation would be

$$
\begin{aligned}
q_x &= -\alpha \frac{\partial}{\partial x} (\rho C_p T) \\
&= -k \frac{\partial T}{\partial x}
\end{aligned} \tag{9.15}
$$

which is known as *Fourier's law of conduction*. Similarly, Eq. (6.15) can apply for time dependent heat conduction when c is replaced by $\rho C_p T$ and D is replaced by α. Thus, just as D represents the magnitude of the mass flux from a unit magnitude mass concentration gradient, so α represents the magnitude of the heat or thermal energy flux from a unit magnitude gradient in energy concentration per unit volume.

The mass diffusivity D is exponentially temperature dependent. However, the term ρC_p in the denominator of α is relatively temperature insensitive, and, therefore, α varies with temperature in roughly the same manner as the thermal conductivity. Because this is not an exponential relationship, thermal diffusivity differs significantly from mass diffusivity in this respect. This arises from the fundamental difference in the molecular mechanisms which the two parameters reflect: the mass diffusivity represents a mass flux resulting from the physical transport of atoms or molecules in space, while the thermal diffusivity represents an energy flux arising from the motion of phonons through a relatively stationary atomic array. Since the phonons are merely wave forms, the atoms vibrate in unison, but are not physically transported.

Coupled Thermal Phenomena

Other phenomena associated with temperature fields can occur which, in some cases, may be quite important to the engineer. Such pertinent phenomena, caused by the interaction of thermal disturbances with other material response modes, result in coupled phenomena. The thermoelastic effect, the various thermoelectric effects, and thermal diffusion are examples of such coupled phenomena.

Thermoelastic Effect

Whenever a solid is subjected to hydrostatic pressure or other stress that results in a change in volume, work is done on it. If this work occurs isothermally, an amount of heat dQ is released which may be calculated from the first and second laws of thermodynamics as

$$dQ = T\left(\frac{\alpha_v}{\beta}\right) dV \qquad (9.16)$$

where α_v is the volume coefficient of thermal expansion, and β is the compressibility of the material. This relationship can be further modified [2] to give an expression for the *adiabatic temperature rise* (occurring without transfer of heat to or from the surroundings) of the material:

$$\left(\frac{\partial T}{\partial p}\right)_S = \frac{T\alpha_v V}{C_p} \qquad (9.17)$$

In many metallic and ceramic materials, this coefficient is very small and, therefore, unimportant from an engineering standpoint. However, in molecular materials that have long coiled molecules, such as elastomers, it can become quite pronounced. A simple demonstration of this effect can be performed with an ordinary rubber band. If the rubber band is rapidly stretched to its capacity and immediately touched to the lip while in this stretched condition, it will feel warm because the internally stored work causes adiabatic heating. Similarly, when the band is rapidly relaxed and touched to the lip again, it will feel cool. This is a manifestation of the thermoelastic effect.

Thermoelectric Effects

Thermoelectric effects are those in which either electric or magnetic fields are generated in a solid as a result of the application of a temperature gradient. A number of such effects are known [3], but they are essentially variations of three basic phenomena that bear the names of Peltier, Seebeck, and Thompson.

The *Seebeck effect*, the basis of the often-used thermocouple, occurs when two electrically and thermally dissimilar metallic conductors are joined at two places, and these junctions are maintained at different temperatures. Under these conditions, an electric current is observed to flow between the junctions. This current depends only upon the natures of the two metals and the temperatures of the two junctions; it is independent of what happens in between as long as the two conductors are homogeneous and isotropic. Of course, the introduction of dislocations due to strains can alter the behavior of the materials and cause spurious currents if other junctions exist. An example of this occurs when a cold worked and annealed sample of the same metals are joined. This particular effect is highly sensitive to many types of chemical and physical environmental changes, and is a major cause of certain types of corrosion phenomena (to be discussed in Chapter 11).

If a current flows through the loops of a thermocouple by means of an external source of electrical potential, a temperature difference will be observed between the two junctions of the couple. This inverse of the Seebeck effect is called the *Peltier effect*. It is reversible in that reversing the sign of the applied external electric field results in a reversal of the induced temperature difference. As above, in the case of isotropic homogeneous materials, this effect is independent of everything except the materials and the temperatures of the junctions.

The third phenomenon, the *Thompson effect*, is associated with the Peltier effect. It is observed that the rate of heat input at one junction does not always equal the rate of heat output at the other junction, indicating an apparent net gain in or loss of heat from the circuit. This phenomenon was explained by noting [2] that when an electric current passes through an iron bar that is simultaneously conducting heat, absorption or generation of heat by the bar occurs, depending upon whether the current flow opposes or coincides with the direction of heat flow. In copper, this effect is inverted. Since lead is essentially unaffected by this phenomenon, it is commonly used as a reference material.

Thermoelectric and thermomagnetic effects are almost entirely associated with metals or other materials that have relatively free electrons in their bonding structures. Molecular solids and strongly bonded ionic and ceramic materials do not have such free electrons and, there-fore, do not exhibit the above-mentioned effects. The simple effects must be corrected when crystal anisotropy, dislocations, etc., are present in the materials used.

Thermal Diffusion

In multicomponent mixtures, the application of large temperature gradients can result in the generation of concentration differences in the various components at the two different temperature sites. This is often called thermal diffusion, or the *Soret effect* [4]. The inverse of this effect (in which concentration gradients produce temperature differences) is also observable, but is generally smaller. Although thermal diffusion is itself rather small, it has been used [4] in nuclear processes to separate isotopic gas mixtures as well as complex mixtures of very similar organic compounds. This phenomenon is unimportant in solids, however. ([4] contains many bibliographic references to works that treat this interesting phenomenon in the kind of detail not possible here.)

Thermal Radiation

To some degree, all materials possess the ability to generate and emit photons from their surfaces as a result of the thermal excitation of their surface atoms. Similarly, all materials will absorb impinging photons to varying degrees and generate phonons within their structure, thus showing an increase in temperature. If all of the incident energy of the impinging photons is absorbed, the body is said to be a *black body*. In actual practice no real materials are true black bodies, and, therefore, a *coefficient of absorptivity* is defined as the ratio of absorbed energy to total energy incident upon the surface. A similar coefficient, called the *emissivity*, is defined in terms of the ability of a real surface to emit electromagnetic radiation in relation to this ability in a purely black body. For a given material [5] at equilibrium conditions, these two coefficients will be identical.

The amount of thermal radiation emitted from or absorbed by a solid object is given by the *Stefan–Boltzmann law*

$$q_r = e\sigma_r T^4 \tag{9.18}$$

where $\sigma_r = 1.355(10^{-12})$ cal/(sec cm^2 °K^4) is the Stefan–Boltzmann constant, and e is the emissivity factor. Table 9-5 contains typical values of emissivity factors for a variety of materials. It is obvious that, as the temperature level increases, radiant energy emission by solids becomes increasingly significant and must be dealt with effectively. For example, in ore refining furnaces a very large proportion of the energy transfer to the ore being smelted occurs by radiation from the furnace walls. This is also a significant source of heat loss from high temperature objects. ([5] and [6] contain excellent analyses of this phenomenon.)

TABLE 9-5

The Total Emissivities of Various Surfaces for Perpendicular Emission*

	$T(°K)$	e	$T(°K)$	e
Aluminum				
Highly polished, 98.3% pure	500	0.039	850	0.057
Oxidized at 1110°F	472	0.11	871	0.19
Al-coated roofing	311	0.216		
Copper				
Highly polished, electrolytic	354	0.018		
Oxidized at 1110°F	472	0.57	871	0.57
Iron				
Highly polished, electrolytic	450	0.052	500	0.064
Completely rusted	293	0.685		
Cast iron, polished	473	0.21		
Cast iron, oxidized at 1100°F	472	0.64	871	0.78
Asbestos paper	311	0.93	644	0.945
Brick				
Red, rough	294	0.93		
Silica, unglazed, rough	1272	0.80		
Silica, glazed, rough	1372	0.85		
Lampblack, 0.003 in. or thicker	311	0.945	644	0.945
Paints				
Black shiny lacquer on iron	298	0.875		
White lacquer	311	0.80	367	0.95
Oil paints, 16 colors	373	0.92–0.96		
Aluminum paints, varying age				
and lacquer content	373	0.27–0.67		
Refractories, 40 different				
Poor radiators	871	0.65–0.70	1271	0.75
Good radiators	871	0.80–0.85	1271	0.85–0.90
Water, liquid, thick layer	273	0.95	373	0.963

*From *Transport Phenomena*, by R. B. Bird, W. E. Stewart, and E. N. Lightfoot, copyright 1962, John Wiley and Sons, Inc. Used by permission.

References Cited

[1] Ziman J., "Thermal Properties of Materials," *Materials, A Scientific American Book*, W. H. Freeman & Co., San Francisco, 1967, p. 119.

[2] Sinnott, M. J., *The Solid State for Engineers*, John Wiley and Sons, Inc., New York, 1961, Chapter 14.

[3] Bridgeman, *The Thermodynamics of Electrical Phenomena in Metals*, Macmillan Company, New York, 1934.

[4] Bird, R. B., W. E. Stewart, and E. W. Lightfoot, *Transport Phenomena*, John Wiley and Sons, Inc., New York, 1962, p. 568.

[5] *Ibid.*, Chapter 14.

[6] Jakob, M., *Heat Transfer, Vol. 2*, John Wiley and Sons, Inc., New York, 1957.

PROBLEMS

9.1 Assume that the thermal conductivity of platinum varies with temperature according to the relationship $k = 1/(a + bT + c/T)$, where T is absolute temperature and a, b, c are empirical constants. Compute both the temperature at which k assumes a minimum value and the minimum value of k by using the following data:

t (°C)	-200	0	600
k (Btu/hr ft °F)	45.0	40.6	47.1

9.2 For copper, compute the value of (a) Gruneisen's constant, (b) $C_p - C_v$ at 500°K, and (c) C_p/C_v at 300°K.

9.3 Compute Gruneisen's constant for lead.

9.4 At 300°K, $C_v = 0.211$ cal/g °C. Using the fact that $MC_p = 5.74 + 0.000617T$, where T is °K, M is molecular weight, and C_p is heat capacity in cal/g °C, estimate α_L at 450°K for Si.

9.5 The velocity of sound in nickel is 16,320 ft/sec. Using Eqs. (9.13) and (9.14) and the fact that nickel crystallizes in a FCC lattice, compute the thermal conductivity of nickel at room temperature (25°C). $T_m = 1455$°C.

9.6 At a mean temperature of 300°K, compute the thermoelastic temperature rise in a sample of manganese if a pressure of 3000 atm is applied adiabatically.

9.7 The velocity of sound in aluminum is 6420 m/sec, and its melting temperature is 659.7°C. Using Eqs. (9.13) and (9.14), estimate the thermal conductivity of aluminum at 25°C, and determine the per cent of deviation of your result from the experimental value tabulated in Table 9-4.

9.8   The polymer polyvinylchloride has the following properties: $C_v = 0.25$ kcal/kg °K, $\alpha_v = 3.0(10^{-4})°K^{-1}$, $v = 1/\rho = 0.71$ cm³/g. Compute the thermoelastic coefficient $(\partial T/\partial P)_S$ for this polymer at 320°K. Compare this coefficient with the thermoelastic coefficient for lead at the same temperature.

9.9   If a temperature gradient of 36°C/mm is applied to a sample of aluminum with a cross-sectional area of 100 cm², what will be the heat flux through it?

9.10   If the resultant heat flow in Problem 9.9 is to occur in a sample of iron which is twice as thick in the direction of heat flow, compute the required cross-sectional area of the sample. (Assume that the temperature *difference* in the direction of heat flow is identical to that in Problem 9.9.)

9.11   Two metal plates, each 1 mm thick with a cross-sectional area of 100 cm², are clamped tightly together to form a composite piece 2 mm thick. One of these plates is nickel and the other is magnesium. If temperatures of 100°C and 50°C are maintained at the two external surfaces of the composite plate, compute the temperature which exists at the junction plane where the two metals contact, and the magnitude of the heat flux through the plates. (HINT: Observe that the same heat flux must pass through each plate. Write Eq. (9.15) for each plate separately and equate.)

9.12   Repeat Problem 9.11, assuming the plates are each 2 mm thick.

9.13   Repeat Problem 9.11, assuming the nickel plate is 2 mm thick and the magnesium plate is 1 mm thick.

9.14   For each situation in the three preceding problems, compute the rate of heat flow in cal/sec.

9.15   Two vertical planes face one another. If one of these is the firebrick ($e = 0.75$, $t = 1800°F$) wall of a furnace, and the other is a plate of tungsten ($e = 0.30$, $t = 5000°F$) serving as a heating element in the furnace, calculate the net radiant heat flux from the heating element to the wall.

9.16   If the wall of the furnace in Problem 9.15 is lined with a 1 mm thick layer of soot (carbon black, $e = 0.96$), compute the surface temperature of the soot on the wall. Assume that the heating element operates at the same temperature, and 2.97 times the net flux occurs.

9.17   The thermal conductivity of the soot in Problem 9.16 is 3.0 Btu/hr ft °F. Calculate the temperature drop across the carbon layer and, hence, the surface temperature of the furnace wall.

10

ELECTROMAGNETIC PROPERTIES OF MATERIALS

In our highly mechanized society, the ability of materials to conduct an electric current or to generate and maintain a magnetic field is of great practical importance. As in the phenomena previously studied, various electromagnetic phenomena arise from the nature of the atoms and the interatomic bonding forces. Although some can be explained in terms of the models of bonding forces already discussed, other more interesting and significant electromagnetic phenomena, such as semi-conduction and certain aspects of magnetization, will require more detailed models.

Electrical Properties

Conductivity

When the symbols are appropriately changed, the mathematical form of Fick's first law of diffusion (Eq. (6.11)) can also be applied to heat conduction and electrical conduction. Thus, replacing J_x by $J_e = i/A$, D by k_e, and $\partial c/\partial x$ by $\partial V/\partial x$, where i is the current in coulombs, A is the cross-sectional conductor area, $\partial V/\partial x$ is the voltage gradient, and k_e is the electrical conductivity, results in Ohm's law

$$J_e = k_e \left(\frac{-\partial V}{\partial x} \right) \tag{10.1}$$

Clearly, k_e in Eq. (10.1) represents the current density or charge flux that would result from a unit voltage gradient. When J_e is expressed in amperes/cm^2 and $\partial V/\partial x$ in volts/cm, the units of k_e are $1/(\text{ohm cm})$. The numerical magnitude of k_e is an approximate indication of a material's ability to act as a conductor, a semiconductor, or an insulator. In general, materials having electrical conductivities greater than about 10^2 (ohm cm)$^{-1}$ are considered conductors, while those with $k_e < 10^{-2}$ (ohm cm)$^{-1}$ are usually classed as insulators. The poorly defined middle range is called the semiconductor class of materials. As will be seen shortly, semiconductors can be much more sharply defined by a consideration other than the magnitude of k_e.

The phenomenon of electrical conduction can arise from two major sources: ionic conduction or electron movements. The transport of electric charges by ionic conduction occurs primarily in ionic solids, ceramic materials, and liquids and glasses. Because ionic conduction is a diffusion phenomenon involving the physical transport of ions within the lattice structure of the solid phase, it is strongly affected by dislocations and imperfections in the crystal structure. The existence of Schottky and Frenkel defects in ionic materials is therefore of considerable importance in providing diffusion sites for ionic conduction.

A simple thermodynamic argument may be used [1] to calculate the number of Frenkel and Schottky defects in a given ionic material. The number n of atoms that are out of proper register with their neighbors, forming the defects in a crystal containing N total atoms having N' interstitial positions, is

$$n = \sqrt{NN'} \, e^{-W/2kT} \tag{10.2}$$

for Frenkel defects or

$$n = N e^{-W/kT} \tag{10.3}$$

for Schottky defects where k is Boltzmann's constant, T is the absolute temperature, and W is the work or energy expenditure necessary to create the vacancy. These relationships are only approximate, as refined calculations of volume changes or the variations of W with temperature may introduce differences of several orders of magnitude. However, the relationships do indicate the exponential nature of the dependence of the number of defects on the temperature. The pronounced dependence of ionic conductivity on temperature is indicated by the data for AgBr in Fig. 10-1.

Since ionic conduction requires the physical transport of ions through the crystal lattice, ionic conductivity is closely related to the

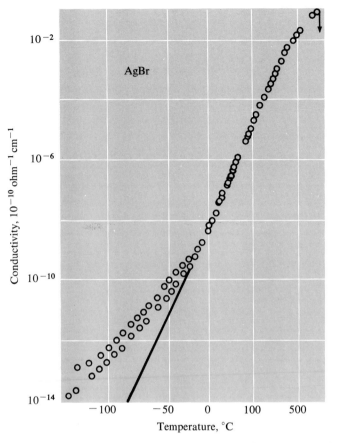

FIG. 10-1 Ionic conductivity of AgBr as a function of temperature. (From W. Lehfeldt, *Zeit. Physik,* **85,** p. 717 (1933). Used by permission.)

ordinary diffusion coefficient. This relation is given by the Einstein–Nernst equation

$$\mu_i kT = qD \tag{10.4}$$

where k is Boltzmann's constant, T is the absolute temperature, q is the ionic charge, D is the diffusivity of the ion in the material, and μ_i is the ionic mobility. The ionic mobility may be thought of as the drift velocity of an ion through the host medium that results from the application of an electric field of unit magnitude.

The ionic conductivity k_i is related to the ionic mobility by

$$k_i = nZe\mu_i \tag{10.5}$$

where $Ze = q$, e is the electronic unit charge, Z is the ionic valence, and n is the concentration of charged particles per unit volume. Combining Eq. (10.4) and Eq. (10.5) and noting Eq. (6.16), the temperature dependence of the electrical conductivity for ionic materials can be observed in the form

$$k_i = k_{i0}e^{-Q/kT} \tag{10.6}$$

where Q represents the diffusional activation energy.

From the experimental evidence available, it appears that, in materials consisting of ions with different valences, the ion possessing the larger charge has the greater mobility. Increasing the disparity in ionic sizes also increases the conductivity of the material. Similarly, the addition of impurity atoms that form mixed crystal or solid solutions almost always increases the conductivity of the ionic material, due to the introduction of crystal imperfections in the formation of solid solutions. Finally, if the lattice structure of an ionic material contains vacancies as a natural result of the ordering of its ions, as in Ag_2HI_4 or AgI, for example, unusually large conductivities due to the ease of ionic motion through the open crystal structure may be observed.

Metallic Solids

Primarily because of the difference in the natures of ionic and metallic bonds, electrical conduction in metallic solids occurs by a much different mechanism than the simple ionic diffusion discussed above. The ionic bonds formed between oppositely-charged ions are very strong, and the bonding electrons are generally tightly bound to a given nuclear core. However, the metallic bond is created when the valence electrons of one nuclear core ion are brought closer to the core of a neighboring atom than to their own in the free atomic state. This allows the valence electrons to move about the structure in a more or less random fashion, much as the molecules of a gas.

This free electron-gas model of the metallic solid, first proposed by Drude [2] and later elaborated and improved upon by Lorentz [3], was successful in accounting for several observable properties in metals. In particular, the Drude–Lorentz theory leads to the relationship between the thermal and electrical conductivities of metals, known as the Wiedman–Franz–Lorentz law[1]

$$\frac{k_t}{k_e T} = 3\left(\frac{k}{e}\right)^2 \tag{10.7}$$

where k_t is the thermal conductivity, k_e is the electrical conductivity,

---

[1]This relation was developed experimentally prior to the Drude–Lorentz theoretical calculation—thus, the different name for this law.

T is the absolute temperature, k is Boltzmann's constant, and e is the unit of electronic charge. Equation (10.7) agrees reasonably well with the experimental facts, indicating that some of the properties of the Drude–Lorentz model are valid and, therefore, useful. However, this model has serious shortcomings if it is used to calculate the temperature variation of k_c and the contribution of the electrons to the specific heat of the metal. These shortcomings were overcome by the advent of the quantum theory, which led to the formulation of a new model of electronic energy distribution in a metallic solid.

This new model is based primarily on the concepts of the Pauli exclusion principle. In an individual atom, only two electrons may possess the same set of quantum numbers (n, l, m). When two or more atoms are brought closely together in a metallic bond, the same rule applies to the entire collection; that is, the valence electrons interacting to form the bond cannot possess precisely the same energy. Because a large number of electrons are involved in the bond, the distribution of energy into discrete levels associated with individual atoms is no longer possible. Rather, those levels that are allowed split and subdivide around the mean atomic level to accommodate the different electrons from contributing atoms. These finely subdivided energy states lie so closely together that they effectively smear out into broad *bands* or *zones* of energy in which the energy content of a given electron may be found.

The energy band formed by these valence electrons is called the *valence band*. The magnitude of the repulsive forces generated when the core ions are brought closely together prohibits the subvalence electrons from interacting to the degree that the valence electrons do. Consequently, these subvalence electrons form much narrower energy bands, and, therefore, these electrons are said to be more tightly bound to their respective nuclei than the valence electrons. This model of the electronic energy distributions in a bulk solid can be applied equally well to any type of solid, regardless of its bonding type. The characteristics of the electronic energies associated with any one of the allowed energy bands determines, in large measure, the relative electrical conductivity of a material.

Band Theory of Conduction

In many instances, interactions between atoms in a solid structure result in an electronic energy distribution in the bands which differs considerably from a corresponding distribution in individual free atoms. Some electrons in the bound state are promoted to higher energy levels than they normally occupy in the free atomic state, while others

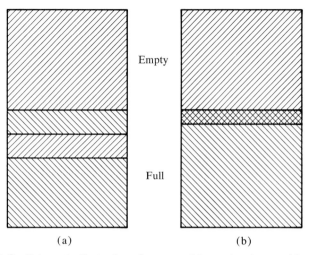

FIG. 10-2 Schematic illustration of two possible overlapping combinations of energy bands. (a) Partially filled lower band overlaps empty upper band. (b) Full lower band overlaps empty upper band.

occupy lower energy levels in the bound state than in the free state. In some substances, therefore, the energy states comprising a band from one type of electron, say a 3*p*, may overlap with those from a different type of electron, say a 4*s*. In free atomic distributions, these quantum energy states would be separated by a range of energies that the electron would not be allowed to possess. The fact that these energy bands do overlap (see the bar graphs in Fig. 10-2) means that an electron from the upper levels of the lower energy band can be promoted to the lower levels of the higher energy band with the acquisition of very little energy. If this higher energy band happens to correspond to a relatively free electron, as in the valence band of the metallic bond, the material will conduct an electric current with relative ease.

The upper energy band in which the electron is energetic enough to be moved from its core atom by an externally applied electric field is called a *conduction band*, as opposed to the lower or *valence band*. In the case of the metallic bond, these two bands overlap to such a considerable extent that the valence electrons are also the conducting electrons.

In many solids, particularly molecular solids and those bound by strong covalent or partially ionic bonds, the valence band is separated from the more energetic conduction band (which is unoccupied in such instances) by a rather large zone of quantum-mechanically-forbidden energy states. The magnitude of this forbidden zone, called the *energy gap*, determines whether the material is a semiconductor or a nonconductor. (Two possible energy gaps are illustrated schematically in Fig. 10-3.)

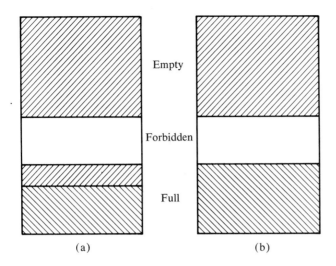

FIG. 10-3   Schematic illustration of two possible energy band combinations with forbidden zones or energy gaps. (a) Partially filled lower band with empty upper band. (b) Full lower band with empty upper band.

Even in the case of metallic conductors, not all of the valence electrons participate in conduction. Only those electrons occupying energy states in the upper portion of the valence-conduction band will be sufficiently energetic to participate in the conduction process. Consider Fig. 10-4, a schematic representation of the energy-density probability

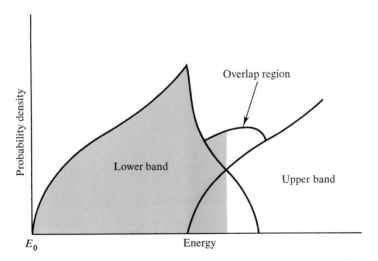

FIG. 10-4   Schematic energy-density probability diagram for the case of overlapping lower and upper bands. The ordinate represents the number of states having the particular energy. E_0 is the quantum mechanical, zero energy of the lower zone.

distribution for the lower and upper energy zones of a metal in which considerable band overlapping occurs. If the degree of overlapping of the two distributions is great, many of the electrons will have sufficient energy for conduction, and the material will be a reasonably good conductor. In monovalent metals, the peak in the distribution curve occurs in the middle of the energy band, leaving numerous energy levels in the high-energy portion of the band to which the electrons may be promoted by the acquisition of energy from an external field; thus, a monovalent metal will be a relatively good conductor. In divalent metals, the lower band is nearly filled, and a small degree of overlapping into the higher energy band occurs. The electrons near the boundary of the two bands share in both regions, decreasing the number of electrons which can conduct an electric current. Consequently, the conductivity in a divalent metal is not as great as in a monovalent metal. Since the lower band does not fill in a linear fashion [2], it will not be completely filled when the upper band begins to fill. In trivalent metals the lower band will be completely filled and the upper one will be partially filled. Thus, a trivalent metal will generally show greater conductivity than a divalent metal.

In the individual atoms of many metallic transition elements, the *d*-levels lie below the *s*-levels. Upon solidification of a metal, the electron density of the broadened *s*-states is lower than that of the overlapping *d*-band, because only two electrons can occupy a given *s*-state, whereas ten may occupy a given *d*-state. The *d*-band electrons are less conductive because there are fewer vacant states to which they may be promoted. On the other hand, the *d*-levels are completely filled in the elements copper, silver, and gold, with each atom possessing a single electron in the appropriate outer *s*-level. This half-filled outer *s*-level creates a half-filled band with many available higher energy positions near its upper end; thus, such metals will be extremely good conductors. In elements situated near copper, silver, and gold in the periodic table, with partially filled *d*-levels that are more energetic than their *s*-levels, *d*-electrons can overflow into the *s*-bands, creating vacancies in the *d*-band to which other electrons may be transferred. The net result of this electron transferral is the lowering of the electrical conductivity and an increase in the specific heat of these metals, in comparison with a good conductor such as copper.

TEMPERATURE AND ALLOYING EFFECTS   The simplified Drude–Lorentz model of the metallic electron distribution was incorrect in its account of the influence of temperature or alloying upon the electrical conductivity of metals. The essential error of this model lies in its incorrect assumption [2] that stationary ions scatter the conduction electrons, imparting frictional resistance to their passage through a crystal so that the mean free path of the electrons is limited to approximately the interionic distance in the crystal. Quantum mechanical arguments have shown [2] that this assumption is incorrect.

According to quantum mechanical laws, an electron possesses certain wave characteristics. An electron wave can be modulated or modified by the influence of electrostatic fields surrounding the ion-core array through which it passes. As the wave approaches a core ion, the ion field distorts it by increasing its frequency, which, in turn, accelerates the electron and increases its kinetic energy. This means that the electron spends a shorter time near the ion than it might under other conditions, resulting in a decrease in the influence of the core ion. In some simple structures, the electrons behave almost as the free particles in the Drude–Lorentz model. However, because the electron wave can adjust itself to the periodic, ion-core field it passes through, it could pass through a perfect crystal without being scattered. That is, quantum mechanically, it is possible for a perfect crystal to exert zero resistance force on the electron and, consequently, have infinite conductivity.

Real crystalline solids, of course, have large numbers of structural imperfections that scatter the electron wave as it passes through the crystal. As the number of deviations from uniform periodicity in the ion-core fields increases, so does resistance to the passage of the electron wave, thereby decreasing the conductivity. The introduction of impurity atoms through alloying or unintentional contamination greatly increases the number and severity of the lattice imperfections, which decreases the conductivity of an alloy in comparison with that of a pure metal.

The influence of increasing temperature on conductivity arises from the same type of phenomenon. As the thermal energy levels of the ion cores increase, they vibrate more rapidly and with greater amplitude about their equilibrium positions. The amplitude of these ion lattice vibrations is directly proportional to the absolute temperature. Since the degree of scattering is proportional to the deviation in ion-field periodicity, the electrical conductivity varies inversely with the absolute temperature. As will be seen shortly, this characteristic of metallic conductors is one of the important criteria that distinguish conductors from semiconductors.

Since the introduction of imperfections into the crystal by alloying leads to a decrease in the electrical conductivity of the alloy over the pure metals, it is to be expected that any mechanical process that increases the number of dislocations will result in a decrease in the electrical conductivity of the material. This is observed in the case of strain hardening, for example, where the cold-worked metals have a higher resistivity (lower conductivity) than annealed samples of the same metal.

Electronic Properties

Semiconduction

Semiconduction is one of the most important electronic properties to be discovered in recent years because it led to the invention of the transistor. The importance of the transistor in technology and in our daily lives is well known.

Semiconductors fall into two broad classes: *intrinsic* and *extrinsic*. Intrinsic semiconductors are materials that are insulators or nonconductors at ordinary temperatures, but become conductive when the temperature is increased by a moderate amount. That is, in contrast to metallic conductors, these materials show an increasing conductivity-temperature relation, a criterion that separates semiconductors from conductors when the two materials have otherwise comparable levels of electrical conductivity. If the conductivity decreases with rising temperatures, a material is classed as a conductor; if the conductivity increases with rising temperatures, it is considered a semiconductor.

Many materials at moderate temperatures are insulators and cannot be converted into conductors solely by increasing the temperature. However, by appropriate alloying with certain additives in very small amounts, some of these materials can be converted to semiconductors; such alloy semiconductors are called extrinsic. (This mechanism will be discussed shortly.)

Intrinsic Semiconductors

In an intrinsic semiconductor, the valence band of electron energies is nearly or completely filled and is separated in energy from the empty conduction band by a relatively small *energy gap* E_g. If the quantity $E_g/kT \simeq 1$, where k is Boltzmann's constant and T is the absolute temperature, the material can function as an intrinsic semiconductor. The reason for this is easily seen if the physical significance of the valence and conduction bands for such semiconductors as silicon or germanium are considered.

Figure 10-5 is a schematic representation of a portion of a silicon crystal. In the normal crystal, the silicon atoms are arrayed in a diamondlike crystal by means of relatively strong tetrahedral covalent bonds. If sufficient thermal energy is added to this crystal, some of the valence electrons that are involved in the formation of the covalent bonds and that are also very near the top of the energy distribution of

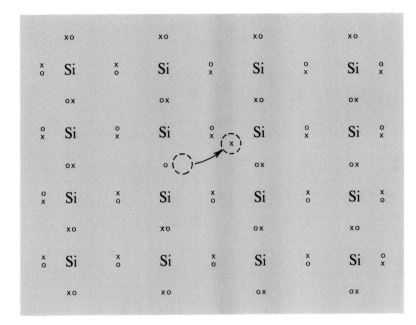

FIG. 10-5 Schematic representation of a thermally-freed electron in a Si crystal. The interstitial electron (enclosed in one of the dashed circles) is a conduction electron.

the valence band acquire enough energy to break loose from the bonding orbitals, becoming free to drift through the crystal under the influence of an external electric field. An electron freed by thermal energy addition creates a vacancy in the electronic energy distribution where it was originally bound, thus becoming an interstitialcy defect. This electron vacancy-interstitialcy pair is completely analogous to the Frenkel defect, discussed previously in connection with ionic conductivity in ceramics and ionic solids. Due to this analogy, Eq. (10.2) will give the number of vacancies or free conduction electrons so generated, provided W is replaced by E_g and $\sqrt{NN'}$ by a proportionality constant. Because the conductivity is proportional to the number density of charge carriers, it follows that the conductivity of an intrinsic semiconductor is related to the temperature by an expression of the form

$$k_e = K \exp\left(\frac{-E_g}{2kT}\right) \tag{10.8}$$

Thus, increasing the temperature results in an increase in the conductivity of the material, rather than a decrease as is characteristic of the metallic conductor.

When the valence electron is thermally activated and removed from the covalent bond to become a free conduction electron, it creates a vacancy or hole in the electronic energy band distribution. In terms of the mechanism of transport of an electric current, such a hole plays the role of a positive charge carrier and can transport charge just as effectively as the freed electron. In order to clarify this, consider the series of schematic orientations of the hole in Fig. 10-6; assume that the hole initially starts with atom A. When an external field is applied as indicated in part (a) of the figure, a force is exerted on the valence electrons of atom B, causing one of them to break loose from its bonding orbital to fill the vacancy in the bonding orbital of atom A. When this occurs (part (b)), the vacancy reappears at atom B, and a similar sequence of events occurs between atoms B and C to transport the hole to atom C (part (c) in the figure). The net results of this process are the transport of valence electrons from C to A and the motion of the hole from A to C. This situation can be viewed, then, as the transport of a positive charge (the hole) in the direction of the applied field. Thus, in semiconductor technology we speak of conduction by holes or by conduction electrons. In many instances, the mobility of the holes μ_p is less than the mobility of the conductance electrons μ_n [4]. This is illustrated for three common semiconductor materials in Table 10-1.

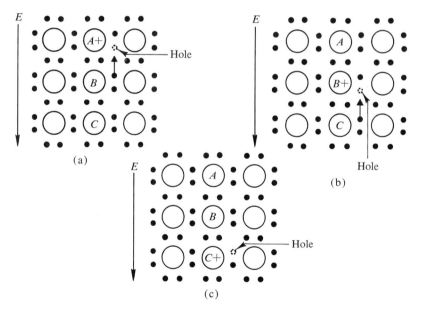

FIG. 10-6 Schematic illustration of conduction by holes in the electron valence band. (From *Principles of Solid State Microelectronics*, by S. N. Levine, copyright 1963, Holt, Rinehart and Winston, Inc. Used by permission.)

TABLE **10-1**

Energy Gaps and Charge Carrier Mobilities for Some Common Semiconductors*

MATERIAL	E_g (electron volts)	μ (holes) (cm^2/volt sec)	μ (electrons) (cm^2/volt sec)
Ge	0.70	1900	3900
Si	1.10	500	1400
GaAs	1.40	680	6800

*From *Principles of Solid State Microelectronics*, by S. N. Levine, copyright 1963, Holt, Rinehart and Winston, Inc. Used by permission.

Extrinsic Semiconductors

Extrinsic semiconductors are essentially very dilute, substitutional alloys in which the solute atoms possess valence characteristics that differ from the solvent material. Depending upon the type of impurity atoms that are introduced into the lattice of the host material by substitution, the alloy can conduct either by a hole or by a conductance electron mechanism. In order to understand these mechanisms, the schematic energy band diagram shown in Fig. 10-7 must be examined. First consider part (a) of the diagram where an atom having fewer valence electrons than the silicon host has been introduced.

Since silicon and germanium are from group IV of the periodic table, they have four valence electrons and crystallize in a diamond cubic lattice. When a group III element, such as aluminum, is substitutionally introduced into the silicon lattice, a mismatch in the electronic bonding structure occurs due to the deficiency of the Al atom in bonding orbitals relative to the Si host. (This is illustrated in Fig. 10-7(a) by the absence of a black cross, representing the electrons, adjacent to the Al atom.) In the sense described above in connection with intrinsic semiconductors, we now have a hole in the electronic structure. In this case, however, there is no corresponding free electron. If an external field is applied to the crystal, one of the neighboring electrons from another tetrahedral bond can acquire sufficient energy to break loose from its bond and be transferred to the vacancy created by the Al atom. This, in turn, causes the hole to move to the position formerly occupied by the electron. In this manner, the hole can move throughout the crystal as a positive charge carrier.

If the situation just described is examined in terms of the band diagram shown in Fig. 10-7(b), we see that the Al atom has provided an energy level that is only slightly higher than the upper limit of the valence band and that is inside the forbidden region. Thus, a nearby valence electron can easily be excited into this intermediate level. Such an energy level is called an *acceptor level*, and the atoms of group III which provide these levels are called *acceptor atoms*. Since the mechanism of charge transport in such an alloy is due to holes in the electronic band structure, the alloy is called a *p-type* (with positive charge carriers)

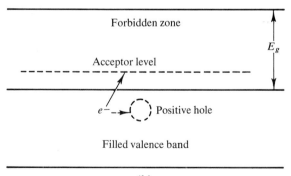

(a)

Empty conduction band

Forbidden zone

Acceptor level

e^- → Positive hole

E_g

Filled valence band

(b)

FIG. 10-7   (a) Schematic illustration of the introduction of an Al^{3+} ion into a Si^{4+} matrix, creating an electron deficiency hole. (b) Corresponding energy band diagram, showing acceptor level with a positive hole in the valence band.

extrinsic semiconductor. The process of adding minute amounts of appropriate alloying agents to create an extrinsic semiconductor is called *doping* a crystal; the element so added is called the *dopant.*

Figure 10-8(a) is a schematic illustration of the addition of a group V element into the Si matrix; phosphorus can be used as a typical example. In this case, all of the normal Si tetrahedral bond orbitals are filled by electrons from the P atom. There remains, however, an additional electron from the P atom which is not used in forming any P–Si bonds. Although this electron is still associated with the parent P atom, it is loosely bound and requires little additional energy to be removed from its parent nucleus and made into a free conductance electron. In terms of an energy band diagram, Fig. 10-8(b), this means that the extra electron occupies an energy level in the forbidden region that is just slightly lower than the boundary of the conduction band. Adding a small amount of energy (much less than E_g) to this electron promotes it to the conduction band. Such an energy level is called a *donor level,* and the atoms of group V which provide these additional electrons are called *donor atoms.* Because the conduction mechanism is due to the extra electrons introduced into the structure, such materials are called *n-type* (with negative charge carriers) *extrinsic semiconductors.*

From an examination of Table 10-2, it is evident that the amounts of additional energy required to produce a charge carrier are generally at least an order of magnitude smaller than the energy gap of the host elements listed in Table 10-1. The concentrations of dopant which must be added to create extrinsic semiconductors are generally very small, as too large a concentration of dopant will destroy the semiconductive

TABLE 10-2

Ionization Energies of Donors and Acceptors*

IMPURITY ATOMS†	IONIZATION ENERGY (eV)	
DONORS	Si	Ge
P	0.044	0.0120
As	0.049	0.0127
Sb	0.039	0.0096
ACCEPTORS		
B	0.045	0.0104
Al	0.057	0.0102
Ga	0.065	0.0108
In	0.160	0.0112

*From *Principles of Solid State Microelectronics,* by S. N. Levine, copyright 1963, Holt, Rinehart and Winston, Inc. Used by permission.
†Atomic size increases downward in table.

property by converting the material to a conductor. In the process of manufacturing semiconductor grade silicon and germanium, extreme care must be exercised to obtain ultrapure materials with relatively

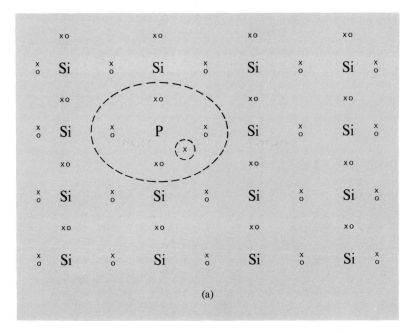

(a)

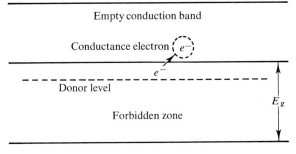

(b)

FIG. 10-8 (a) Schematic illustration of the doping of a Si^{4+} matrix with P^{5+} atoms. (b) Corresponding energy band diagram, showing position of donor level. Large dashed circle around P atoms in part (a) indicates that the fifth electron (x enclosed in smaller dashed circle) is still weakly attached, although not bound, to the parent P nucleus.

perfect crystal structures before tiny amounts (in parts per million) of an appropriate dopant are added.

In addition to the elemental semiconductors Si and Ge discussed above, numerous compounds, such as GaAs and PbS, also exhibit semiconducting properties. The same concepts regarding donor or acceptor dopants apply, except that elements from different groups of the periodic table must be used. In the case of such III-V compounds as GaAs, substitution of group II elements for the group III member results in the production of a *p*-type semiconductor, whereas substitution of a group VI element for the group V member results in an *n*-type semiconductor.

The p-n Junction

If only individual semiconductors were available to the engineer, they would represent a relatively unimportant curiosity with little practical value. However, the combination of an *n*-type semiconductor with a *p*-type material through a rather sharply defined interface results in a device of extreme utility and importance in electronic engineering. It is, therefore, important to study the properties of the boundary region or *junction* between these two different types of semiconductors.

Figure 10-9 is a schematic representation of a *p-n* junction. The plus signs represent the holes or positive carriers in the *p*-region, and the minus signs represent the conductance electrons in the *n*-region. Such a device is generally made by growing half of the crystal with the addition of one type of dopant impurity, and completing crystal growth with the addition of the other type. The junction or boundary between the *n*- and *p*-regions cannot be distinguished optically in this type of crystal growth.

If a battery or other external source of potential is applied to the

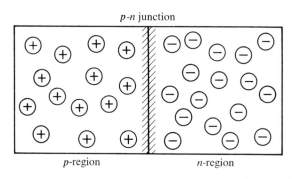

p-n junction

p-region   *n*-region

FIG. 10-9   A schematic *p-n* junction. Plus signs represent holes in the *p*-region; minus signs are electrons in the *n*-region.

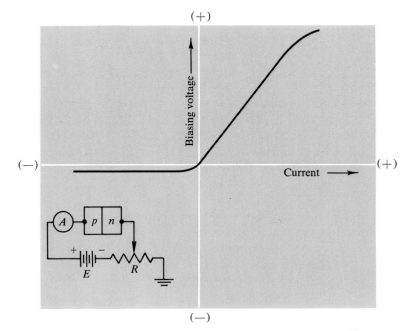

FIG. 10-10   Forward biasing of *p-n* junction by an external source. The curve illustrates both positive and negative biasing.

p-n junction, the type of current flow response observed depends upon the polarity (orientation of positive and negative poles) of the applied field. If the external field is applied with the positive pole in contact with the *p*-region and the negative pole in contact with the *n*-region, there is a considerable flow of current. In Fig. 10-10, this *forward biasing* is indicated by a positive potential. Here the flow of current occurs because the positive pole siphons off electrons from the *p*-region, creating more holes and driving existing holes toward the junction boundary. Simultaneously, the negative pole injects electrons into the *n*-region, forcing existing conductance electrons toward the junction. The two opposite types of carriers meet at the junction and recombine with one another, producing a continuous flow of current.

If the polarity of the external field is reversed, the very small flow of current indicates a high resistance or low conductivity. This occurs because the negative pole injects electrons into the *p*-region, annihilating positive charge carriers within it, while the positive pole sucks electrons from the *n*-region, again removing charge carriers. Thus, the concentration of charge carriers near the junction is significantly decreased, and very little current is able to flow. This property of passing a large current for positive field polarity, but a negligible current for negative field polarity is known as *rectification*. Clearly, if an alternating field is applied to the *p-n* junction, current will flow only when the field cycle corresponds to positive polarity, thus rectifying or converting the input alternating field into an output current that is positive only.

Rectification is one of the important uses of junctions in electronic applications. For example, some silicon rectifiers are capable of handling over a hundred amperes of forward current and of withstanding reverse voltages as high as 600 volts. At the other end of the scale, tiny rectifiers for use in computers operate at power levels of a few milliwatts (1 milliwatt $= 10^{-3}$ watts) and are capable of field reversal and current stoppage in a nanosecond (10^{-9} sec). Many other types of *diode* or single junction devices are manufactured [5].

Transistors

A transistor is obtained when two *p-n* junctions occur sequentially in the same crystal of the material. These two junction devices can be made with either a *p-n-p* or an *n-p-n* sequence of regions. They are of great value in electronic applications since they can serve as amplifiers, allowing fluctuations in tiny currents to control fluctuations in much larger currents. Transistors are practical because they require very low levels of operating power when compared with conventional vacuum tubes and because they can be manufactured in extremely tiny sizes. The latter feature has permitted the ultraminiturization of many electronic devices no doubt familiar to the reader.

Figure 10-11 is a schematic representation of an *n-p-n* transistor biased by external batteries in order to permit the flow of a substantial

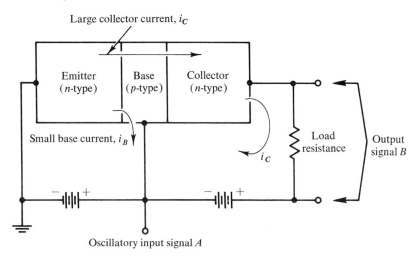

FIG. 10-11   Biasing of an *n-p-n* transistor to amplify a base input signal. Oscillations in i_B are duplicated in i_C. Since $i_C/i_B \gg 1$, the signal emerging at B is amplified relative to the signal at A.

current from the first n-region (called the *emitter*) through the p-region (called the *base*) to the other n-region (called the *collector*). The reason for this pattern of current flow becomes clear upon examining the biasing arrangements for the two different junctions. The n-p junction between the emitter and base regions is biased in the forward direction, so that electrons are forced toward the junction by the negative pole of the external field. Similarly, the holes are generated by appropriate biasing at the base and forced toward the emitter-base junction. The base-collector junction, however, is biased in the negative or reverse direction, so that no current can flow from the base to the collector. Since the base region is very thin, the great majority of electrons that arrive at the emitter-base junction under the action of the external biasing field are moving with sufficient drift velocity to overshoot this narrow base region before they have time to recombine with holes there. Consequently, these electrons are trapped in the unfavorably biased field of the collector and, since they cannot return to the base, they are carried onward toward the positive terminal at the other end of the collector. In this manner, the base pumps the electrons from the emitter to the collector, thus creating a relatively large current flow.

Because of the mechanism described above, it should be evident that the relative biasing potential across the emitter-base junction is very important in determining the rate at which electrons are pumped from the emitter to the collector. Therefore, if the biasing voltage for the base terminal fluctuates, the current flowing from the collector will fluctuate sympathetically and, in this manner, the device serves as an amplifier. Even though the current which flows in the base terminal as an input signal is of a very small magnitude when compared with the large current flowing through the emitter-collector path, a certain per cent of fluctuation in the base current also causes the collector current to fluctuate in a similar manner. When this large, fluctuating current is permitted to pass through an external resistance, the ohmic voltage[2] developed across the resistance likewise fluctuates. Thus, the magnitude of the fluctuation in the base biasing voltage and, thus, in the current is greatly amplified by the action of the transistor. Obviously, the same operation could be obtained with a p-n-p transistor by appropriate changes in the mode of biasing.

Organic Semiconductors

Although organic molecular crystals (more completely described in Chapter 12) generally involve bonding by the covalent mode, the con-

---

[2]Ohmic voltage is developed in the resistance R by the current I, $\Delta V = IR$.

cepts of the band theory can be applied to these materials, as well as to conventional metallic semiconductors. It has been found [6] that certain organic materials, notably phthalocyanine and anthracene, exhibit semiconductive behavior. Even though they possess a narrower band structure than conventional semiconductor materials, reasonably good predictions [6] of charge carrier mobilities for organic materials have been made on the basis of band theory.

At the present time, organic semiconductive properties are significant because they help to explain various biological characteristics of materials [6]. A detailed description of the semiconduction mechanism in organic materials is complicated by the addition of such effects as multiple bond resonances, aperiodicities in molecular crystal structures, and distributions in molecular weights. (A definitive discussion of this interesting phenomenon can be found in reference [6].) The possibility that organic materials possessing semiconductive and perhaps even metallic characteristics will be developed in the future [2] is very exciting.

Superconductivity

As the temperature approaches the absolute zero in certain materials, electrical conductivity becomes very large. This phenomenon, known as *superconduction*, generally occurs rather abruptly at a well-defined temperature. A number of different types of substances exhibit this property, although the monovalent metals and those with ferromagnetic or antiferromagnetic properties (to be discussed later in the chapter) generally do not. For many years, the compound niobium-tin, which has a superconductivity critical temperature of 18°K, was thought to be the upper-limiting example of a superconductive material. Recently, however, alloys of niobium, aluminum, and germanium have been found [2] which exhibit critical temperatures of 20°K. The possibility exists [7] that superconductivity can be obtained at ordinary room temperatures by the appropriate design of certain organic materials.

Since it has been found that strong magnetic fields can disrupt superconductivity, it is believed that the phenomenon is related to the magnetic properties of matter [8]. The magnetic fields that cause this superconductive disruption are functions of the temperature level. The critical temperature for superconductivity is found experimentally to be directly proportional to the Debye characteristic temperature of the material. (At the present time, however, the phenomenon is not thoroughly understood.)

Magnetic Properties

Magnetic Terminology

Magnetism and magnetic phenomena have been known for centuries and, from a practical standpoint, form the basis of a large and very important segment of modern technology. Undoubtedly the proper theoretical explanation of magnetic phenomena lies in the quantum mechanical analysis of multiatom systems. To date, however, such an analysis has proved to be too difficult to perform and, therefore, no generalized theoretical explanations of these phenomena exist. Almost all of our knowledge concerning magnetism and magnetic materials is based upon experimental studies.

A number of terms are frequently used to describe the various types of observed magnetic behavior. After these terms have been defined, a simple physical model of atomic magnetic fields will be presented to partially account for them.

The most familiar materials that exhibit magnetic properties are iron and some of its alloys, which form the basis of permanent and electromagnets. Some materials, notably iron, will become more magnetic when exposed to a weak magnetic field. They may also exhibit significant residual magnetic behavior after an externally applied magnetic field has been removed. Such materials are said to be *ferromagnetic*. Above a certain relatively well-defined temperature, called the *Curie temperature*, however, ferromagnetic materials lose their ferromagnetism and no longer retain their magnetic behavior, even when subjected to large external fields.

When subjected to an externally applied magnetic field, a number of materials exhibit a property called *paramagnetism*—a relatively weak magnetic behavior that varies in proportion to the applied field, but is lost entirely if this external field is removed. Ferromagnets above the Curie temperature become paramagnets.

All known materials exhibit *diamagnetism* to some degree; that is, they develop a magnetic field which opposes the applied external field. Even para- and ferromagnetic materials are diamagnetic, although this phenomenon is usually masked by the more noticeable characteristics of these materials and can easily be overlooked. In any event, the effect of diamagnetism is generally too small to be of any practical engineering value.

Three other terms, antiferromagnetism, ferrimagnetism, and helimagnetism are occasionally used to describe certain types of magnetic behavior. Before these terms can be defined, however, it is necessary to develop a model of the magnetic state.

Magnetic Domains

The concept of magnetic domains was introduced by Weiss in 1907, long before quantum theory was conceived. At the time of its introduction, the domain concept represented a remarkable speculative leap which departed radically from conventional thinking [9]. According to this concept, ferromagnetic materials below their Curie temperatures are composed of large numbers of small, permanently magnetized regions called *domains*. The overall magnetic strength of the material is the algebraic sum of these little magnets. If the magnetic axes of the domains are randomly oriented, the magnetism of the entire material will be small, or zero. The application of an external field of moderate strength can result in the alignment of the magnetic domain vectors in a direction that is, in most instances, parallel to the direction of the applied field, thus magnetizing the material.

Although there was no theoretical or experimental evidence of the existence of domains at the time Weiss' concept was introduced, direct experimental evidence has since been obtained that demonstrates their reality. Domains are generally small areas, ranging between 0.01 cm to 0.1 cm across. Since regions of these proportions are extremely large in comparison with atomic and molecular dimensions, it is difficult to understand how their origin is related to atomic and molecular forces. Weiss accounted for this by introducing a second radical concept—the molecular field. He assumed that the atoms of certain materials were themselves tiny magnets, maintained in some degree of alignment below the Curie temperature by means of a magnetic field that was an intrinsic property of the material. He explained the Curie temperature by assuming that above this temperature the energy of thermal agitation is sufficient to disrupt the intrinsic aligning power of the molecular field and convert the ferromagnetic material to a paramagnetic one.

Although, as previously mentioned, Weiss introduced his hypotheses before the quantum mechanical model of the atom was developed, both of his hypotheses proved to be essentially correct when the quantum theory was applied, revealing Weiss' tremendous insight.

Molecular Origin of Magnetism

The fact that some elements exhibit intrinsic magnetism and others do not can be explained in terms of their atomic electron distributions. Since moving electric charges create magnetic fields, the orbital motion of electrons about an atomic nucleus (in terms of the Bohr–Sommerfeld visualization) gives rise to a net atomic magnetic field. Two effects associated with electron movement contribute to the magnetic field:

the angular momentum of gross orbital motion of an electron about the nucleus, and the angular momentum of its spin about its own axis. The resultant, magnetic moment vector associated with these two effects is directed along the axis of rotation of the electron. Because of the Pauli exclusion principle, electrons occupy orbits in antiparallel pairs; that is, their magnetic moment vectors are oriented antiparallel to one another. This results in a cancellation of magnetic and angular momenta, leaving a zero resultant. When an external magnetic field is applied to the atom, either or both of these paired momentum vectors may be slightly disrupted. Diamagnetism will result if the gross orbital pairing is disrupted. If the spin pairing is also disrupted, a weak paramagnetism will result. The latter effect generally occurs in metals where the electronic structuring for the valence electrons is much looser than it is in ionic or covalent materials.

In certain transition metals and rare-earth elements, the close energy spacing of some of the subvalence orbitals that are only partially filled permits the repulsive forces between the two electrons to overpower the energy-reducing tendency of antiparallel spin vector pairing. As a result, some orbitals possess single, unpaired electrons rather than antiparallel, oriented pairs. Since these unpaired electrons exist in energy states that do not normally enter into valence bonding, the net magnetic moments that they generate tend to be permanently associated with the atoms, resulting in ferromagnetism.

The interactions between individual ferromagnetic atoms are relatively short range in nature, extending, at most, to one or two layers of nearest neighbors. Due to such interactions between adjacent layers, each affecting only nearest neighbors, however, the overall effect of this short-range coupling is a long-range ordering of magnetic moments to form macroscopically large magnetic domains. Thus, although Weiss' picture of a long-range molecular field that couples each individual atomic magnet to every other one with equal strength is not true in detail, it is nearly correct if the combined effects of the short-range coupling forces are considered.

In some materials, as manganese fluoride, for example, the reduction of thermal vibrations by decreasing the temperature increases the short-range coupling effect to such an extent that each atomic magnet forms an antiparallel pair with one of its neighbors, and all gross magnetism is lost. This phenomenon is known as *antiferromagnetism*, and the temperature at which the transition from para- to antiferromagnetism occurs is called the *Néel temperature*.

Another interesting effect arises from the short-range coupling of atomic magnets. In some materials the coupling forces are of such a magnitude that two of three neighboring atomic magnets will be parallel and the third antiparallel. The net result is a weak magnetic moment due to the single unpaired magnet. This arrangement, known as *ferrimagnetism*, is not to be confused with ferromagnetism, where the large majority of the atomic magnets align parallel with one another to form domains. In reality the mineral magnetite, commonly known as lodestone, is ferrimagnetic rather than ferromagnetic. Some complex oxides of iron and various rare-earth elements, known as rare-earth iron garnets, possess pronounced ferrimagnetic characteristics. A typical example of such a material is a yttrium iron garnet (known as YIG) which, like numerous other similar materials, is used to make tiny magnetic core elements for high speed digital computer memory units. Such iron garnets are used in these units because of the extreme rapidity with which they can be magnetized and demagnetized.

Although the short-range coupling forces that favor antiparallel alignment extend beyond the range of immediate nearest neighbors, it is evident that not all atomic magnetic moments can be aligned antiparallel to one another; some of the second and third nearest neighbors will be parallel. An interesting arrangement of atomic magnetic moments occurs in the stable helical spiral observed in manganese oxide below $-189°C$. The helix turns at about $129°$ per plane of atoms, making five revolutions in fourteen planes. This phenomenon, known as *helimagnetism*, appears to be an intermediate manifestation between antiferromagnetism (a helix with a turn angle of $180°$ per plane) and ferromagnetism (a helix with a turn angle of zero degrees).

Domain Boundaries: The Bloch Wall

The state that exists at a boundary between adjacent magnetic domains is somewhat similar to the arrangement of the helix in helimagnetism. The ordering of adjacent atomic magnetic moments twists or spirals into a new orientation over the space of a few atomic planes. However, the rate of twist is much smaller than that observed in the helimagnetic helix. The transition zone, illustrated schematically in Fig. 10-12, is known as a Bloch wall (after its discoverer); it is generally several hundred atomic planes thick [9].

Increasing the number of domains of a material shrinks the external, magnetic force line loops and reduces the energy of the external field, while it simultaneously increases the short-range coupling energy. Therefore, the distribution of domains in a magnetic material is the result of the equilibrium between these opposing effects that represent the lowest total energy of the overall material. Experimental proof that magnetic domains actually exist can be obtained through the use of the Bitter technique [9]. In this, colloidal suspensions of magnetic, iron oxide particles are spread over the surface of a polished magnetic

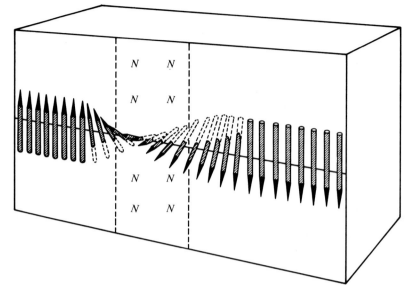

FIG. 10-12   Schematic illustration of atomic magnetic moments in a Bloch wall. (From *Introduction to Solid State Physics*, by C. Kittle, copyright 1953, John Wiley and Sons, Inc. Used by permission.)

material. If the sample is viewed through a microscope, the magnetic, iron oxide particles can be seen congregating along the boundaries of the magnetic domains of the material and clearly delineating them. Figure 10-13 is a photomicrograph showing magnetic domain boundaries.

Magnetization Curves: Hysteresis

In dealing with magnetic materials in a quantitative fashion, it is necessary to have experimental data available in the form of a magnetization curve—that is, one which describes the response of the material to an applied magnetic field. In order to understand the various aspects of magnetization curves and their significance in technical applications, several magnetic field quantities must be defined.

Magnetic fields are polar in nature; that is, lines of constant, magnetic field intensity radiate outward from the *pole*, or center of magnetic intensity in a spherically symmetric pattern. A *field line of force* is defined as one line of field intensity on each square centimeter of the surface of a sphere one centimeter in radius which surrounds a pole of unit strength. A unit pole is one of such strength that it experiences a force of one dyne when placed a distance of one centimeter from another unit pole. Since the surface area of a sphere of radius 1 cm is 4π cm^2, the strength of a unit pole is 4π lines.

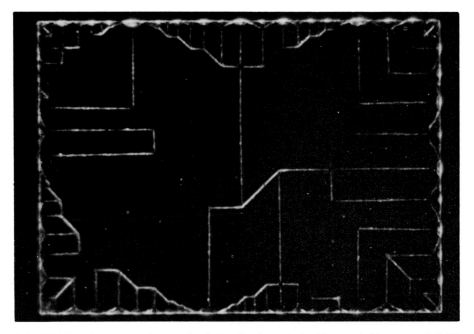

FIG. 10-13   Bitter photomicrograph of domains in a single Ni crystal. (Courtesy of R. W. DeBlois, The General Electric Company, Schenectady, New York. Used by permission.)

All real magnetic materials are dipolar in nature. The intensity of magnetization or the magnetic dipole moment per unit volume for these materials is given the symbol I; thus, the strength of the intrinsic magnetic field of the material is $4\pi I$ lines/cm^2. The magnetic unit lines/cm^2 is called a *Gauss;* the intrinsic magnetism of the material is, therefore, $4\pi I$ Gauss. (This intrinsic magnetism B_i is sometimes called ferric induction.)

The magnetic force H applied to a material is generally produced by passing a current i (measured in amperes) through a coil of wire having n turns per centimeter. When these units are used for i and n, the units of H are defined as oersteds and H is related to i and n by $H = 0.4\pi n i$.

The total magnetic field induced in the material by the external field H and the intrinsic strength, or ferric induction, is the magnetic induction B, given by

$$B = H + 4\pi I = H\left(1 + \frac{4\pi I}{H}\right) = \mu H \qquad \textbf{(10.9)}$$

The quantity most frequently used is the ratio B/H, which is defined as the magnetic permeability μ. This quantity represents the slope of a straight line through the origin and any given point on the magnetization curve.

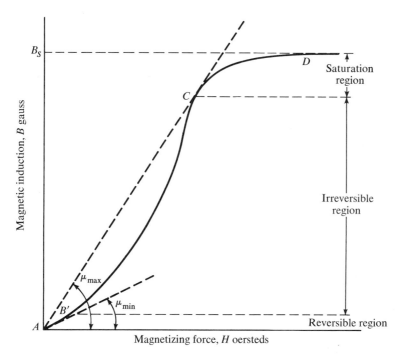

FIG. 10-14 Schematic magnetization curve. $\mu = \mu_{min}$ for the reversible AB portion. μ increases to μ_{max} for the irreversible region. For the saturation region, B rapidly approaches the saturation value B_s.

A typical magnetic induction response curve for an increasing field H, as shown in Fig. 10-14, illustrates several features of the magnetization process for a ferromagnetic material. Perhaps the most obvious feature is the variability of μ. The curve can be divided roughly into three parts. The first part, indicated by the AB' portion of the curve, corresponds to small reversible amounts of magnetic induction. Here displacements of magnetic moment vectors from their equilibrium positions in the material are small, and no gross deformation of domain boundaries is observed. If the applied force H is removed or reversed, the induced field B is likewise removed or reversed, a process analogous to elastic deformation.

Near part B' of the magnetization curve, the permeability begins to increase with increasing H until it reaches a maximum at point C. The $B'C$ portion of the curve corresponds to large irreversible shifts in the domain boundaries, which are analogous to plastic mechanical deformations. In this region of the magnetization curve, many of the magnetic moment vectors that were originally aligned in an opposite direction to the applied field are realigned with it, changing the position of the Bloch walls and realigning the domain boundaries.

At point C, the permeability begins to decline rapidly—a manifestation of the finite number of magnetic moment vectors the material possesses. When all of these moments are aligned parallel to the field H, no amount of increase in H can bring about a further increase in B and the material is said to be magnetically saturated. The CD portion of the curve in Fig. 10-14 is the saturation region of the magnetization curve, and the level of B corresponding to point D is the saturation magnetization B_s.

The magnetization curve for a paramagnetic material with a relatively low permeability would be similar to the AB' portion of Fig. 10-14. The corresponding curve for a diamagnetic material would be linear, with a negative permeability corresponding to the opposing induced field.

Because of the nonlinearity of the magnetic induction response curve for ferromagnetic materials due to the irreversible domain boundary shifts, the curve can not be repeated in the reverse. That is, if the field H is increased to the saturation value and then decreased, the response field B will follow a different curve back. A complete cyclic oscillation in H is illustrated schematically in Fig. 10-15. The arrowheads on the curve indicate the direction change of the field H.

The $AB'CD$ portion of the curve in Fig. 10-15 is the initial magnetization curve shown in larger detail in Fig. 10-14. As the field H is reduced to zero, the induced field B decreases somewhat, but is not eliminated. Rather, due to the irreversible part of the initial magnetization, many of the domains remain in their changed states. Some of the reversible component is recovered, and the material arrives at zero applied field with a residual induced field strength B_r, called the *residual induction* or *remanence*. In order to remove the residual induction, a negative field of magnitude H_c, known as the *coercive force*, must be applied. If the driving field H is increased negatively at this point, saturation is reached when the negative magnitude of H becomes sufficiently large; the saturation field is then $-B_s$. At this point, if the field H is reversed so that it increases positively, the reversible magnetization component will be recovered, leaving the residual remanence due to the irreversible domain rotations. Once again at zero H, the field $B = -B_r$, and application of $H = +H_c$ is required to drive B to zero. Increasing H to its saturation level now drives B to $+B_s$, bringing us back to point D on the curve.

Continued cycling of the driving field H will result in retracing the above-mentioned looped or *hysteresis curve*. The area that this curve encloses is very important in determining the value of the material for applications involving oscillating magnetic driving fields. As the domains rotate and their boundaries shift, energy from the applied field H is irreversibly consumed and, therefore, cannot be fed into the induced field B. This lost energy, known as the *hysteresis loss*, is measured by the area of the hysteresis curve. In addition to hysteresis losses,

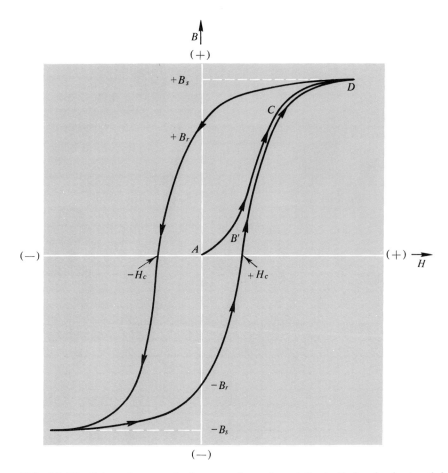

FIG. 10-15 Schematic magnetization curve for cyclic variation in H, showing hysteresis loop.

materials placed in alternating magnetic fields lose energy due to the generation of small circulating internal currents, called *eddy currents*, that result from the movements of the magnetic moments in a conductive medium. If paths are available for current flow, eddy currents can become large, resulting in considerable ohmic heating and power loss. (Anyone who has placed his hand on a large transformer covering realizes the amount of heat these currents can generate.)

If the magnitude of H_c is relatively small and that of B_r is relatively large in comparison with that of B_s, the material is said to be *magnetically soft*. That is, it is easily saturated, retains a large fraction of its saturation magnetism upon removal of the external field, and requires little field reversal to demagnetize. Magnetically soft materials tend to be physically soft as well; annealed iron is a good example of magnetic softness. The coupling between physical and magnetic softness or hardness is associated with the ease or difficulty with which the magnetic domains rotate to saturate the material magnetically. If few dislocations or crystal impurities are present in a material, the domains rotate much more freely than if the reverse were true. Since the absence of dislocations makes the material softer or more ductile, the relationship above follows.

A material with a very large H_c value is said to be *magnetically hard*. Once it has been magnetized to its saturation value, such a material requires considerable force to rotate its domains into a demagnetized position. In addition, a magnetically hard material also tends to be physically hard, as in the case of strain-hardened iron. The relationship between physical and magnetic hardness is the same as that between physical and magnetic softness discussed above. When a material is strain-hardened, large numbers of dislocations are introduced into the crystal structure, resulting in severe distortion of the lattice. This greatly restricts the movement of atoms in the structure, and these atoms are not free to rotate in their domains when an external field is applied. In order to saturate the material magnetically, most of the domains must be rotated and, consequently, a large external field is required. Because of this restriction on domain movement, a magnetically hard material will have very large hysteresis losses when subjected to an oscillating magnetic field.

In choosing materials for transformer cores and other devices that are subjected to oscillating magnetic fields, it is desirable to have a soft magnetic material with a large remanence. This means that the material must exhibit low saturating and demagnetizing fields, and also that it must lose relatively little input energy through hysteresis. Soft iron meets these requirements and is therefore used in electromagnets or transformer cores. However, if large pieces of soft iron are used as solid cores, the high electrical conductivity of iron causes extensive eddy current heating. Consequently, these cores are frequently made of large numbers of thin iron layers, laminated together with an electrical insulating material such as lacquer or varnish to separate each layer from its neighbors and reduce the available current flow path. In this way the magnetic properties are effectively utilized, while the eddy currents are decreased.

The magnitude of the eddy current losses increases markedly with the increasing frequency of oscillations of the applied field. Thus, even extremely fine subdividing of the iron core elements cannot overcome this problem if the frequency of the driving field becomes large. For this reason, materials other than ordinary ferromagnetics must be used in those special electronics applications where very high field frequencies are encountered. Ferrite materials, which are ferrimagnetic, are frequently used for such applications because of their high electrical resistivity. This high electrical resistivity, which may run as much as 10^{11} times that of metallic conductors, essentially eliminates eddy current formation. However, ferrimagnetic materials are not able to generate as large values of saturation magnetization as ferromagnetic materials and, therefore, are not used in such devices as generators, motors, and power transformers, where large induced magnetisms are required. On the other hand, they are extremely useful in such low field, high frequency devices as the "flyback" transformers of television picture-tube scanning circuits, various microwave elements, and computer memory core units.

Since the state of purity of a crystal structure, as reflected by dislocations and impurity atoms, markedly affects the magnetic character-

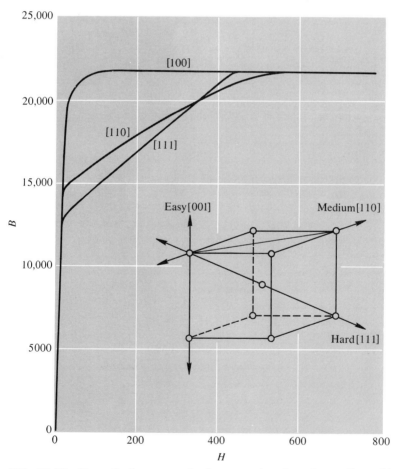

FIG. 10-16 Magnetization curves for iron as a function of crystallographic orientation. (From *The Solid State for Engineers,* by M. J. Sinnott, copyright 1961, John Wiley and Sons, Inc. Used by permission.)

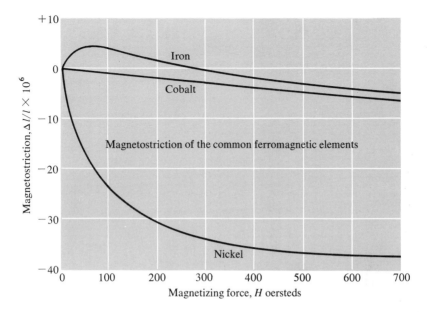

FIG. 10-17   Magnetostrictive strain for common ferromagnetic elements. (From *The Solid State for Engineers,* by M. J. Sinnott, copyright 1961, John Wiley and Sons, Inc. Used by permission.)

istics of ferromagnetic materials, it is not surprising that pure materials show marked anisotropy of magnetic properties. That is, different crystallographic directions (varying with the crystal) exhibit differing degrees of magnetic softness or hardness. In ordinary BCC iron, for example, the [100] direction is the easiest to magnetize, the [110] direction is of medium difficulty, and the [111] direction is the most difficult (see Fig. 10-16).

Magnetostriction is an interesting and sometimes important effect associated with the anisotropy of magnetization of a crystal lattice. Depending upon the orientation of magnetic domains relative to various crystallographic principal directions, the application of an external magnetic field may result in an elastic deformation in the material. This deformation, called magnetostriction, can be positive or negative and large or small, depending upon the particular alloy and the field and domain orientations. Magnetostriction data for three common ferromagnetic elements are shown in Fig. 10-17.

Magnetostriction can cause considerable stress to develop in massive machine parts, sometimes resulting in severe vibrational problems in large rotating electrical machinery. However, in some devices such as transducers, this property is used to convert electrical signals into mechanical deformations, or conversely. The magnetostriction effect assumes three general forms: a simple longitudinal strain, the *Joule effect;* a volume change, the *Barrett effect;* and a torsional strain, the

Wiedemann effect. The Joule effect is generally the most important of the three, but all are merely manifestations of the same phenomenon.

Powdered Metal Magnetic Cores

In order to reduce eddy current losses in cores, a core is often made from finely ground powders of ferromagnetic oxides compounded with a resinous binder material that serves as an electrical insulator. This compact can then be fashioned into shapes that perhaps could not be fabricated otherwise. As the magnetic material is finely subdivided (particle sizes range in practice from 0.00005 to 0.1 in.), the permeability of this composite decreases from that of the basic bulk material, but such permeability also becomes more and more constant. Thus, it is possible to form cores that have constant permeability in a varying field and in which losses due to eddy currents are reduced to a minimum.

Because such varied shapes can be created by powder compacting, it is possible to fabricate complicated shapes for permanent magnets from hard and brittle materials that could not otherwise be cast or machined into these configurations. This positive result must be balanced against the decrease in remanence that results from a decrease in permeability.

Ferrites

Ferrites are ferrimagnetic materials which are becoming increasingly useful in high frequency applications. The basic characteristics of ferrites which make their magnetic properties attractive arise from their molecular constitution. Ferrites are generally mixed oxides of iron and one of a number of rare-earth metals. The particular crystal structure of these materials, called a *spinel structure*, is a close-packed arrangement of oxygen ions with the metallic ions located in the interstices. There are twelve interstitial positions in this close-packed array: eight of these have a coordination number of 4, while the remaining four have a coordination number of 6. Thus, a metal ion can experience different environments, depending upon which interstitial location it occupies. In the normal spinel structure, the rare-earth metal ions occupy one of the four interstices having $CN = 6$, while the iron ions occupy the other type of interstices. In some structures, iron atoms can also occupy some empty $CN = 6$ interstices, giving rise to the so-called *inverse spinel structure*. (The actual unit cell of the spinel structure is rather complex and involves many atoms.)

Although iron has been used as the example of a ferromagnetic ion in the preceding discussion, spinel structures containing Al, Mg, Mn, Fe, Co, and Ni exist and are generally of the inverse type. It is possible to form solid solutions of these different spinel structures in which the individual spinel arrangements of the separate constituents are preserved [10].

The magnetic properties of the spinel structure do not arise from the ferromagnetism of the metal ions, but from their placement within the lattice. If the electron spin vectors of the included metal ions are parallel, the spinel structure will be ferro- or ferrimagnetic, depending upon the metal ion distribution. The electron spin vectors will cancel if they are antiparallel, resulting in antiferromagnetism for the overall structure. (This occurs when Zn or Cd ions occupy tetrahedral positions in the structure.)

Because the spinels are oxide structures, they are ceramic materials with extremely high electrical resistivity. This makes them useful when low eddy current losses are a desirable feature. In addition, spinels generally have a considerably lower density than solid metallic materials, making them especially attractive in modern, low weight applications. Spinel materials can also be ground into a powder base, mixed with binders, and pressed into appropriate forms. It is possible, then, to vary the composition of the mix among several different ferrite materials to produce a composite material having almost any desired combination of properties.

References Cited

[1] Sinnott, M. J., *The Solid State for Engineers*, John Wiley and Sons, Inc., New York, 1961, Chapter 16.

[2] Ehrenreich, H., "The Electrical Properties of Materials," *Materials, A Scientific American Book*, W. H. Freeman & Co., San Francisco, 1967, p. 128.

[3] Sinnott, *op. cit.*, Chapter 15.

[4] Levine, S. N., *Principles of Solid State Microelectronics*, Holt, Rinehart and Winston, Inc., New York, 1963, p. 30.

[5] *Ibid.*, p. 103.

[6] Gutmann, F., and L. E. Lyons, *Organic Semiconductors*, John Wiley and Sons, Inc., New York, 1967.

[7] Little, W. A., *Sci. Ameri.*, **212**, p. 21 (Feb./1965).

[8] Sinnott, *op. cit.*, p. 367.

[9] Keffer, F., "The Magnetic Properties of Materials," *Materials, A Scientific American Book*, W. H. Freeman & Co., San Francisco, 1967, p. 161.

[10] Sinnott, *op. cit.*, Chapter 18.

PROBLEMS

10.1   From Eq. (10.1) and the common form of Ohm's law $(-\Delta V = iR; i = \text{current}, R = \text{resistance})$, derive the result $R = L/Ak_e$, where L is the conductor length, A its cross-sectional area, and k_e the electrical conductivity of the material.

10.2   A nichrome wire must have a resistance of 100 ohms. If it is 0.001 in. in diameter, how long must it be $(\rho_e = 1/k_e = 10^{-4}$ ohm cm$)$?

10.3   A 6 in. filament wire must be designed for a heater application in which the available voltage drop to create the heat is 110 volts. If the material chosen is tungsten $(\rho_e = \rho_{20} [1 + a(t - 20)]$, where $\rho_{20} = 5.6 \times 10^{-6}$ ohm cm, $a = 0.0045°C^{-1}$, $t = °C$, and emissivity $= 0.30)$, and it is to operate at no more than 80% of its melting temperature $(3{,}400°C)$, compute the minimum diameter of the heater wire which can be used. (HINT: Assume the wire radiates heat to its surroundings at a net flux given by Eq. (9.17) if T is the wire temperature. The rate of heat production is $q = i^2 R$, where $i = -\Delta V/R$ is the current generated by a voltage $-\Delta V$ applied across the resistance R.)

10.4   If the only tungsten wire available has a diameter twice that required for Problem 10.3, what length of heater wire would allow it to work in the same way?

10.5   The ionic mobilities of K^+ ions in solid KCl at 250°C and 727°C are $0.688(10^{-4})$ cm^2/volt sec and $8.2(10^{-3})$ cm^2/volt sec, respectively. The ionic mobility of Na^+ ions in solid NaCl at 550°C is $5.5(10^{-10})$ cm^2/volt sec. Assuming that the temperature dependence of diffusivity is the same for both materials, estimate the ionic conductivity of Na^+ ions in NaCl at 250°C. (HINT: Assume $Q(KCl) = Q(NaCl)$ in Eq. (6.16).)

10.6   The diffusion coefficients of K^+ ions in solid KCl are $0.708(10^{-5})$ cm^2/sec at 1000°K and $0.523(10^{-5})$ cm^2/sec at 477°C. Compute the ionic mobility of K^+ ions in solid KCl at 250°C, 300°C, and 350°C.

10.7   Using the data in Problem 10.6, compute the activation energy for the K^+ ion movement in KCl.

10.8   Silicon has a density of 2.40 g/cm^3. Phosphorus is added to make it an n-type semiconductor, with $k_e = 1.0$(ohm cm)$^{-1}$ and $\mu_n = 1750$ cm^2/volt sec. Compute the concentration of charge carriers/cm^3.

10.9   Compute the number of Si atoms per charge carrier in Problem 10.8.

10.10   Germanium has $\rho_e = 1.8$ ohm cm and an electron hole concentration of $2.0(10^{15})$ holes/cm^3. Compute μ_p for the holes in this semiconductor material.

10.11   Using the E_g values from Table 10.1, compute the change in concentration of Frenkel type charge carriers in Ge and Si at 400°C relative to 300°C.

10.12   The resistivity of silver is given by $\rho_e = 1/k_e = 1.59(10^{-6})\,[1 + 0.0038(t - 20)]$ ohm cm, where t is in °C. Compute the electrical resistance of a silver wire, 1000 ft long and 0.001 in. in diameter at 250°F.

10.13   In a certain application, an electrical element must be used that will experience a static load of 1000 lb$_f$, have a resistance no less than 0.0001 ohm/in., and operate success-fully under strain-hardened conditions at 425°C. For safety, the stress level must be at least 0.75 times the tensile strength of the material. Five materials have been selected as possible candidates. Using the following physical property data for these five, select the material which could be used to satisfy all the above conditions. What is the possible range of wire diameters for this material?

MATERIAL	T_{melt}(°C)	$10^{-4}\,\sigma_t$(psi)	$10^6\,\rho_e$(ohm cm)	ρ(g/cm^3)
A	1539	3.07	9.7	7.87
B	650	0.933	4.5	1.74
C	1650	6.0	56.7	4.54
D	1450	12.0	50.0	7.75
E	660	5.34	2.67	2.80

11

THE RESPONSE OF MATERIALS TO CHEMICAL ENVIRONMENTS

The chemical environment in which a material is used may be passive and have little effect on its service life, or it may be hostile and severely limit or destroy its usefulness. In selecting a material for a particular application, after all mechanical, thermal, and other properties have been considered, the engineer must then examine the environment in which this material will operate. For example, a certain aluminum alloy may be an excellent process piping material from the point of view of cost, strength, weight, and thermal stability, but if the process involves such highly basic liquid solutions as NaOH, the aluminum would be an extremely poor choice of material. Aluminum reacts rapidly with basic solutions to form $Al(OH)_3$, and the entire system would dissolve in a very short time. This dissolution or deterioration of a material by its chemical environment is called *corrosion*.

In this chapter, the basic principles of corrosion and some of the methods of controlling it will be examined. In general, corrosion is the destruction or deterioration of a material by chemical reaction with its environment. It will be observed, however, that the interaction of mechanical stresses with such chemical action may accelerate the destruction process.

The engineer must learn how and why corrosion occurs in order to control and manipulate its mechanisms to serve his purposes. Although the wise application of fundamental knowledge can greatly reduce the problem of corrosion, much of what must be done is more in the spirit of art than science.

Corrosion is a tremendously costly phenomenon; the annual cost in the United States alone runs into the billions of dollars. It can cause the unexpected and undesired shutdown of plants, unsightly deterioration of visible parts, and loss of valuable products through leakage and failure of equipment. Corrosion also creates serious health and safety hazards, and pollutes and contaminates vital substances such as water, foods, and drugs.

Because corrosion can take many forms, it may be discussed from a number of different points of view. Here, however, we will divide all corrosion processes into two major categories: *wet corrosion* and *dry corrosion*. Wet corrosion is any process in which the liquid phase plays an essential part, either as a direct reactant or as a reaction medium. Dry corrosion is any process in which the liquid phase is absent; here the corroding agents are generally gases or vapors. For example, consider the difference in the action of chlorine gas on titanium and on steel. Titanium is severely attacked by dry chlorine gas in the absence of water, but, by contrast, the dry gas essentially has no effect on steel. However, the presence of even small amounts of moisture, such as the ordinary humidity in the atmosphere, will cause severe attack and deterioration of steel by the chlorine.

The quantitative effects of corrosion are measured by the amount of weight loss occurring as a result of the deterioration of a material over a given period of time. The nearly universal method of expressing this is

$$\text{mpy} = \frac{534W}{\rho At} \tag{11.1}$$

where mpy is the abbreviation for mils per year, W is the weight loss in mg that occurs in a period of t hours, ρ is the mass density of the material in g/cm^3, and A is the area of the test specimen in square inches.

Galvanic Corrosion

Although the preceding definition of wet corrosion includes direct chemical attack by the liquid and all other processes in which the presence of a liquid phase is essential for destruction, the vast majority of wet corrosion situations involve electrochemical or galvanic action in which the liquid plays the role of a reaction medium. Therefore, in order to understand wet corrosion, it is essential that the fundamentals of electrochemistry be examined first.

Electrode Potentials

When placed in aqueous solutions, different metals dissolve and enter the solution as ions in different ways. The measure of ionization tendency is the free energy difference between a metal atom and its metal ion. Whenever a metal atom ionizes and leaves the surface of the solid to become a positively charged ion in the solution, it leaves the excess electrons that it gave up in the ionization process behind in the solid. This is illustrated by the chemical equation for the ionization of zinc

$$Zn \leftrightharpoons Zn^{++} + 2e^- \qquad (11.2)$$

where e^- represents the liberated electron. Since this reaction is reversible (as indicated by the double-headed arrow), any change in environmental conditions that alters the concentrations of either the reactants or products will cause the reaction to shift in the opposite direction. This happens in accordance with the law of mass action. Initially, the absence of Zn^{++} ions in the solution causes dissolution to occur readily—that is, the free energy driving force is large. However, as the concentration of the ions in solution increases, the free energy difference between metal and ion decreases. Eventually a condition of equilibrium is reached in which the free energy difference vanishes. Here the metal will possess a net negative charge due to the presence of the ionized electrons. This negative electric potential E is related to the concentrations of the reduced (metallic) and oxidized (ionic) species in the solution by the Nernst equation [1]

$$E = E_0 + \frac{RT}{nF} \ln\left(\frac{a_{ox}}{a_{red}}\right) \qquad (11.3)$$

where R is the gas constant, T is the absolute temperature, n is the number of electrons transferred in the reaction, F is Faraday's constant (96,500 coulombs/equivalent),[1] and a_{ox} and a_{red} are, respectively, the activities of the oxidized and reduced species in the reaction. The activity of any specie in a solution is related to its concentration by $a = \gamma c$, where c is the concentration and γ is an activity coefficient close to unity for all but highly nonideal solutions. In electrochemical systems, γ is generally considered to be unity and neglected, so that essentially $a = c$. The term E_0, called the *standard electrode potential*, corresponds to the condition $a_{ox} = a_{red}$. The potential E is related to the free energy change ΔG of the electrochemical dissociation reaction by the equation $-\Delta G = nEF$, where the symbols have the above meanings.

Because of the nature of thermodynamic functions, an absolute value for E_0 cannot be assigned and, therefore, values are determined

---

[1]An equivalent weight in electrochemical reactions is $1/n$ times the molecular weight.

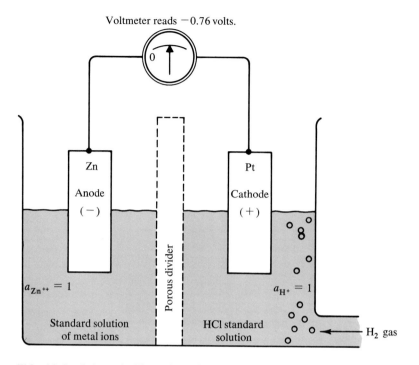

Voltmeter reads -0.76 volts.

FIG. 11-1 Schematic illustration of a standard electrochemical cell for the reaction $Zn + 2H^+ = Zn^{++} + H_2(g)$.

relative to an arbitrarily chosen standard. The standard used is the hydrogen electrode, which consists of a plantinum wire immersed in a 1N hydrochloric acid solution[2] under a blanket of hydrogen gas at one atmosphere pressure and 25°C. For these conditions, Eq. (11.3) reduces to $E = E_0$, where E_0 is arbitrarily chosen to be zero. The standard electrode potential of any other metal can be determined by measuring the potential difference between this standard hydrogen electrode and a standard electrode of the metal in question immersed in a solution of its own ions of unit activity. (Under these conditions, the logarithmic term in Eq. (11.3) will vanish.) This process is accomplished in an electrochemical cell like the one illustrated schematically in Fig. 11-1.

Galvanic Couples or Cells

The coupling of two dissimilar electrodes through an external electrical circuit (wire) and an electrolyte (ionic solution) is called a

---

[2]A 1N HCl solution contains 1 mole of HCl per mole of water and has $a_{H^+} = 1$. This is called a normal solution or a solution of unit H^+ ion activity.

galvanic couple or *cell*. The electrode that supplies electrons to the wire and positive ions to the electrolyte is conventionally called the *anode*, while the other electrode is the *cathode*. The electrical signs used to designate the electrode potentials of the anode and the cathode depend upon whether the work of a physical chemist [2] or an electrochemist [1] is consulted. The physical chemist labels the anode positive and the cathode negative, whereas the electrochemist reverses this sign order. (A similar dilemma in sign convention is encountered in electrical problems with regard to positive current flow versus electron flow.) Here, the electrochemical labeling of the anode as negative will be followed. This seems reasonable since the anode will be negatively charged because the oxidation reaction that supplies positive ions to the electrolyte leaves the excess electrons in the anode.

The difference between these physical and electrochemical systems also results in a difference in the sign of $\Delta G = -nEF$; depending upon the system used, a minus sign must be inserted or deleted so that ΔG will be negative when ionization reaction occurs in the forward direction. In all dealings with corrosion cells, however, it is the difference in the standard electrode potentials which is important, and so this sign convention is not of critical interest here.

The standard cell reaction indicated in Fig. 11-1 can be written

$$Zn = Zn^{++} + 2e^- \tag{11.4}$$

$$2e^- + 2H^+ = H_2 \tag{11.5}$$

$$\overline{2H^+ + Zn = Zn^{++} + H_2} \tag{11.6}$$

The two reactions given by these equations are called *half-cell* reactions. Since Eq. (11.5) is the standard system for which $E_0 = 0$, it follows that the total cell reaction given by Eq. (11.6) has the same standard cell potential as the half-cell reaction in Eq. (11.4). This value of E_0 is therefore assigned to Eq. (11.4) as the standard half-cell reaction potential. If different metals are coupled with the hydrogen electrode in a standard cell, a half-cell potential characteristic of the particular metal will be measured. When all of these standard potentials are arranged in order of increasing potential, the standard electromotive series of metals is obtained (see Table 11-1).

Examination of the data in Table 11-1 provides insight into the process of corrosion. Any metal near the bottom is anodic relative to hydrogen or to any metal above this metal in the table. The further apart two metals are in the table, the greater the electrode potential difference between them when they are coupled in a cell. For example, consider the metals copper and iron.

Copper is cathodic with respect to iron. Therefore, if these two

TABLE 11-1

Standard Electrode Potentials*

METAL REACTION	E_0 (VOLTS)
$Au = Au^{3+} + 3e^-$	+1.498 (cathodic)
$O_2 + 4H^+ + 4e^- = 2H_2O$	+1.229
$Pt = Pt^{++} + 2e^-$	+1.20
$Pd = Pd^{++} + 2e^-$	+0.987
$Ag = Ag^+ + e^-$	+0.799
$2Hg = Hg_2^{++} + 2e^-$	+0.788
$Fe^{+3} + e^- = Fe^{++}$	+0.771
$O_2 + 2H_2O + 4e^- = 4OH^-$	+0.401
$Cu = Cu^{++} + 2e^-$	+0.337
$Sn^{4+} + 2e^- = Sn^{++}$	+0.15
$2H^+ + 2e^- = H_2$	0 (arbitrary reference)
$Pb = Pb^{++} + 2e^-$	−0.126
$Sn = Sn^{++} + 2e^-$	−0.136
$Ni = Ni^{++} + 2e^-$	−0.250
$Co = Co^{++} + 2e^-$	−0.277
$Cd = Cd^{++} + 2e^-$	−0.403
$Fe = Fe^{++} + 2e^-$	−0.440
$Cr = Cr^{3+} + 3e^-$	−0.744
$Zn = Zn^{++} + 2e^-$	−0.763
$Al = Al^{3+} + 3e^-$	−1.662
$Mg = Mg^{++} + 2e^-$	−2.363
$Na = Na^+ + e^-$	−2.714
$K = K^+ + e^-$	−2.925
$Rb = Rb^+ + e^-$	−2.925
$Li = Li^+ + e^-$	−2.959

*Values are at 25°C and produce normal ionic activity when referenced to a normal hydrogen electrode.

metals are coupled in the cell illustrated schematically in Fig. 11-2, a potential difference will be observed where the copper electrode is the cathode. The reaction taking place at the cathode will vary with the acidity or alkalinity of the electrolyte solution. In an acidic solution, the reaction that occurs at the cathode of the cell is Eq. (11.5), where H_2 gas is *plated out*. If the solution is neutral or basic (that is, with a high OH^- concentration), the reaction which tends to occur at the cathode is

$$O_2 + 2H_2O + 4e^- = 4OH^- \qquad (11.7)$$

In this reaction, dissolved oxygen reacts with the water of the aqueous electrolyte and the electrons that arrive from the cathode, forming more hydroxyl ions and increasing the alkalinity of the solution. The potential developed by the cell is dependent upon which reaction predominates at the cathode. The value is calculated as the difference between the cathodic electrode potential and the anodic electrode potential (the values in Table 11-1).

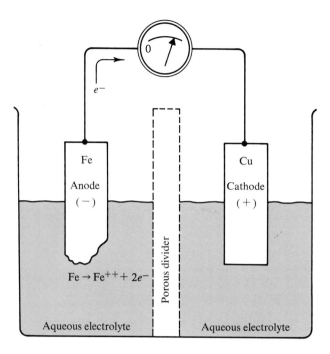

FIG. 11-2   Cell reaction for an Fe–Cu couple, where Fe is the anode and Cu is the cathode.

Thus, we have seen that three essential components must be present simultaneously for electrochemical or galvanic corrosion to occur: (1) an anode-cathode couple, (2) an electrical connection between the anode and cathode (such as a wire), and (3) an electrolyte. When these three elements are present, current flows and corrosion occurs; when any one of these is removed, the current cannot flow and corrosion cannot occur.

Because the anodic reaction is one of oxidation, the anode will suffer physical deterioration. Thus, if the system in Fig. 11-2 is allowed to run for a sufficient time, the lower portion of the iron electrode is eaten away, breaking the circuit and halting the flow of current.

Electrode Kinetics

Thus far, the equilibrium behavior of electrochemical systems has been examined on the basis of thermodynamic arguments. However, corrosion is a nonequilibrium or kinetic process, and thermodynamic considerations cannot reveal the rate at which corrosion will occur or whether it can occur at all. If anything, thermodynamic arguments have somewhat of a negative influence. That is, if the thermodynamic potentials are unfavorable for corrosion, it will not occur; if they are favorable, corrosion may or may not occur, or it may occur at a negligible rate. Thus, to discover more about the rate of the corrosion reaction, the kinetic details of electrodes must be examined.

POLARIZATION   In an electrochemical cell, at the instant the anode electrically couples with the cathode, the potential difference is the equilibrium thermodynamic value as calculated from the standard electrode potentials. As soon as current is drawn from this cell, however, its overall potential begins to decrease as a result of various phenomena that occur at the individual electrodes, causing them to approach each other in potential. This process, called *electrode polarization*, is defined [1] as the displacement of the electrode potential from its thermodynamic equilibrium value as a result of a net current flow. The magnitude of the polarization is expressed in terms of the *overvoltage* η. In general, two different mechanisms can create electrode polarization: *concentration polarization* and *hyrdogen overvoltage*.

CONCENTRATION POLARIZATION   As current begins to flow in a cell, ions must enter the solution at the anode and leave at the cathode. When concentration gradients resulting from the relatively slow diffusion process cause the ionic concentration in the immediate vicinity of the electrode to become unfavorable for the particular electrode reaction, the overall reaction is retarded and the potential changes. At the anode in Fig. 11-2, for example, as the concentration of Fe^{++} ions in the immediate neighborhood of the electrode increases due to their inability to diffuse away fast enough, the reaction slows down and the potential of the electrode becomes less anodic, thus polarizing the anode. The voltage difference between the polarized condition and the equilibrium value is the overvoltage. Similarly, a depletion of H^+ or O_2 ions (depending upon the pH of the system)[3] at the cathode slows down the reaction at the cathode and causes the potential to become less cathodic. Again, the difference between the polarized and equilibrium potentials is the overvoltage.

For an electrode in which concentration polarization is the only operative mechanism [1], the overvoltage may be calculated from the equation

$$\eta_c = \frac{RT}{nF} \ln \left(1 - \frac{i}{i_L} \right) \qquad (11.8)$$

---

[3]pH is a logarithmic measure of a H^+ ion concentration. Thus, if $a_{H^+} = 10^{-7}$, $\log_{10} a_{H^+} = -7$. pH $= -\log_{10} a_{H^+}$. Therefore, if $a_{H^+} = 10^{-7}$, which is the concentration of H^+ ions in distilled water, then pH $= 7$. Basic solutions have pH > 7, while acid solutions have pH < 7.

where i is the current drawn from the cell; R, T, n, and F have the same meanings as in Eq. (11.3); and i_L is the limiting, diffusion current density given by [3, 4]

$$i_L = \frac{DnFc_B}{x} \qquad (11.9)$$

where D is the diffusion coefficient of the reacting ions in the electrolyte, c_B is the concentration of these ions in the bulk solution (that is, some distance away from the electrode), n and F have the same meanings as above, and x is the thickness of the diffusion layer.[4] Since x depends upon the shape of the particular electrode, the geometry of the system, and the degree of solution agitation, it is extremely difficult to calculate and, except for very simple systems, must be determined by experimental measurements. This, however, does not entirely limit the usefulness of these results when we consider the fact that Eq. (11.9) does indicate the factors that influence i_L. For example, agitation decreases x and, therefore, increases i_L, which, in turn, decreases η_c. Similar qualitative observations can be drawn concerning the other variables.

Concentration polarization is essentially unimportant until i approaches i_L rather closely. When this condition is realized, the overvoltage will increase sharply.

HYDROGEN OVERVOLTAGE   The *hydrogen overvoltage* phenomenon is a second form of electrode polarization occurring at the cathode. As hydrogen ions are reduced at the cathode surface according to Eq. (11.5), a sequence [5] of separate events must occur sequentially. The mechanism illustrated in Fig. 11-3 is a simplified version of the many possible, more complex mechanisms [5].

The first step involves diffusion of the H^+ ions from the bulk solution to the cathode surface. (In most situations this is not the limiting step.) In the second step, the ion is adsorbed on the surface where it combines with an electron from the interior of the cathode, thus forming an atom of neutral hydrogen. This hydrogen atom diffuses along the surface of the electrode until it encounters another adsorbed hydrogen atom with which it combines to form a molecule of diatomic hydrogen—the third step in the process. The fourth and final step is the nucleation of the molecular hydrogen to form a bubble of H_2 gas, which detaches itself from the surface and escapes. If steps three and four occur at a slower rate than the first two steps in the reaction, the surface of the cathode becomes covered with an adsorbed layer of hydrogen. This, in turn, precludes additional H^+ ions from being adsorbed on the surface, slowing down the rate of hydrogen reduction, reducing the current flow, and making the cathodic potential more

---

[4]A diffusion layer is the region immediately surrounding an electrode through which the ions must diffuse before their concentration is essentially c_B.

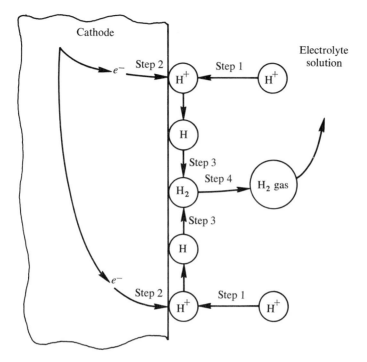

FIG. 11-3 Simplified schematic representation of the step sequence involved in H⁺ ion reduction at the cathode surface.

anodic. The decrease in potential from the equilibrium value due to this mechanism is the hydrogen overvoltage.

Table 11-2 contains values of hydrogen overvoltage for a selected number of metals [6]. From the column headed $E_0 + \eta_{H_2}$ in this table, it is clear that the hydrogen overvoltage for a number of metals anodic relative to hydrogen under equilibrium conditions is sufficient to render them cathodic. This means that these metals would not react with an acid solution, as might be anticipated solely on the basis of their E_0 values. The low value of $E_0 + \eta_{H_2}$ for zinc indicates that it will react slowly when first placed in an acid solution. Initially, this is observable; once the reaction begins, however, the rising bubbles of hydrogen sweep the surface above them clean, exposing fresh surface to the acid and creating a vigorous reaction.

ACTIVATION POLARIZATION The process of polarization in the reduction of hydrogen at the cathode, described above, is an example of *activation polarization*. In general, activation polarization refers to electrochemical reactions that are controlled by a slow step in the reaction sequence [1], such as the nucleation of the hydrogen gas bubble

TABLE **11-2**

Hydrogen Overvoltage for Selected Metals

METAL	η_{H_2} (VOLTS)*	E_0 (VOLTS)	$E_0 + \eta_{H_2}$ (VOLTS)
$Pt = Pt^{++} + 2e^-$	0.12	1.20	1.32
$Ag = Ag^+ + e^-$	0.29	0.80	1.09
$Cu = Cu^{++} + 2e^-$	0.25	0.34	0.59
$2H^+ + 2e^- = H_2$	—	—	—
$Pb = Pb^{++} + 2e^-$	0.60	−0.13	+0.47
$Sn = Sn^{++} + 2e^-$	0.50	−0.14	+0.36
$Ni = Ni^{++} + 2e^-$	0.25	−0.25	0
$Cd = Cd^{++} + 2e^-$	0.50	−0.40	+0.10
$Fe = Fe^{++} + 2e^-$	0.27	−0.44	−0.17
$Zn = Zn^{++} + 2e^-$	0.70	−0.76	−0.06

*η_{H_2} data, taken from reference [6], correspond to metals being immersed in normal acid solutions at a current density of 0.1 amp/cm^2.

or the formation of the hydrogen atom by electron transfer in the case above.

The current density of a corrosion cell of moderate current density level is related to the overvoltage of the electrode by the *Tafel equation* [1]

$$\eta_a = \pm \beta \ln \left(\frac{i}{i_0} \right) \tag{11.10}$$

where η_a is the activation polarization overvoltage, i is the cell current density, β is the slope of a semilogarithmic plot of η_a vs i, and i_0 is an empirically measured constant called the *exchange current density* (see Table 11-3 for selected values). As the value of i decreases toward zero, the Tafel equation no longer applies. The deviation point is then a polarization about 50 mV more active than the corrosion potential[5] of the cell. For most electrochemical reactions, the magnitude of β, or the *Tafel constant*, ranges between 0.05 and 0.15 volts [1], the most common value being approximately 0.1 volt.

MIXED-POTENTIAL THEORY   Generally both types of electrode polarization described above are operative; activation polarization is the controlling mechanism at low reaction rates, and concentration polarization is the controlling mechanism at high ones. The total polarization

---

[5]The corrosion potential of a cell is that potential at which the rates of oxidation and reduction are exactly equal. This potential is determined from the mixed-potential theory, to be described shortly.

TABLE **11-3**

Exchange Current Densities for Selected Metals*

REACTION	ELECTRODE	ELECTROLYTE	$i_0(amp/cm^2)$
$2H^+ + 2e^- = H_2$	Al	2N H_2SO_4	10^{-10}
$2H^+ + 2e^- = H_2$	Au	1N HCl	10^{-6}
$2H^+ + 2e^- = H_2$	Cu	0.1N HCl	$2(10^{-7})$
$2H^+ + 2e^- = H_2$	Fe	2N H_2SO_4	10^{-6}
$2H^+ + 2e^- = H_2$	Hg	1N HCl	$2(10^{-12})$
$2H^+ + 2e^- = H_2$	Hg	5N HCl	$4(10^{-11})$
$2H^+ + 2e^- = H_2$	Ni	1N HCl	$4(10^{-6})$
$2H^+ + 2e^- = H_2$	Pb	1N HCl	$2(10^{-13})$
$2H^+ + 2e^- = H_2$	Pt	1N HCl	10^{-3}
$2H^+ + 2e^- = H_2$	Pd	0.6N HCl	$2(10^{-4})$
$2H^+ + 2e^- = H_2$	Sn	1N HCl	10^{-8}
$O_2 + 4H^+ + 4e^- = 2H_2O$	Au	0.1N NaOH	$5(10^{-13})$
$O_2 + 4H^+ + 4e^- = 2H_2O$	Pt	0.1N NaOH	$4(10^{-13})$
$Fe^{3+} + e^- = Fe^{++}$	Pt	—	$2(10^{-3})$
$Ni = Ni^{++} + 2e^-$	Ni	0.5N $NiSO_4$	10^{-6}

*From J. O'M. Bockris, *NBS Circular 524*, United States Government Printing Office, Washington, D.C., 1953, pp. 243–62.

overvoltage of any electrode is therefore the sum of the contributions due to these two effects

$$\eta_t = \eta_a + \eta_c \qquad (11.11)$$

Concentration polarization does not affect the rate of the reaction during the process of anodic dissolution, because the presence of a piece of solid metal provides an essentially infinite supply of metal atoms at the reaction site. For the dissolution reaction at the anode, therefore, Eq. (11.11) becomes

$$\eta_t = \eta_{diss} = \beta \ln\left(\frac{i}{i_0}\right) \qquad (11.12)$$

In the case of the cathodic reaction, however, depletion of H^+ ions can cause influential concentration polarization, especially if the rate of reduction of hydrogen ions approaches the limiting, diffusion current density i_L. Therefore, for the cathodic reduction reaction, Eq. (11.11) must be written

$$\eta_t = \eta_{red} = -\beta \ln\left(\frac{i}{i_0}\right) + \frac{RT}{nF} \ln\left(1 - \frac{i}{i_L}\right) \qquad (11.13)$$

The formal statement of the mixed-potential theory may be summarized [1, 7] as follows:

1. Any electrochemical reaction system can be represented in terms of two or more half-cell, oxidation-reduction reactions.

2. In any isolated sample undergoing an electrochemical reaction there can be no net accumulation of electrical charge.

This theory becomes useful when the behavior of a *mixed electrode* is considered. (A mixed electrode is an isolated metal electrode in simultaneous contact with two or more oxidation-reduction systems.) Consider the anodic and cathodic reactions in a system composed of an isolated piece of iron immersed in a sulfuric acid solution.

The anodic reaction can be visualized as an iron electrode in equilibrium with its own ions. This reaction is characterized by an equilibrium potential E_0, taken from Table 11-1, and by a corresponding exchange current density. The cathodic reaction is the reduction of hydrogen ions on an iron surface in acid solution; it is characterized by an E_0 value of zero and by an exchange current density (see Table 11-3 for selected values). These two situations are represented schematically in Fig. 11-4, where Eq. (11.12) and Eq. (11.13) are plotted separately for each electrode.

In the actual situation, the iron does not contact a 1N solution of its own ions; instead it is in an acid solution that is relatively dilute in Fe^{++} ions. The total electrode, which must assume some uniform

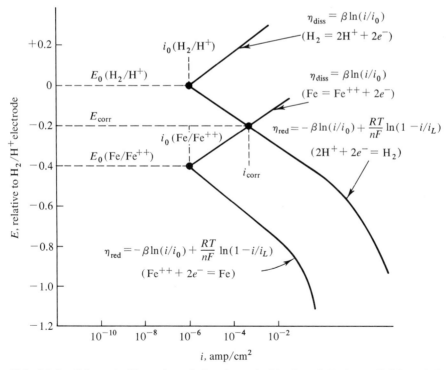

FIG. 11-4   Schematic illustration of the electrode kinetics of Fe in an H_2SO_4 solution.

potential because of the second hypothesis of the theory, cannot be at either of the half-cell standard potentials and, therefore, must lie somewhere in between the two. The only electrode potential that simultaneously satisfies Eq. (11.12) for the anode and Eq. (11.13) for the cathode is given by the intersection of the two curves that represent these equations. This point is indicated in Fig. 11-4 by the potential E_{corr} and the current density i_{corr}, called the *corrosion potential* and the *corrosion current* of the cell, respectively. At this particular set of conditions, the rate of dissolution of the iron exactly balances the rate of reduction of the hydrogen ions.

This examination of mixed electrode kinetics leads to some rather interesting conclusions. To illustrate the importance of kinetic factors on the corrosion rate, consider placing iron and zinc in HCl. Since zinc has a more anodic value of E_0 than iron, it might be expected to corrode more quickly. This is not true, however, because zinc surfaces have a lower value of i_0 for H^+ reduction in HCl than iron surfaces [1], which makes $i_{corr}(Zn/HCl)$ less than $i_{corr}(Fe/HCl)$. That is, iron actually corrodes faster in HCl than does zinc. This fact clearly illustrates the error made by equating relative values of E_0 with relative rates of corrosion.

PASSIVITY   A very interesting phenomenon occurs in some metals when they are surrounded by strongly oxidizing solutions. If a sample of iron is placed in concentrated (70 $w/0$)HNO$_3$, no reaction is observed: the iron seems to be inert or *passive* in the concentrated HNO$_3$ solution. If an approximately equal volume of water is added to this solution while the passivated iron is still present, the passivity of the iron is retained and no reaction occurs. If the iron is then scratched with a stirring rod or agitated in a way that it strikes the container walls, a violent reaction occurs with profuse liberation of nitrogen oxide gas. The same violent reaction is observed if the iron is placed in the dilute acid initially.

Clearly, the concentrated nitric acid did something to the iron to render it passive that the dilute acid did not do. This passivity is evidently the result of the formation of some kind of film on the surface of the iron that isolates it from the acid and, therefore, from further reaction. In this case, the film appears to be very fragile, for merely scratching the surface of the iron initiates a violent reaction.

Many metals other than iron exhibit passivity. Some passivity is only temporary and easily destroyed, as in the Fe-HNO$_3$ system; some is relatively permanent, as in stainless steels containing more than eighteen per cent Cr. Figure 11-5 is a schematic representation of the anodic polarization behavior of a metal undergoing an active-passive

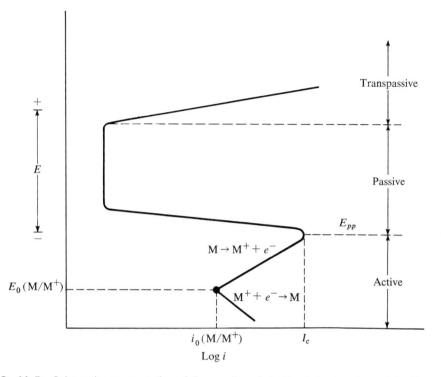

FIG. 11-5   Schematic representation of the anodic polarization behavior of a metal with an active-passive transition at E_{pp}.

transition. This metal exhibits typical Tafel behavior at electrode potentials not too far removed from $E_0(M/M^+)$. However, at a potential E_{pp}, called the *primary passive potential*, which is less anodic than E_0, the metal suddenly undergoes a transition to an extremely low value of corrosion current density. It retains this current density with increasing cathodic potential until, at some higher potential (generally significantly higher than E_{pp}), the corrosion current density suddenly begins to increase at a very rapid rate. The region between E_0 and E_{pp} is the *normal active state*, that between E_{pp} and the next break point in the curve is the *passive state*, and that above the second break point is the *transpassive region*. Apparently, in the transpassive region, the oxidizing power of the solution destroys the passive film formed at E_{pp}. The current I_c that exists at E_{pp} is known as the *critical anodic current density for passivity*.

Because of the phenomenon of passivity, many metals do not behave in practice as their standard electrode potentials indicate. It has been necessary, therefore, to develop an alternate series for ordering the anodic or cathodic natures of various metals. This *Galvanic series* (see Table 11-4) is based entirely upon experience with various metals in different environments. In selecting materials to operate in corrosive

TABLE **11-4**

Galvanic Series of Metals and Alloys in Seawater

	Platinum
	Gold
NOBLE OR CATHODIC	Graphite
	Titanium
	Silver
	⎡ Chlorimet 3 (62 Ni, 18 Cr, 18 Mo)
	⎣ Hastelloy C (62 Ni, 17 Cr, 15 Mo)
	⎡ 18-8 Mo stainless steel (passive)
	⎢ 18-8 stainless steel (passive)
	⎣ Chromium stainless steel 11-30% Cr (passive)
	⎡ Inconel (passive) (80 Ni, 13 Cr, 7 Fe)
	⎣ Nickel (passive)
	Silver solder
	⎡ Monel (70 Ni, 30 Cu)
	⎢ Cupronickels (60-90 Cu, 40-10 Ni)
	⎢ Bronzes (Cu-Sn)
	⎢ Copper
	⎣ Brasses (Cu-Zn)
	⎡ Chlorimet 2 (66 Ni, 32 Mo, 1 Fe)
	⎣ Hastelloy B (60 Ni, 30 Mo, 6 Fe, 1 Mn)
	⎡ Inconel (active)
	⎣ Nickel (active)
	Tin
	Lead
	Lead-tin solders
	⎡ 18-8 Mo stainless steel (active)
	⎣ 18-8 stainless steel (active)
	Ni-Resist (high Ni cast iron)
	Chromium stainless steel, 13% Cr (active)
	⎡ Cast iron
	⎣ Steel or iron
	2024 aluminum (4.5 Cu, 1.5 Mg, 0.6 Mn)
ACTIVE OR ANODIC	Cadmium
	Commercially pure aluminum (1100)
	Zinc
	Magnesium and magnesium alloys

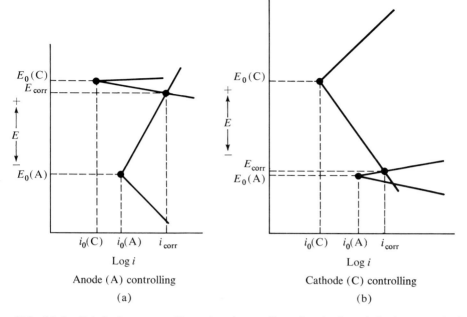

FIG. 11-6 Polarization curves illustrating the anodic and cathodic polarization control of corrosion current.

service conditions, the Galvanic rather than the standard EMF series must be used because it incorporates the observed effects of passivity.

POLARIZATION CONTROL   Figure 11-4 is an example of a system in which both the anode and the cathode undergo polarization; that is, β is approximately the same for each electrode in terms of the Tafel equation. In some situations, however, one electrode or the other exhibits very little polarization, and this will be manifested in the polarization curve by a very small Tafel constant. Under such conditions, illustrated schematically in Fig. 11-6, the electrode with the large Tafel constant is said to be the controlling electrode. That is, the reactions occurring at that particular electrode are the ones that will limit the corrosion current.

Origin of Potential Differences on Metal Surfaces

When two galvanically different metals are electrically coupled in the presence of an electrolyte, a corrosion current flows and one of the metals is preferentially deteriorated or corroded. The origin of this action is easily understood in view of the preceding theoretical discussion. The presence of dissimilar metal couples is not necessary, however. Samples of single metals placed in contact with a corrosive electrolyte readily corrode, indicating that anodic and cathodic areas both exist on the single metal surface itself. It therefore becomes necessary to know the origin of these cathodes and anodes.

ENVIRONMENTAL FACTORS   From the preceding sections, many of the environmental factors that produce relative anodic and cathodic areas on the surface of a single piece of metal can be deduced. For example, it is known from the Nernst equation that the concentration of oxidized and reduced species in a solution influences the electrode potential relative to the standard value. Therefore, it should not be surprising to learn that differences in concentration at two points on a metal surface will result in differences in electrode potential at the same two points. For whatever reason, as soon as a potential difference exists, the region with the more negative potential is anodic relative to the more positive area, and a cell has been established.

EXAMPLE 11.1

The cell illustrated in Fig. 11-7 contains two Fe electrodes immersed in $FeCl_2$ solutions of differing concentrations. The solution on the left has an activity $a_{Fe++/Fe} = 10^{-5}$, while that on the right has a similar activity of 10^{-1}. Estimate the potential difference that will develop in the cell.

Solution:

For this calculation, $R = 8.32$ joules/mole °K, $F = 96,500$ coulombs/mole, $T = 298°$K, and $n = 2$. Equation (11.3) then becomes $E = E_0 + 0.0285 \log_{10} (a_{Fe++/Fe})$. For the left electrode $E = -0.5825$ volts, and for the right electrode $E = -0.4685$ volts. Thus, the left electrode is strongly anodic relative to the right one. The cell potential is $-0.4685 - (-0.5825) = +0.114$ volts, and a corrosion current can be expected to flow.

From this example, it is clear that any mechanism resulting in a difference in electrolyte concentration between two regions of a metal surface will establish a concentration difference cell where corrosion will probably occur. Such concentration cells can be created by the partial shielding of one area from a source of metal ions or by flow velocity differences in the solution over two different parts of the surface so that the diffusion layer thickness is affected to varying degrees.

The cathode reaction Eq. (11.7) indicates another important source

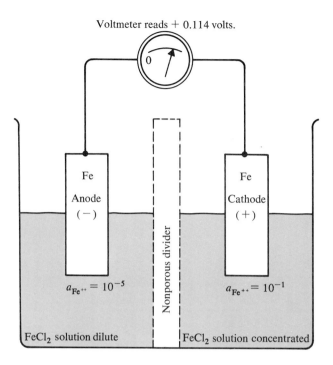

Voltmeter reads + 0.114 volts.

Fe
Anode
(−)

Fe
Cathode
(+)

$a_{Fe^{++}} = 10^{-5}$

Nonporous divider

$a_{Fe^{++}} = 10^{-1}$

FeCl$_2$ solution dilute

FeCl$_2$ solution concentrated

FIG. 11-7   Illustration of a concentration difference cell. An anode forms in the dilute concentration region, while a cathode forms in the high concentration region.

of potential difference on the surface of a single metal. If the concentration of dissolved oxygen is decreased for some reason near one area of this surface, the reaction rate of the oxygen with water and its ability to combine with electrons to form hydroxyl ions will both be decreased, thus polarizing the cathodic reaction and making it more anodic. Consequently, an area where this oxygen concentration is decreased relative to its surroundings becomes a relatively anodic region, and a cell has been established again. Such a cell, called an *oxygen concentration cell*, is a very important source of corrosion cells in single metal corrosion.

Oxygen concentration cells are very easily established. Deposition of dirt or scale on the surface of an otherwise clean metal shields the area under the deposit from access to dissolved oxygen. This creates an anode that, in turn, results in the anodic deterioration of this surface *under the deposit*. In this manner, the areas underneath rivets, bolt heads, washers, etc., become anodes for oxygen concentration cells; even breaks in paint or other protective coatings create these cells. The process of developing oxygen concentration cells by restricting the supply of dissolved oxygen to some portion of a metal surface is called *differential aeration*. A very common example of this is the formation of a spot of rust beneath a drop of water when it is placed on the surface of a piece of iron.

The presence of dissolved oxygen in an otherwise nonoxidizing acid solution such as HCl creates corrosion cells. For example, under normal conditions copper is passive in an HCl solution because its electrode potential is positive and therefore cathodic relative to hydrogen evolution. However, if the acid solution is strongly aerated with oxygen, the cathodic hydroxyl iron reaction given by Eq. (11.7) can readily occur and the copper will then be strongly attacked by the acid.

Stress is another environmental factor effective in producing galvanic couples on the surface of a single metal. Whenever a metal is cold-worked, the region that undergoes the greatest amount of plastic deformation also stores the greatest amount of strain energy, and is therefore in a higher energy state than its immediate neighbors—the requirement for an anode. Thus, locally work-hardened areas of a metal surface will be slightly anodic relative to annealed areas or the areas which have been work-hardened to a lesser degree.

A good example of this is seen in the rusting of an ordinary iron wire nail. Nails are made by wire drawing processes that severely work-harden the wire involved. However, the shearing process that cuts the nail from the main wire and forms its point work-hardens the point more than the shank of the nail. Similarly, the operation that flattens the wire to make the nail head also work-hardens this part of the nail more than the shank. Therefore, the point and the head of a nail are anodic relative to its shank. This fact is easily demonstrated by placing a nail in a gel composed of the chemical indicators phenolphthalein and potassium ferricyanide, with NaCl added to serve as the electrolyte (see Fig. 11-8). The phenolphthalein turns pink in the presence of OH^- ions, revealing the location of cathodes where the oxygen-water reaction in Eq. (11.7) occurs. The ferricyanide ions react with any Fe^{++} ions present to form bluish ferrous ferricyanide, revealing the location of anodes where the iron dissolution reaction occurs.

METALLURGICAL FACTORS   We know that the metal in steel is not pure iron, but a complex mixture of ferite and cementite. These two alloys of carbon with iron have different electrode potentials which create a myriad of tiny microcells in the material itself that are intimately connected electrically. The presence of an electrolyte is all that is necessary to establish a galvanic cell, one reason why ordinary steel rusts so rapidly in a moist atmosphere. Obviously, steel is not unique in this aspect; any alloy of two or more elements will possess similar, internal galvanic systems.

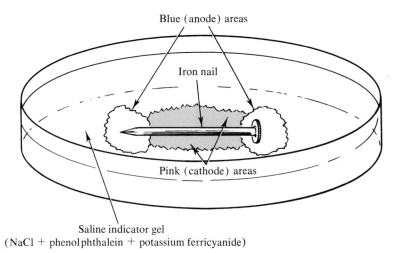

Blue (anode) areas

Iron nail

Pink (cathode) areas

Saline indicator gel
(NaCl + phenolphthalein + potassium ferricyanide)

FIG. 11-8 Schematic illustration of an experiment to demonstrate the existence of stress-caused, anode-cathode couples in an ordinary iron wire nail.

Even in a pure unalloyed metal such as zinc, myriads of tiny galvanic couples exist in the surface by virtue of its structure. Grain boundaries and other crystalline imperfections represent strained areas possessing higher energy than their neighboring atoms due to the strain energy of their imperfections. These areas, therefore, are slightly anodic relative to the bulk grains, and again galvanic microcells are established by the score.

Locally strained regions surrounding impurity inclusions, surface cracks, and any other source of crystalline irregularity also introduce tiny anodes into the surface of the metal. It is evident, then, that all metal surfaces—with the possible exception of carefully grown, perfect single crystals—are covered with a network of tiny anodes and cathodes which can be activated upon contact with a suitable electrolyte.

Ringworm corrosion is an interesting form of corrosion resulting from galvanic couples established by metallurgical operations. In many operations involving steel, the metal is subjected to large, local temperature gradients. In the flanging of a steel pipe by forging operations, for example, the end of the pipe is heated locally to very high temperatures, resulting in a variance of phase structure along the pipe. Often the section of the pipe that only receives moderate heating (that is, the portion slightly removed from the hot end) becomes spheroidized. If the entire section is not subsequently heat-treated to produce a uniform grain structure, the spheroidized section will corrode preferentially, producing a ring of corroded metal around the pipe. Because of this ring-shaped region, the name ringworm corrosion was developed.

Other Forms of Corrosion

Galvanic corrosion is only one of many characteristic types of attack known to occur [8]. Although most of these different forms of corrosion are electrochemical in nature, their separate classifications help engineers to keep their different characteristics clearly in mind.

Uniform Attack

The examples of corrosion previously described have been localized attacks at anodic areas. However, it is a common experience to observe an unprotected and exposed iron structure that is rusting away more or less uniformly. Such *uniform attack* is characterized by electrochemical or chemical reactions that proceed uniformly over the entire surface of the material exposed to the corrosive medium.

Although uniform attack on a tonnage basis is the greatest cause of metal destruction, particularly of steel, it is relatively easy to control and not an especially difficult problem technically. By proper selection of materials and use of protective coatings, as well as other techniques to be discussed later, this form of corrosion can essentially be eliminated.

Crevice Corrosion

A particularly destructive form of localized, electrochemical corrosion is *crevice corrosion*. This is generally associated with small volumes of stagnant liquid such as might form in shallow holes, under gasket surfaces, and in crevices under bolt and rivet heads. Two characteristically different types of crevice corrosion may be observed: *gasket corrosion* and *filiform corrosion*.

GASKET CORROSION   To occur, gasket or ordinary crevice corrosion requires the presence of a crevice that is wide enough to admit liquid but narrow enough to allow this liquid to remain stagnant. Therefore, this type of attack is generally reserved for openings a few mils or less in width. Fibrous gaskets, which act as wicks to adsorb an electrolytic solution and to maintain its contact with the metal surface, provide nearly perfect locations for crevice corrosion.

While both metal ion and oxygen concentration cell formations undoubtedly contribute to gasket corrosion, neither is the primary mechanism responsible for it [8]. Rather, it has been postulated

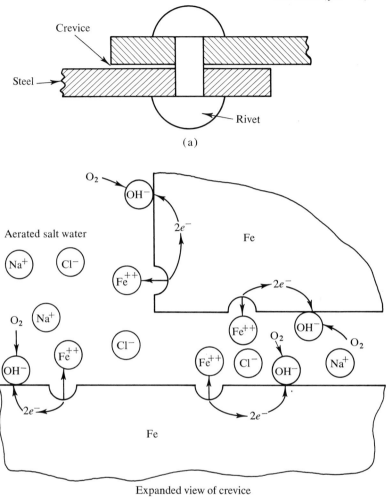

FIG. 11-9   Schematic illustration of the initial stages in crevice corrosion.

[9, 10, 11] that a self-catalytic mechanism involving the metal ion hydrolysis of water and acid formation is responsible (see Figs. 11-9 and 11-10).

In the crevice of the riveted section of steel plate illustrated in Fig. 11-9, the reactions taking place are the one in Eq. (11.7) and

$$Fe = Fe^{++} + 2e^- \tag{11.14}$$

It is assumed that this crevice is operating in aerated seawater (pH = 7), so that no particular problems should be anticipated from

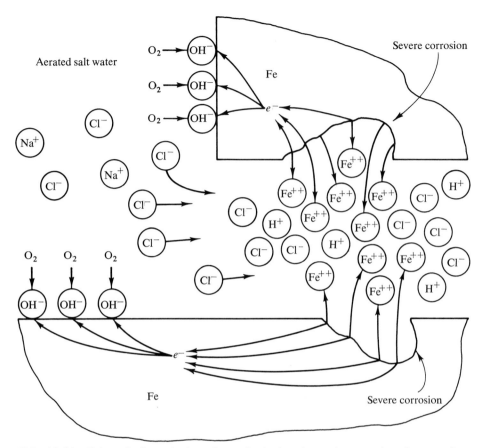

FIG. 11-10   Enlarged schematic view of a crevice in the advanced stages of crevice corrosion attack.

acids or bases. Initially, the anode and cathode reactions occur uniformly; however, the supply of oxygen in the crevice rapidly becomes depleted, halting the oxygen reduction there. Since the area of the crevice is negligibly small when compared with the total area of the metal exposed to the seawater, the decrease in oxygen reduction that occurs here is negligible in comparison with the total reaction. Therefore, the rate of metal oxidation in the crevice does not change appreciably, resulting in an increase in the metal ion concentration of the solution in the crevice and causing a similar local increase in positive charge. The condition of conservation of charge, discussed earlier in connection with the mixed-potential electrode theory, requires that negative ions diffuse into the crevice to neutralize this excess positive charge. Cl^- ions are more mobile than OH^- ions and, therefore, diffuse into the crevice more rapidly. However, $Fe^{++}(Cl^-)_2$ (as well as many other heavy metal salts) hydrolizes in water according to the reaction

$$Fe^{++}(Cl^-)_2 + 2\,H_2O = Fe^{++}(OH^-)_2 + 2\,H^+Cl^- \qquad \text{(11.15)}$$

and the $Fe^{++}(OH^-)_2$ formed is precipitated from the solution due to its insolubility. This removes Fe^{++} ions from the solution, depleting their concentration and accelerating the ionization reaction in Eq. (11.14). Furthermore, the presence of H^+Cl^- greatly increases the acidity of the solution in the crevice and accelerates the reaction even more. Thus, as indicated in Fig. 11-10, the rate of corrosion in the crevice accelerates relative to the corrosion outside.

Because of the necessity of conservation of charge, the electrons that accumulate as a result of Fe dissolution flow through the metal to its large exposed surface, where OH^- ions can easily be formed with the oxygen from the aerated salt water. Thus, the surface regions become cathodes and do not experience appreciable deterioration. In other words, the large exposed areas of a metal are cathodically protected at the expense of the small area under the crevice. The severity of this type of corrosion is great, as illustrated by the stainless steel sample in Fig. 11-11.

Unfortunately, stainless steel, which depends upon surface passivity for its effectiveness, is especially susceptible to crevice corrosion because of the high Cl^- ion concentrations that develop, destroying the passive film, in turn, rapidly accelerating such corrosion. The concentration of Cl^- ions has been observed [8] to increase by as much as a factor of 10, and the pH of the solution in the crevice has been known to drop from the neutral value of 7 to as low as 2; these factors indicate the severity of the conditions that may develop in a crevice.

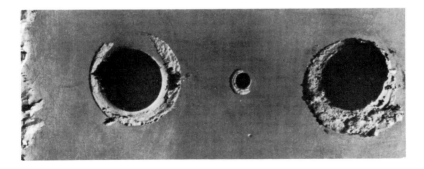

FIG. 11-11 Photograph of severe crevice corrosion under the washers on a riveted stainless steel plate immersed in sea water. (Courtesy of the International Nickel Company. Used by permission.)

FILIFORM CORROSION   A form of crevice corrosion that has a different appearance from that of gasket corrosion may be frequently observed on the surfaces of enameled or lacquered food and beverage containers. When these are exposed to the atmosphere after their coatings develop pinhole breaks near the edges, this corrosion appears as a network of tiny trails or filaments that somewhat resemble the trails of tiny bugs or worms. The filaments, which form intricate patterns as they intersect and interact with one another, are the reddish-brown color characteristic of iron rust and have the blue-green head that is characteristic of Fe^{++} ions.

This form of crevice corrosion occurs primarily between 65 and 90 per cent relative humidity and is rarely observed at humidities below this range [8]. At higher than 90 per cent humidity, the attack becomes so pronounced that it assumes the form of large blisters; although this type of corrosion is generally not harmful to the structure of the metal, it does greatly detract from its appearance and is therefore undesirable from the standpoint of the marketability of food and beverage containers. No completely satisfactory method of controlling this corrosion has been developed, although storing cans in low humidity environments and using films that exhibit very low water permeability do help [8].

The mechanism by which this process is believed to occur [8, 12] is illustrated schematically in Fig. 11-12. Nearly all films are at least partially permeable to the oxygen and the water vapor that diffuse through them from the atmosphere to the metal beneath. Thus, if a pinhole or crack occurs in the film where corrosion can start (near a sharp edge on a can, for example), a continuing supply of O_2 and H_2O can be fed to the corrosion site. When Fe dissolves in a small drop of water condensed from vapor at the head of a filament, a local region of low oxygen concentration is created. The decreased water concentra-

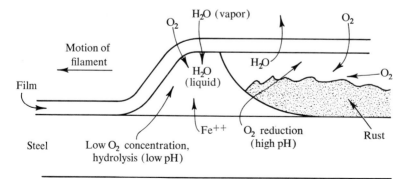

FIG. 11-12   Schematic representation of the postulated mechanism of filiform corrosion.

tion which results drives water from the air and through the film by osmosis, thus keeping the active head supplied with water. The dissolved O_2 reacts with the water and with electrons from the metal, producing OH^- ions that react with the Fe^{++} ions to precipitate insoluble $Fe(OH)_2$. This, in turn, depletes the Fe^{++} concentration and dissolves more Fe, perpetuating the reaction. As the insoluble hydroxide precipitates, the decreased concentration of dissolved salts increases the water concentration in the inactive tail, driving some of the water out of the crack through the film by osmosis. The O_2 that has diffused through the film into this region and the O_2 which has diffused longitudinally up the tail of the filament react with the $Fe(OH)_2$ to form $Fe(OH)_3$, according to the equation

$$4\,Fe(OH)_2 + 2\,H_2O + O_2 = 4\,Fe(OH)_3 \qquad \textbf{(11.16)}$$

(The resultant ferric hydroxide is the familiar reddish-brown rust.)

Active corrosion is restricted to the filament head, where hydrolysis of the corrosion products creates an acidic environment. Thus, filiform corrosion can be considered a self-propagating crevice and is, therefore, a proper form of crevice corrosion. (The peculiar growth pattern in which the corroding head does not spread out laterally, and the strange interactions [8, 12] between filaments have not yet been fully explained.)

Pitting Corrosion

Like crevice corrosion, pitting is an extremely severe form of localized corrosion. In this type of attack, holes that may be very small (pinhole size) or very large form in the metal surface, and generally they penetrate completely through the metal. If these pits are minute they may be covered over by corrosion products, often making this form of corrosion difficult to discover before part failure occurs. The local severity of the corrosion frequently results in sudden and unexpected service failure.

Pitting may require a relatively long initiation period, but, once started, it becomes autocatalytic (like the crevice corrosion mechanism discussed above) and its rate accelerates rapidly. Pitting generally occurs in the direction of gravity on the lower surface of a piece of equipment. It is rarely observed [8] progressing in an upward direction on an upper surface. (The mechanism of the pitting process for an iron surface is illustrated schematically in Fig. 11-13.)

Once the pit is initiated (usually a random process caused by momentary local increases in corrosion rate [13]) the mechanism is essentially the same as the mechanism for crevice corrosion. The concentration of Fe^{++} ions in the pit and the depletion of dissolved

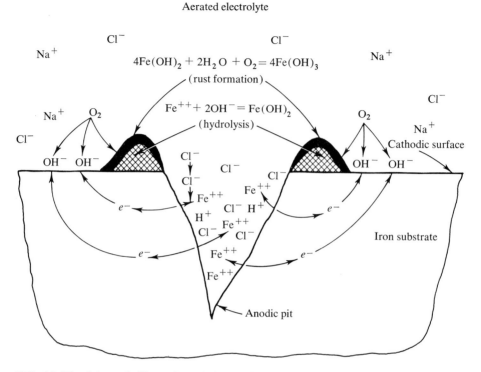

Aerated electrolyte

FIG. 11-13   Schematic illustration of the mechanism of pit formation on an Fe surface.

oxygen result in an influx of Cl^- ions from the solution. The process of hydrolysis converts the $Fe^{++}(Cl^-)_2$ to $Fe(OH)_2$, which is insoluble, and simultaneously creates an acid environment in the pit. In turn, the rate of corrosion is accelerated, rendering the process autocatalytic. Since the electrons migrate through the metal to the exposed surfaces in order to reduce O_2, these surfaces become cathodic and are spared from deterioration. Thus, the pit deepens, cutting its way rapidly through the metal. The increased Cl^- concentration in the pit increases the solution density and the force of gravity keeps the Cl^- ions in place, explaining why pits are normally found only on the lower portions of process equipment.

Another important variable in corrosion problems is exhibited in the pitting process, where polarization curves determine the corrosion current density i_{corr}. Since this is a current *density*, the smaller the area of the anode, the greater both the actual current and the rate of dissolution of the metal will be for a given value of i_{corr}. Thus, in pitting, the detrimental effect of a small anodic area coupling with a large cathodic area further accelerates the penetration rate of the pit. (The influence of relative electrode size will be considered later in connection with a discussion of corrosion prevention methods.)

Intergranular Corrosion

Because the grain boundary regions in metals can be activated to a much greater than normal degree under certain conditions, they can corrode severely while cathodically protecting the bulk grains. This phenomenon is called *intergranular corrosion*. For example, it is known that the grain boundaries of brass are rich in zinc, leading to the possibility of this type of attack. Stainless steels with high carbon contents are likewise susceptible to intergranular attack if heated to a high enough temperature to allow chromium carbide ($Cr_{23}C_6$) precipitation at the grain boundaries—a frequent occurrence during welding operations.

Type 304 stainless steel (eighteen per cent Cr, eight per cent Ni; called 18-8 stainless) generally contains between 0.06 and 0.08 per cent C, while the minimum threshold concentration of C for precipitation of $Cr_{23}C_6$ is about 0.02 per cent. Therefore, if the temperature of an 18-8 stainless steel is raised to the 950–1,450°F range, carbon is mobilized for diffusion and readily diffuses to the grain boundaries. On the other hand, Cr is much less mobile than C and tends to retain its present position. Since the grain boundaries provide suitable nucleation sites, C diffuses there to combine with Cr and form the carbide $Cr_{23}C_6$. This is highly insoluble in the solid solution matrix and, therefore, precipitates. Consequently, the grain regions very near the grain boundaries become severely or completely depleted in Cr, and a small band near the boundaries is depassivated or *sensitized* [11]. Because of the depletion of Cr, the metal of this region has a more anodic potential than the bulk grain, and a galvanic couple is established. The net result is a rapid, severe attack in this region, but little or no attack on the grains that are cathodically protected.

The chromium carbides are realtively immune to corrosion attack. Thus, when intergranular corrosion occurs, the attack results in deep narrow trenches on either side of the carbides, as illustrated schematically in Fig. 11-14. This results in a separation of grains, causing the alloy to lose its strength; such failure of sensitized stainless steel by intergranular corrosion attack can be sudden and catastrophic.

The failure of welds by intergranular *weld decay* is a result of the same mechanism. When metals are welded, the metal immediately adjacent to the weld zone is heated to very high temperatures, and a zone of metal somewhat removed from the actual centerline of the weld is heated to the sensitizing temperature range, thus permitting carbide precipitation at the grain boundaries of this metal zone. If the entire part is not subsequently solution treated to dissolve the carbides, this sensitized band of metal will undergo intergranular attack at a rapid rate when the metal is exposed to a corrosive environment, and the part will fail at its welds.

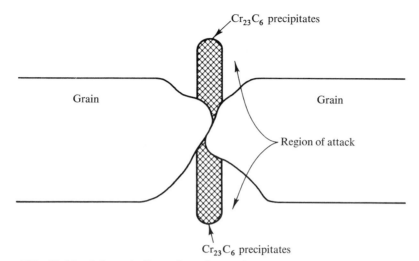

FIG. 11-14 Schematic illustration of corrosion trenches in the intergranular corrosion of a sample of 18-8 stainless steel.

Intergranular corrosion of austenitic stainless steels can be minimized by the following methods: (1) high temperature solution heat treatment following welding or other sensitizing operations, (2) addition of elements that are stronger carbide formers than Cr, and (3) lowering the carbon content of the steel below the $Cr_{23}C_6$ threshold of approximately 0.02 per cent.

The high temperature solution treatment, called *quench-annealing*, consists of heating the metal to about 2000°F and following this by rapid water quenching. The rapid quenching is important to the success of the operation because the entire structure will be sensitized if slow cooling is employed. This process is not practical for a large or massive part such as a storage tank, however, and other methods must be used.

Columbium, tantalum, and titanium are the carbide forming elements usually added to 18-8 stainless steels to minimize intergranular attack. Since titanium is generally lost during welding operations [14], it is not a good additive for welding rods. In general, the addition of an amount of columbium equal to approximately ten times [14] its carbon content renders a stainless steel immune to intergranular attack. Special grades of extra low-carbon (ELC) stainless steel are also used to limit this type of corrosion attack.

Stress Corrosion

A significant number of corrosion failures arise from the phenomenon called *stress-corrosion cracking*, defined [15] as the combined action of stress and corrosion that leads to the cracking or embrittle-ment of a metal. This general definition includes three major classes of failure: (1) cracking where crack progression is at least partially due to electrochemical dissolution of the metal at the crack tip (the case generally implied in practice), (2) hydrogen embrittlement, and (3) liquid metal embrittlement. Characteristically, only tensile stresses are damaging [16] for all three of these classes of failure, and the corrosion environment-alloy combination—no environment will crack all alloys, and no alloy will crack in all environments—is specific where each is observed. (This phenomenon is of especial interest to corrosion scien-tists because the complete details of its mechanism are not yet fully known [8, 16].)

STRESS-CORROSION CRACKING   The problem of stress-corrosion cracking can be illustrated by two classical examples: *season cracking of brass* and the *caustic embrittlement* of steel. The term season cracking of brass originated when it was observed that brass cartridge cases developed cracks at the point where the case is crimped to the slug during periods of heavy rainfall in the tropics. Although this phenome-non eventually became associated with the rainy season, seasonality actually has nothing to do with it; in fact, the phenomenon is caused [8] by the corrosive action of ammonia on the stressed cartridge casing. The season only contributes large quantities of moisture which, in turn, cause the decomposition of organic matter necessary to release the true corrosive agent—ammonia.

Caustic embrittlement refers to the failure of steel that is often observed in explosions of riveted boilers in steam-driven locomo-tives [8]. The ruptured steel exhibits white deposits containing caustic, or NaOH, in the cracks at the rivet holes. Brittle failure in the presence of caustic naturally resulted in the term caustic embrittlement.

The mechanisms of stress-corrosion cracking are complex and appear to involve an interplay of metal, interface, and environmental properties. Many different theories have been proposed [16], most of which fall into two main categories: those viewing stress-corrosion cracking as an electrochemical process proceeding normally to the prevailing tensile stress of the material [17, 18], and those viewing the process as a mixed, corrosion-mechanical failure [19, 20, 21]. All stress-corrosion cracking theories admit the necessity of the electrochemical corrosion process in either generating or propagating the crack. Once a corrosion pit or another type of surface deterioration has occurred, it serves as a stress raiser (see the discussion of mechanical fatigue in Chapter 8). Stress-corrosion cracks often begin [8] at the base of a corrosion pit.

Once the crack is initiated, the small radius at its terminus is an extremely efficient stress concentrator. (In one study, the mechanical

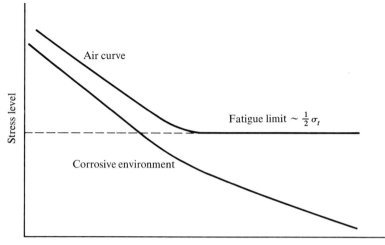

FIG. 11-15   Schematic fatigue curves for steel in a corrosive environment and in air.

jumping of the crack could actually be audibly discerned by the pinging noises it made [22].) Corrosion is also important in advancing the crack, as has been demonstrated [21] by alternately applying and removing an external potential which converts the metal into a cathode. Each time the cathodic protection is applied, the crack ceases to advance; each time it is removed, the crack begins to move again. This has proved to be the case for several cycles of current application.

The latter observation leads to a method of distinguishing between stress-corrosion cracking and failure by ordinary mechanical fatigue. Figure 11-15 is a schematic set of fatigue curves, comparing the behavior of steel in air and steel in a corrosive environment. The upper curve shows the typical limiting *fatigue stress*, which is roughly half the tensile strength of the metal in steels. The lower curve shows the typical behavior of a steel in a stress-corrosion environment. One of the striking features of the lower curve is the absence of any apparent fatigue stress limit; this curve, which always lies below the air curve, continues to decrease with an increasing number of cycles. The relative position of this curve is affected by the corrosion rate [16], but not by the tensile strength of the material. Therefore, the effective life of a steel part subjected to stress-corrosion fatigue cannot be increased by increasing the tensile strength of the steel used. In general, the only measures that increase the effective life of a part are those that reduce the corrosion rate; in fact, increasing the strength of a steel can sometimes be detrimental [8] in corrosive environments. High-strength steels are usually more brittle than low-strength, ductile steels, and a stress-corrosion crack, once formed, may actually propagate more rapidly through a brittle material than through a more ductile material of lower strength.

HYDROGEN DAMAGE   Three different types of hydrogen damage [8] will be included in this category: (1) hydrogen blistering, (2) hydrogen embrittlement, and (3) decarburization. (The latter is actually a form of dry corrosion.)

Hydrogen blistering results from the diffusion of atomic hydrogen into steel, because molecular H_2 (large, by atomic standards) cannot penetrate the crystal structure of the metal. Most of the atomic hydrogen diffuses through the steel, escaping on the other side of the metal as molecular H_2 gas. (Consider the schematic representation of this process in Fig. 11-16.) However, some of the atomic hydrogen will find its way into a void or microcrack in the steel wall (defects most common in rimmed steels [8]), where it can combine to form molecular hydrogen which is then trapped in the void or crack because it cannot diffuse through the steel. Rupture of the steel is guarantied when this occurs, since the equilibrium pressure for molecular hydrogen in small bubbles or holes may be several hundred thousand atmospheres. (Figure 11-17 is a photograph of a carbon steel section which has failed in this manner.)

Hydrogen embrittlement is a phenomenon where atomic hydrogen reacts with steel to form a more brittle material, which then fails by brittle cracking (under the internal pressure of hydrogen gas, entrapped

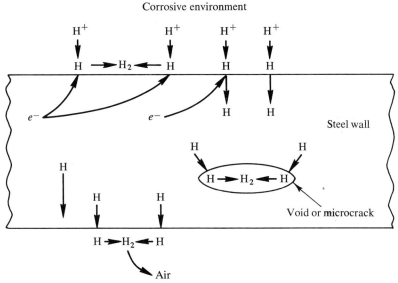

FIG. 11-16   The proposed mechanism for hydrogen blistering.

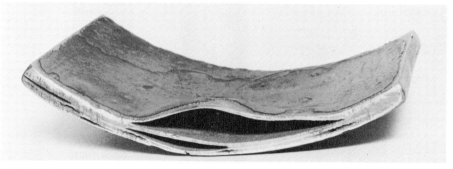

FIG. 11-17 Cross-section of a carbon steel plate, showing a large hydrogen blister after two years of exposure to a petroleum process stream. (Photograph by Imperial Oil Limited, Canada. From *Corrosion Engineering*, by M. G. Fontana and N. D. Greene, copyright 1967, McGraw-Hill Book Co., Inc. Used by permission.)

by the process just described). The details involved in the mechanism of this particular hydrogen damage are not well understood [8, 16].

Hydrogen embrittlement can be distinguished from ordinary stress-corrosion cracking [23] by the behavior of a hydrogen-damaged metal under cathodic polarization. The rate of the formation of atomic hydrogen is influenced by cathodic polarization; clearly, ordinary stress-corrosion cracking is occurring in a metal if an increase in the cathodic reaction process can stop the cracking. However, if increasing the rate of cathodic reaction shortens the life of a material, then its failure is caused by hydrogen embrittlement. Figure 11-18 is a plot of the average failure times of a particular high-strength alloy steel as a function of externally impressed current density [23]. The normal, unassisted cracking time here is about 100 minutes. Anodic polarization decreases this time markedly, indicating that the stress-corrosion cracking process is involved. The application of slightly cathodic currents (those less than 1 ma/in.^{2}) results in a much longer life for the sample. However, application of larger cathodic currents causes a rapid decrease in the life of the sample, and currents greater than 3 ma/in.^{2} reduce its effective life below the unpolarized value. Such behavior is clearly indicative [23] of hydrogen embrittlement failure.

A third process that involves the destruction of steel through hydrogen activity is *decarburization*. In this process, the carbon in steel reacts with hydrogen to form methane gas (CH_4), thus removing the carbon and causing the steel to lose its tensile strength. The inverse of this process, *carburization*, is the breakdown of hydrogen-hydrocarbon gas streams, with the subsequent absorption of carbon by the steel. Although not as destructive to the properties of the steel as decarburization, carburization can produce some undesired degree of embrittlement through the formation of excess carbides in the iron structure.

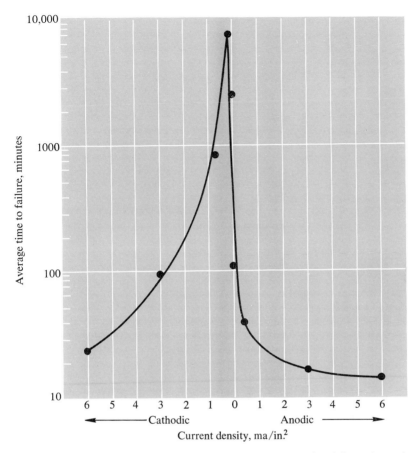

FIG. 11-18   Effect of a cathodic protection current on the failure time of MoV steel in an oxygenated three per cent NaCl solution at pH 6.5. (From H. J. Bhatt and E. H. Phelps, *Corrosion*, **17**, 430*t* (1961). Used by permission.)

LIQUID METAL EMBRITTLEMENT   In the nuclear industry, liquid metals are extensively used as high temperature, heat transfer media. This type of metal can seriously attack its container if an incorrect choice of container material is made. Such a corrosive attack is more physical than chemical in nature [24]; four attack types are identifiable: (1) solution of the structural metal, (2) diffusion of liquid metal into solid metal, (3) intermetallic compound formation, and (4) mass transfer resulting from either composition or thermal gradients.

Since liquid metals are most often used in heat exchangers, the last two types of attack are the most troublesome. In liquid metal heat exchangers, the greatest resistance to heat flow lies in the walls of the tubes through which the liquid metal flows. Consequently, large thermal gradients and stresses exist in this region, and the formation of brittle intermetallic compounds, either on the surface or in the substrate, can result in the brittle cracking of the tube walls.

A few of the problems involved in containing liquid metals should be noted briefly. 18-8 stainless steels are successful in containing relatively high temperature Na, K, and NaK alloys; however, the presence of oxygen greatly increases the oxidation corrosion power of these metals, so that operation under an inert gas atmosphere is desirable [24]. Li is more aggressive in its attack than the other alkali liquid metals. Carbon steel and five per cent Cr steels contain liquid Hg well. Plain carbon steel and cast iron contain liquid Mg, but austenitic stainless steels are attacked rather severely by this liquid metal. No known metal or alloy [24] is immune to attack by liquid Al. Although Ni, Monel metal, and cupronickel alloys resist stress cracking in liquid Pb, Bi, Sn, and their alloys, stress-corrosion cracking is a common occurrence in most metals that are exposed to these molten metals.

Corrosion by fused salts such as NaOH appears to be a combination of electrochemical attack and the liquid metal attack phenomena just introduced [24]. Ni and Ni-based alloys seem to have the best corrosion resistance in such applications.

Erosion Corrosion

A particularly significant type of corrosion involves the coupling of the flow or movement of the corrosion medium relative to the metal surface that has been electrochemically attacked. *Erosion corrosion* is defined as [8] an acceleration in the rate of deterioration due to the relative movement of the corrosive fluid past the metal surface. In most cases where this type of attack is significant, the relative motion is quite rapid and the effects of mechanical wear and abrasion are appreciable. Erosion corrosion is characterized by the appearance on the metal surface of grooves, valleys, waves, and rounded holes, which usually occur in a directional pattern that reflects the flow direction of the corrosive fluid. Figures 11-19 and 11-20 illustrate typical examples of erosion corrosion.

In many instances, failure due to erosion corrosion occurs relatively soon after a part is placed in service. This type of failure is generally unexpected, because corrosion testing is performed under static conditions in which the influence of the velocity of the medium is ignored. Such metals as stainless steel and aluminum, which depend upon the formation of a passive film for their anticorrosion effectiveness, are especially susceptible to this form of attack. The motion of the fluid wears or erodes the passivating film and sweeps it away; thus, the fresh metal surface is exposed to the fluid, accelerating the corrosion and

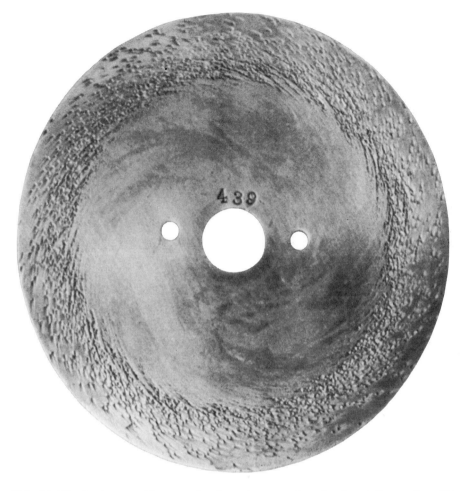

FIG. 11-19 A photographic example of erosion corrosion due to relative motion. This admiralty brass disc was spun in corrosive liquid at a high speed. The outer parts with higher velocities were severely attacked. (Courtesy of the International Nickel Company. Used by permission.)

greatly increasing its severity. Soft metals are more vulnerable than hard metals because their surfaces are more easily damaged physically by erosive action.

Three general types of erosion corrosion attack are important: (1) the erosion of surfaces by impingement, or the rapid flow of corrosive liquid streams, (2) cavitation damage, and (3) fretting corrosion.

IMPINGEMENT AND VELOCITY EFFECTS   Usually the mechanism of erosion corrosion is an acceleration of the normally occurring electrochemical action that results from the physical removal of protective

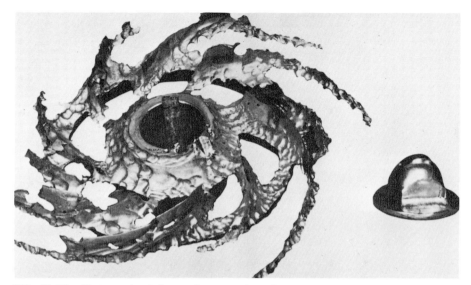

FIG. 11-20   Photograph of the erosion corrosion of a stainless alloy pump impeller after three weeks of service. (From *Corrosion Engineering* by M. G. Fontana and N. D. Greene, copyright 1967, McGraw-Hill Book Co., Inc. Used by permission.)

films by the rapidly moving stream of corrosive liquid. This exposes fresh metal to the galvanic action of the solution, thus perpetuating the corrosion. The presence of entrained air bubbles or suspended solids in the corrosive liquid stream (as in slurries) greatly increases the erosion process. Figure 11-21 shows the influence of velocity on the corrosion of stainless steel by a sulfuric acid-ferrous sulfate slurry [8]; under stagnant conditions, such a steel is completely inert or passive to this environment, but it is severely attacked under flow conditions.

CAVITATION DAMAGE   An especially severe form of erosion corrosion called *cavitation* can be observed in pump impellers, hydraulic turbines, ship propellers, and other devices where very high velocities and low pressures are simultaneously encountered. The mechanism by which cavitation damage occurs is related to vapor pressure considerations.

If the pressure of a liquid is reduced below its equilibrium vapor pressure, the liquid will vaporize or boil. Fluid mechanical considerations of conservation of energy in a moving stream indicate that pressure must decrease when velocity is locally increased, all other factors being equal. Thus, if the velocity behind a moving impeller or turbine blade becomes sufficiently large, the pressure behind this blade will decrease; if this pressure becomes lower than the vapor pressure of the liquid, vaporization will occur. As the rotating impeller blade encounters regions of lower fluid velocity (and consequently higher pressure), the bubble of vapor collapses; thus, bubbles formed during one part of the cycle collapse during another. It has been estimated that the shock wave pressures [8] caused by collapsing vapor bubbles can be as large as 60,000 psi. Such forces can produce plastic deformation in many metals, as indicated by the existence of slip lines found in pump parts subjected to cavitation. Cavitation damage can occur on high speed rotating machinery, where the service failure is often violent and catastrophic.

FRETTING CORROSION   When materials under load are subjected to the vibration and relative slipping of contacting parts, the area between these mating parts often develops corrosion pits or grooves surrounded by corrosion products. This corrosion, called *fretting*, is often accompanied by considerable abrasive action which increases its severity.

Fretting corrosion is particularly damaging because it often occurs where tight pressfit joints are desired—that is, in the areas of contact between bearings and shafts, sleeves, etc. Fretting can result in the loosening of parts due to the abrasion from oxide debris, in the seizing and galling of parts, and even in fatigue fracture caused by excessive strain and stress increases from corrosion-initiated pits. In order for

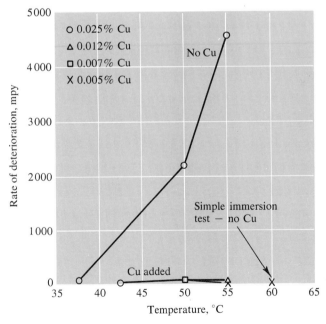

FIG. 11-21   The effects of temperature and velocity on the corrosion of stainless steel by an H_2SO_4–$FeSO_4$ slurry. Without flow, the stainless steel is inert up to 60° C. (From *Corrosion Engineering* by M. G. Fontana and N. D. Greene, copyright 1967, McGraw-Hill Book Co., Inc. Used by permission.)

fretting to be a significant form of corrosive attack, the interface between parts must be under load, the part under attack must be subjected to vibration or repeated relative motion between the mating surfaces, and the load and relative interface motion must be sufficient to cause slip or plastic deformation of the surface metal.

Other Types of Corrosion

Other forms of corrosion also occur, but only two additional types will be mentioned here: selective leaching and biological action. *Selective leaching* is the preferential removal of one or more of the alloying agents in a metal by the action of an electrolyte. In brasses containing large amounts of zinc such as common yellow brass (70 per cent Zn, 30 per cent Cu), for example, the zinc is sometimes preferentially removed from the brass, leaving a spongy, weak matrix of copper—a process known as *dezincification*. The mechanism of dezincification, then, involves three steps [8]: (1) dissolution of the brass, (2) solution of the zinc ions, and (3) redeposition of the weak copper matrix on the brass substrate. Since the addition of zinc to copper to form brass lowers the corrosion resistance of the copper, removal of the zinc by dezincification will increase the corrosion resistance of the copper. As noted above, however, the copper matrix that remains does not have a desirable strength. (This type of failure can be observed in brass boiler or heat exchanger tubes.)

Another form of selective leaching is the *graphitization* of gray cast iron. This material consists of a matrix of steel and iron interspersed with a network of carbon in the form of graphite. When gray cast iron is in the presence of corrosive media, its iron matrix is preferentially attacked because it is anodic to graphite. After this matrix dissolves, only a porous mass of rust and graphite remains. This material has the appearance of the graphite and is soft and crumbly. Because of the external appearance of this end-product, the process was mistakenly named graphitization; it usually occurs rather slowly, without dimensional changes, making it difficult to detect prior to failure.

Various types of *biological corrosion* can occur. Actually, in the sense that it has been considered thus far, biological corrosion is not really a corrosion at all; rather it is the deterioration of metal surfaces by processes which occur either as the direct or indirect result of the action of biological organisms.

Because microorganisms can exist under extremely varying conditions, biological activity may influence corrosion in a wide selection of environments—soils, natural waters, and petroleum products, for example. Biological organisms participate in the corrosion process either by ingesting a certain substance as part of their metabolic activity, or by depositing it in the form of a waste product scale on the surface. In general, the destruction caused by the ingestion of chemical substances is associated with *microorganisms*, while the deposition of scales or other waste products is associated with *macroorganisms* such as barnacles, mold, and fungi. Deterioration processes associated with microorganisms often involve the production of corrosive metabolic by-products in otherwise noncorrosive environments.

Microorganisms are generally classified according to their ability to grow in the presence or absence of oxygen. Bacteria requiring oxygen in their life processes are *aerobic*. *Anaerobic* bacteria, prevalent in wet or soggy soils and in marshes, need no oxygen to exist. The most common anaerobic bacteria reduce sulfate ions to sulfide ions by the schematic reaction

$$SO_4^{--} + 4\,H_2 = S^{--} + 4\,H_2O$$

The hydrogen necessary to this reaction is either obtained from organic matter present in the soil or cathodically produced during electro-chemical corrosion. In either case, the presence of the S^{--} ions formed by these organisms markedly affects the electrochemical corrosion reactions occurring on iron surfaces buried in the soil. The S^{--} ions react with the Fe^{++} ions to precipitate FeS, thus removing ferrous ions from galvanic reactions at the metal and stimulating an anodic dissolution reaction.

Aerobic bacteria utilize dissolved oxygen to form sulfuric acid through the oxidation of sulfur or sulfur-bearing compounds by the reaction

$$2\,S + 3\,O_2 + 2\,H_2O = 2\,H_2SO_4$$

thus increasing the acidity and corrosiveness of the environment. Aerobic bacteria thrive in low pH surroundings and require sulfur or sulfur-bearing materials to sustain life. They are consequently prevalent in oil and sulfur fields, and in soils around sewage disposal pipes that discharge sulfur-bearing effluents.

In addition to the fact that they foul a metal surface by their presence, fungi, molds, and scaly organisms such as barnacles serve as sources of oxygen concentration cells and create crevices for crevice corrosion. These actions destroy the heat transfer ability of heat exchanger surfaces, and reduce the effectiveness of streamlining (as in the case of ship hulls which become coated with heavy barnacle encrustations and require increased power to operate). Many fungi and molds assimilate organic matter and produce large amounts of organic acids as by-products, thus increasing the corrosiveness of their environment.

Most biologically caused or assisted corrosion can be effectively prevented by the use of paints or coatings containing substances that are toxic to biological organisms, coupled with the use of externally applied, cathodic potentials. In the case of aqueous organisms in closed systems, for example, the addition of toxic agents to the water can effectively control their growth.

Dry Corrosion

The preceding discussion has been concerned almost exclusively with situations in which the presence of a liquid electrolyte is essential to the mechanism of corrosion. Since virtually every metal and alloy reacts with air to form an oxide, at ambient or higher temperatures the oxidation of metal surfaces in the absence of liquid electrolytes is an important form of corrosion which must be considered, especially in high temperature material applications.

Mechanisms

Although the oxidation of a metal such as $Fe + \frac{1}{2}O_2 = FeO$ appears to be the simple reaction of metal with oxygen to form metal oxide, this is actually an electrochemical process similar to the examples of aqueous corrosion previously considered. Consequently, the oxidation reaction can be written as the sum of two half-cell reactions:

$$2\,Fe = 2\,Fe^{++} + 4e^- \qquad\qquad (11.17)$$

$$O_2 + 4e^- = 2O^{--} \qquad\qquad (11.18)$$

$$\overline{2\,Fe + O_2 = 2\,FeO} \qquad\qquad (11.19)$$

Metal oxidation generally occurs at the metal-oxide interface, while oxygen reduction usually takes place at the scale-air interface.

To some extent, all metal oxides conduct both ions and electrons. Thus, the electrochemical process above can proceed without an external conductor (the wire in a galvanic cell) or a liquid electrolyte because the oxide scale itself performs both functions. The metal-oxide interface becomes a local anode and the scale-air interface becomes a local cathode. The oxide layer in between the two serves as an ionic conductor or electrolyte, an electron conductor or electrical connection between the electrodes, and a diffusion barrier through which the two

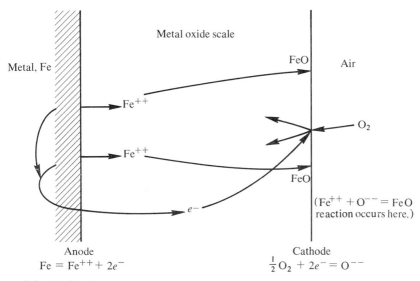

Metal oxide scale

Metal, Fe

FeO

Air

Fe^{++}

O_2

Fe^{++}

FeO

$(Fe^{++} + O^{--} = FeO$
reaction occurs here.)

e^-

Anode
$Fe = Fe^{++} + 2e^-$

Cathode
$\frac{1}{2}O_2 + 2e^- = O^{--}$

FIG. 11-22 Schematic representation of the electrochemical mechanism of dry oxidation corrosion acting on a metallic Fe surface in air.

charge carriers must migrate [24]. Figure 11-22 is a schematic representation of this mechanism.

The conductivity of oxides may occur by either *n*- or *p*-type semiconduction because all metal oxide structures are nonstoichiometric compounds [24]; that is, their actual oxide composition differs from that indicated by their molecular formulae. Some metal oxides possess a metal ion deficiency, while others are deficient in oxygen ions—a situation illustrated schematically in Figs. 11-23 and 11-24. Figure 11-23 represents a nonstoichiometric lattice of ZnO which is characteristic of oxygen deficient or metal-rich oxides; CdO, TiO_2, Ta_2O_5, Al_2O_3, SiO_2, and PbO_2 are other typical examples. The condition of overall electrical neutrality in ZnO requires the presence of excess electrons in the structure to offset the interstitial Zn^{++} ions. Because these excess electrons serve as electronic current carriers, ZnO behaves as an *n*-type semiconductor.

Figure 11-24 illustrates NiO, a *p*-type semiconducting oxide with a metal ion-deficient lattice; metal ion vacancies are indicated by the small boxes in the figure. In order for overall electrical neutrality to prevail, each of these electron vacancies or holes must combine with an adjacent Ni^{++} ion to form a Ni^{3+} ion. Since electronic conduction occurs by the movement of the positively charged holes associated with these ions, NiO is a *p*-type semiconductor. Ionic transport is provided by the diffusion of the vacancies. Typical *p*-type oxides are FeO, Cu_2O, Cr_2O_3, and CoO.

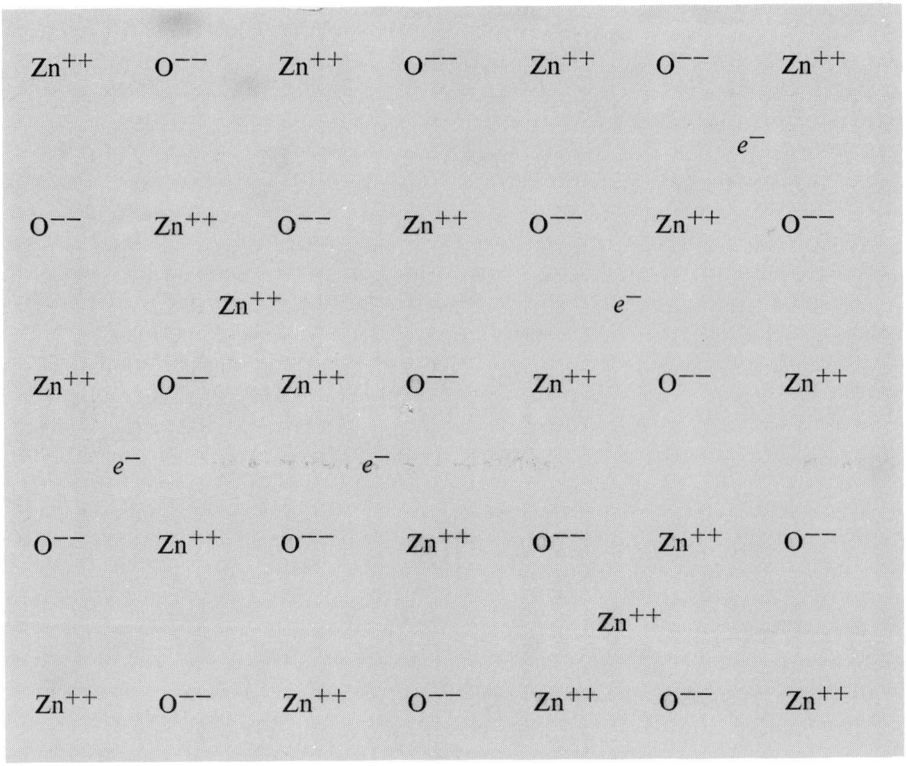

Zn^{++}	O^{--}	Zn^{++}	O^{--}	Zn^{++}	O^{--}	Zn^{++}

e^-

O^{--}	Zn^{++}	O^{--}	Zn^{++}	O^{--}	Zn^{++}	O^{--}

Zn^{++} e^-

Zn^{++}	O^{--}	Zn^{++}	O^{--}	Zn^{++}	O^{--}	Zn^{++}

e^- e^-

O^{--}	Zn^{++}	O^{--}	Zn^{++}	O^{--}	Zn^{++}	O^{--}

Zn^{++}

Zn^{++}	O^{--}	Zn^{++}	O^{--}	Zn^{++}	O^{--}	Zn^{++}

FIG. 11-23 A nonstoichiometric ZnO lattice, with interstitial Zn^{++} ions and electrons. This is an *n*-type semiconductor.

Many metals form a number of thermodynamically stable binary oxides; iron, for example, forms FeO, Fe_2O_3, and Fe_3O_4. In most cases, all of the stable oxides are present in the scale of a metal, and they appear in the order of increasing oxygen content when passing from the metal surface to the oxide-air interface. For iron above 560°C, the observed phase sequence is $Fe/FeO/Fe_3O_4/Fe_2O_3/O_2$. The rate of ionic diffusion through each layer will determine its thickness [24].

Scales formed on metals such as Fe, Ni, Cu, and Cr grow primarily from the air-oxide interface; their rate of growth is therefore limited by the rate of metal ion diffusion through the increasing layer. The condensation of vacancies (due to atomic volume differences between the metal and oxide atoms) near the metal-oxide interface often results in the formation of numerous large voids there, causing the scale to be porous, noncoherent, and structurally weak. This permits oxygen to diffuse through the gas phase in the porous interstices, reducing the resistance to growth that the increasing thickness of the scale would otherwise offer. Such scales are nonprotective of the substrate metal.

Ni^{++}	O^{--}	Ni^{+++}	O^{--}	Ni^{+++}	O^{--}	Ni^{++}
O^{--}	Ni^{++}	O^{--}	$\square$	O^{--}	Ni^{++}	O^{--}
$\square$	O^{--}	Ni^{+++}	O^{--}	Ni^{++}	O^{--}	Ni^{+++}
O^{--}	Ni^{++}	O^{--}	$\square$	O^{--}	Ni^{++}	O^{--}
Ni^{+++}	O^{--}	Ni^{++}	O^{--}	Ni^{+++}	O^{--}	Ni^{++}

FIG. 11-24   A nonstoichiometric NiO lattice, with metal ion deficiency vacancies indicated by $\square$. This is a p-type semiconductor.

Oxidation Rate Equations

It has been seen that the anodic process in electrochemical wet corrosion results in the deterioration of the anode surface, which is why rate laws for aqueous corrosion processes are expressed in terms of the rate of metal weight loss. On the other hand, dry oxidation corrosion is generally characterized by the growth of an adherant scale on a material surface, and oxidation rate equations are usually expressed in units of weight *gain* per unit area per unit time. Five empirical variations of weight gain with time are used to describe the kinetics of oxidation corrosion reactions [24]: (1) the linear law, (2) the parabolic law, (3) the cubic law, (4) the logarithmic law, and (5) the inverse logarithmic law.

In some materials such as Na, K, Ta, and Cb, the transport of reactant ions occurs at a much faster rate than the actual chemical reaction, which then becomes the limiting step. Such materials follow the linear rate expression

$$W = k_L t \qquad (11.20)$$

where W is the weight gain per unit area per unit time, t is time, and k_L is the linear rate constant—that is, the molecular rate constant for the limiting chemical reaction.

If ion diffusion is the sole controlling step in the oxidation process, pure metal oxidation should follow the parabolic relation

$$W^2 = k_p t + C \qquad (11.21)$$

TABLE 11-5

Parabolic Rate Constants for Zn and its Alloys*

$(T = 390°C; p(0_2) = 1 \text{ atm})$

MATERIAL	$k_p(\text{g}^2/\text{cm}^4 \text{ hr})$
Zn	$8(10^{-10})$
Zn + 1.0 a/o Al	$1(10^{-11})$
Zn + 0.4 a/o Li	$2(10^{-7})$

*From Gensch, C., and K. Hauffe, *Zeit. Phys. Chem.*, **196**, p. 427 (1950).

where W and t have the significance in Eq. (11.20), C is a constant, and k_p is the parabolic rate constant. Metals such as Fe, Co, and Cu follow this rate equation relatively well. Since it is characteristic of thick, coherent, oxide-scale forming metals, Eq. (11.21) is fairly commonly observed.

In some restricted circumstances, such as the short time exposure of zirconium, a cubic relation is observed

$$W^3 = k_c t + C \tag{11.22}$$

where W, t, and C have the above connotations, and k_c is the cubic rate constant. This irrational behavior is probably produced by a combination of limiting ionic diffusion through the scale and complications arising from porosity and structural defects in the scale [24].

Some metals such as Al, Fe, and Cu oxidize at *ambient* conditions to form thin oxide films, according to the logarithmic relation

$$W = k_e \log (Ct + A) \tag{11.23}$$

or the inverse logarithmic relation

$$\frac{1}{W} = C - k_i \log (t) \tag{11.24}$$

where C and A are constants, and k_e and k_i are the corresponding logarithmic rate constants.[6] Although the exact mechanism that gives rise to the logarithmic variation of W with t is not known, the postulation has been made [24] that it results from the promotion of ionic transport across the scale by electrical fields within very thin oxide layers.

One characteristic of a metal that oxidizes according to a logarithmic rate expression is that its oxide film approaches an apparent limiting thickness with increasing time. For example, aluminum that is oxidizing in air at ambient temperatures behaves logarithmically, and

---

[6]Generally, if Eq. (11.23) fits the data, so will Eq. (11.24).

TABLE **11-6**

Parabolic Rate Constants for Ni and Cr-Ni Alloys*

w/0 Cr	k_p(g^2/cm^4 sec)
0	3.8(10^{-10})
0.3	15(10^{-10})
1.0	28(10^{-10})
3.0	36(10^{-10})
10.0	50(10^{-10})

*From Wagner, C., and K. E. Zimens, *Acta Chem. Scand.*, **1**, p. 547 (1947).

the film growth stops after a few days of exposure—accounting for its excellent atmospheric, oxidation resistance and making it a good construction material.

Parabolic and logarithmic rate equations are preferable to linear relations when materials will be used in high temperature applications. Tables 11-5, 11-6, and 11-7 contain parabolic rate constants derived from experimental data for a number of alloys.

In some metals that exhibit linear rate equations, the reaction occurs with rapid exothermic oxidation at the metal-oxide surface. This causes a continuous increase in the rate constant, resulting in a chain reaction. Increases in the reaction rate liberate large amounts of heat which, in turn, accelerate the reaction rate, and so on. Such a process, termed *catastrophic oxidation*, can actually lead to ignition of the surface in extreme cases (as is possible with columbium). Alloys containing even small amounts of molybdenum or vanadium often exhibit catastrophic oxidation, thus limiting their high temperature utility in oxidizing atmospheres [25]. The presence of high chromium and nickel contents in alloys will retard the effect of catastrophic oxidation of the other elements [24].

TABLE **11-7**

Parabolic Rate Constants for Ni in Differing Atmospheres*

($T = 1000°C$)

ATMOSPHERE	k_p(g^2/cm^4 sec)
pure O_2	2.5(10^{-10})
$O_2 + Li_2O$ vapor	5.8(10^{-11})

*From Pfeiffer, H., and K. Hauffe, *Zeit. Metalkunde*, **43**, p. 364 (1952).

Corrosion Control

STAINLESS STEEL

That which is understood can often be controlled. If an engineer uses the basic principles of electrochemistry and oxidation kinetics intelligently, many of the difficulties caused by corrosion in poorly designed systems can be avoided. This may result in considerable economic savings—highly desirable from an engineering standpoint; since the complete elimination of all corrosion is neither a desirable nor a practical goal, the ultimate decision as to which control measures should be employed is an economic one. For example, it may be more economical to periodically replace certain equipment than to construct it from materials which will not corrode. It is also often possible to effect large savings in maintenance and replacement costs by the proper design and selection of materials. Without regarding their economic features, several methods which are at the engineer's disposal in controlling or eliminating corrosion will now be discussed.

Materials Selection

Corrosion can be most obviously and most commonly controlled by choosing materials that minimize the undesirable effects of the corrosive environment.

Although most of the previous corrosion principles have been discussed in terms of metals, the student should not have the impression that a panacea for all corrosion problems is the utilization of non-metallic materials which are insusceptible to galvanic attack. In some instances, of course, this is a possible solution, but in most cases it is not because few nonmetals possess the desirable combination of the strength, ductility, and fabricability of steel. Polymeric materials, for example, are much weaker, softer, and less resistant to strong inorganic acids than metals. They are also highly susceptible to swelling and attack by numerous solvents (as will be seen in the next chapter), and are usually severely limited by their temperature properties, making their possible use at much above ambient temperature rare. While ceramics are exceptionally resistant to corrosion, possess very good, high temperature properties, and can be strong, hard materials, they are generally quite brittle, with very low tensile strengths, and become fragile under impact loads. In view of these examples, it can be seen why nonmetallic materials are mainly employed in corrosion control in the forms of liners, gaskets, and coatings.

Stainless steel is a very often misunderstood and misused corrosion-resistant metal. Actually, stainless steel is a generic name for a large class of chromium-bearing steel alloys, some 30 of which are presently marketed. This addition of chromium to steel increases its resistance to attack by oxidizing acids such as nitric acid. However, this material is less resistant to chloride-containing mediums and more susceptible to stress-corrosion cracking than an ordinary, low-carbon structural steel. Also, stainless steels depend upon surface passivity for their corrosion resistance and are, therefore, much more vulnerable to crevice corrosion, intergranular corrosion, and pitting than ordinary steels which do not contain chromium. Therefore, the indiscriminant use of stainless steel alloys to remedy corrosion problems can often lead to expensive and disastrous failures. Another common misconception [26] is that since stainless steels are nonmagnetic alloys, they are superior to magnetic alloys in their corrosion resistance; *no* correlation is known to exist between magnetic susceptibility and corrosion resistance [26].

In general, some classes of metals perform better under certain types of environmental conditions than others; thus, there is usually a metal "of choice" for each condition. Several examples follow:

Stainless steels are well suited for oxidizing acid solutions such as those produced by nitric acid, while nickel and its alloys work well in basic solutions such as those produced by caustic (NaOH). Monel metal is resistant to hydrofluoric acid—a medium which is extremely corrosive to a number of normally corrosion-resistant materials, including glass. A whole series of Hastelloys has been developed that shows excellent resistance to hot hydrochloric acid. Lead is very resistant to dilute H_2SO_4 under stagnant conditions, but is easily susceptible to erosion corrosion in such media under nonstagnant conditions. Because of its logarithmic oxidation rate, aluminum is very good for atmospheric exposure applications where nonstaining characteristics are desired. Tin is almost always chosen as the container or piping material for very pure distilled water since it is completely inert to water. Because it does not appear to possess a transpassive region, titanium is useful in very strong, hot, oxidizing solutions. Ordinary structural steel is suitable for containing concentrated sulfuric acid, but is attacked by dilute or aerated acid.

Perhaps the most perfect corrosion-resistant metal is tantalum, which resists most acids at all concentrations and temperatures. Because it very nearly parallels the corrosion-resistance of glass [26], tantalum is often used as a plug material by manufacturers of glass-lined equipment. It is, however, susceptible to attack by hydrofluoric acid and strong caustic solutions.

To summarize: nickel, copper, and the alloys of these metals are used in reducing or nonoxidizing environments; chromium-containing alloys are used in oxidizing conditions; and Titanium and its alloys are used for extremely severe oxidizing conditions.

Alteration of Environment

Environment is especially important in determining the severity of galvanic corrosion. Since significant influences of cation and oxygen concentrations on corrosion reaction rate are now clear (from the above sections), any change in either of these variables that decreases the corrosion current can be viewed as beneficial in controlling the corrosion rate of a part. An example of a very old and effective control measure is the deaeration of water for boilers. This type of protective measure is not universally desirable, however, because systems depending upon oxidative passivation for their protection require oxygen to be present in the water; deaeration would, of course, be detrimental in this regard. If the concentration of a corrosive agent in solution can be decreased, the effect is usually beneficial in reducing corrosion. The removal of chloride ions is especially helpful in situations where such ions accelerate the corrosion.

When possible, decreasing temperatures reduce corrosion because only rate processes with rather pronounced temperature dependencies are involved. Care should be exercised, however, to be sure that the particular type of corrosion being combated will be decreased by reducing the temperature; raising the temperature of seawater toward its boiling point, for example, decreases the solubility of its oxygen and reduces its corrosiveness.

Decreasing the velocity of a corrosive medium is helpful in reducing erosion corrosion, especially when the danger of cavitation exists. In cases where surface passivation is important, however, it is generally desirable to avoid stagnant solutions which might invite scale desposition.

The addition of inhibitors to the electrolyte is another common way of limiting corrosion by the manipulation of environment. Inhibitors are essentially retarding catalysts or reaction poisons; the use of rust inhibitors in automobile radiators is a typical example of this practice. Some inhibitors are adsorbed onto the surface to form a protective film, thus sealing the surface and prohibiting contact with the electrolyte. Other types, such as arsenic and antimony ions, specifically act to retard the evolution of hydrogen at the cathodic areas; such inhibitors are very effective in acid media, but are not especially useful in basic media. Another type of inhibitor, the scavanger, reacts with the corrosive agent (particularly in the case of oxygen) and removes it from the solution. This method of control is somewhat analogous to deaeration and is likewise detrimental when passivation due to oxidation is important; it is also ineffectual if oxidation is not the controlling mechanism of the corrosion. A large number of specific corrosion inhibitors for different metals and environments is known [26].[7]

Coatings

The three general types of coatings used in reducing metal corrosion are: metallic, ceramic and nonmetallic inorganic, and organic.

METALLIC COATINGS   Metallic coatings, produced by the surface application of a thin layer of metal different from the one to be protected, are most often used to separate the environment from the substrate material. However, some metal coatings are intentionally applied to serve as anodes and corrode in place of the underlying substrate metal. Such coatings are called *sacrificial anodes*. Numerous techniques [26], varying in expense, are used to apply coatings; some can produce very thin uniform layers, while others result in much thicker coverings.

The effectiveness of metal coatings varies with the nature of the coating method used. A sprayed coating of aluminum on stainless steel, for example, provides protection from atmospheric oxidation at temperatures up to 1500°F. High-strength aluminum alloys are often coated with a pure aluminum skin, providing a corrosion barrier that protects the underlying alloy from the stress corrosion to which it is normally susceptible.

In some instances, a break in a coating may actually accelerate the corrosion process—a factor which must be considered when a coating material is chosen. For example, steel may be coated with tin to make tinplate, or with zinc to produce galvanized iron; although these two metals react very differently to breaks in their surface coatings, each metal in its particular application is preferable to the other.

If the surface coating of tinplate is scratched, the tinplate will corrode more rapidly than a piece of unprotected steel with a similar scratch. There are two reasons for this: tin is cathodic to iron, inducing an undesirable galvanic couple; the exposed iron anode is small, while the tin cathode is large, creating a very unfavorable electrode area ratio. (This problem of relative electrode size was mentioned earlier in context with crevice and pitting corrosion, where small anodic areas connected with large cathodic regions were observed to corrode at accelerated rates.)

------

[7]Table 6-1 in [26] contains an extensive list, together with a bibliography of 77 references pertaining to corrosion inhibitors. The same list is also found in an article by M. Brooke, *Chem. Eng.*, p. 230 (December 1954).

The reverse situation is obtained with zinc-coated or galvanized iron; here the zinc coating, anodic relative to the iron, is intentionally sacrificed. Furthermore, if the surface coating is scratched, the anode/cathode area ratio is the reverse of the tinplate situation above: the small scratch is cathodic, while the large unbroken surface of zinc is anodic. Very small corrosion currents will result, causing corrosion to occur at a slow, generally uniform rate.

At first, the above considerations might seem conclusive enough to eliminate tin plating as a protective method. However, there are many applications where the other properties of tinplate are preferable to those of galvanized iron. For example, "tin" cans (which are really made of tinplate) are used extensively in the food packaging industry because tin is nontoxic.

NONMETALLIC INORGANIC COATINGS   In some applications, it is desirable to coat the steel with a nonmetallic inorganic material such as ceramics. (Common examples of the application of ceramic coatings to achieve smoothness, durability, and beauty are the porcelain coatings of household appliances, bathtubs, and sinks.) In addition, an inorganic nonmetallic coating is often used where a hard, slick surface that can resist a wide variety of severely corrosive materials is desirable. A typical example of this is the coating of steel with glass. Glass-lined steel vessels are becoming increasingly important in the chemical processing industry because they are easy to clean when sticky materials such as latex are handled, they are inert to a great variety of corrosive materials, and they rarely, if ever, contaminate the contained product. The last property is especially important in the food and drug industries.

ORGANIC COATINGS   Many organic materials, particularly polymers, are very resistant to corrosive environments. Perhaps the most widely used organic coating technique is the application of paints, varnishes, and lacquers. These materials provide a thin, tough, usually quite durable barrier which protects the substrate metal from contact with corrosive environments. In addition to their usefulness in combating corrosion, these materials also fulfill esthetic values.

Like any other protective method, paints are not a panacea for corrosion. If a coat of paint is applied to a severely attacked portion of substrate metal and the paint film cracks, rapid pitting and perforation will probably result on the small metal area exposed to the corrosive environment.

In the majority of cases where paints are inadequate corrosion retardants, the fault is generally either due to improper surface preparation (resulting in peeling paint) or the poor choice of a primer coat (resulting in the nonadherance of subsequent coats). Failure to prepare a surface and choose a primer properly will result in inadequate protection, regardless of the attributes of the top coat of paint that is applied later.

Paints are not the only organic coating materials. Many types of plastic or polymeric films may also be used to prevent corrosion. The lining of steel pipe with chemically resistant polymers is becoming a common practice in some industries, and many items are encapsulated in films of protective plastic. To reduce erosion corrosion, a particular type of resilient rubber is used to line and coat slurry handling machinery, especially pumps. Other examples of organic coatings include tar and asphalt, which are frequently used to coat buried pipelines in an attempt to eliminate both bacterial corrosion and contact with corrosive water in the soil.

Design

The proper design of equipment can be just as important in reducing corrosion as proper material selection. Protection against corrosion should begin on the drawing board [27]; the modern engineering designer therefore considers materials selection and corrosion suitability simultaneously with mechanical, thermal, electrical, and economic property requirements.

Although the solution of specific corrosion problems may require the special consideration of an expert, many common corrosion problems can be eliminated or greatly reduced by adhering to the following [26, 27] general design rules:

1. In designing equipment, allow an additional thickness of metal over the mechanical strength requirement because of the penetrating nature of corrosion. (This is especially necessary in pipes and tanks.)
2. Welding rather than riveting tanks and other containers will greatly reduce opportunities for crevice corrosion.
3. If riveting is absolutely necessary, choose a rivet material that is cathodic relative to the plates being joined. If galvanically dissimilar metals must be bolted together, use insulating washers and gaskets to avoid electrical contact between the metals.
4. Where possible, choose galvanically similar materials throughout the entire structure. In general, try to avoid combinations of environmental conditions or galvanically dissimilar metals which will enhance corrosion.

5. Design tanks and other containers so they can be completely drained and easily cleaned. Avoid situations where stagnant pools of corrosive materials can accumulate after a tank is emptied and configurations which will encourage concentration differences in corrosive solutions.
6. Allow for easy replacement of components in which failure due to corrosion is anticipated. (These components usually exist in pumps, stirrers, and other parts that move in or through corrosive media.)
7. Avoid sharp bends in piping systems and sharp corners where flow will occur; such arrangements promote erosion corrosion.
8. Design heating systems that avoid localized hot spots. These severe, local nonuniformities in temperature distribution encourage comparable localized increases in corrosion rate with attendant localized attack. Hot spots also promote local stresses caused by thermal expansion, which may enhance stress corrosion and lead to failure.
9. Avoid crevices in the design of equipment supports, and allow for proper drainage spaces around tank bases. Flat-bottomed tanks permit dirt, water, and other undesirable materials to accumulate under their surfaces and where their edges contact floors, etc. A tank with a curved bottom, elevated by the proper supports, will eliminate these problems and be easier to maintain.
10. Exclude air entrainment in liquid systems. This consideration is especially important around the inlets to pumps, tanks, and agitators. (Of course, the exclusion of air in passivating metals is not desirable.)
11. Choose the location for a plant with regard to such natural weather phenomena as prevailing winds and tidal flow patterns. This is important in reducing corrosion problems [27].

Cathodic and Anodic Protection

Since the vast majority of corrosion problems are galvanic in nature, a knowledge of the electrochemical properties of the reacting system can convert desirable parts into nondeteriorating cathodes through the application of external cathodic or anodic currents.

CATHODIC PROTECTION   The most common form of external, electrical corrosion control is *cathodic protection*. In this method, a cathodic potential is impressed on the part to be protected, converting it to a cathode. This may be accomplished by the use of an external power

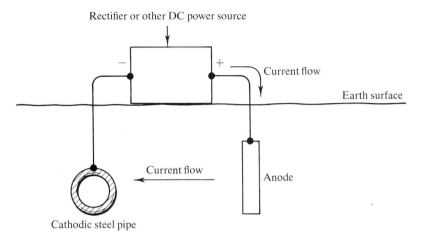

FIG. 11-25   Schematic arrangement of an external power source in the cathodic protection of buried steel pipe.   Current flow convention is standard electrical convention. Negative terminal of external power supplies electrons to the pipe, making it cathodic and suppressing the reaction $Fe = Fe^{++} + 2e^-$.

source or by proper galvanic coupling with a more anodic metal. Figures 11-25 and 11-26 illustrate the cathodic protection of an underground pipe by the first and second methods, respectively. (In Fig. 11-26, the scrap of magnesium is a sacrificial anode.) Cathodic protection is also achieved in this manner when iron is galvanized by coating it with zinc.

In practice, magnesium is most frequently used as a sacrificial anode due to its very negative potential and high current output. The anode materials most widely employed in this impressed current method are steel, graphite, and silicon-iron [26].

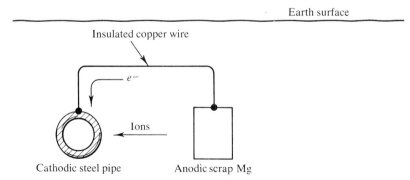

FIG. 11-26   Schematic arrangement of a sacrificial scrap Mg anode in the cathodic protection of buried steel pipe. Mg corrodes, protecting the cathodic steel pipe.

High yield-strength steels [16] and zirconium alloys used in nuclear reactor applications are very susceptible to hydrogen embrittlement. If such alloys are to be applied in corrosive environments, the engineer must be aware of the potential hazards of cathodic protection measures.

ANODIC PROTECTION   *Anodic protection* is a relatively new form of corrosion prevention. It is based on the formation of a protective, passive film on metal surfaces by externally impressed anodic currents [26]. Many metals that display active-passive transitions, such as Ni, Fe, Cr, Ti, and their alloys, may be passivated by the proper application of controlled anodic currents which will reduce their corrosion rate.

Anodic protection requires low currents that are calculable on the basis of the known, appropriate, passive polarization properties of a metal. This protective method is applicable in very corrosive environments [26].

References Cited

[1] Fontana, M. G., and N. D. Greene, *Corrosion Engineering*, McGraw-Hill Book Co., Inc., New York, 1967, Chapter 9.

[2] Brophy, J. H., R. M. Rose, and J. Wulff, *The Structure and Properties of Materials, Volume II, Thermodynamics of Structure*, John Wiley and Sons, Inc., New York, 1964, Chapter 10.

[3] Glasstone, S., *Introduction to Electrochemistry*, D. Van Nostrand Company, Princeton, N. J., 1942.

[4] Potter, E. C., *Electrochemistry*, Cleaver-Hume Press Ltd., London, 1956.

[5] Fontana and Greene, *op. cit.*, Chapter 2.

[6] Jastrzebski, Z. D., *Nature and Properties of Engineering Materials*, John Wiley and Sons, Inc., New York, 1959, Chapter 9.

[7] Wagner, C., and W. Trand, *Zeit. Electrochem.*, **44,** p. 391 (1938).

[8] Fontana and Greene, *op. cit.*, Chapter 3.

[9] Schafer, G. J., and P. K. Foster, *J. Electrochem. Soc.*, **106,** p. 468 (1959).

[10] Schafer, G. J., J. R. Gabriel, and P. K. Foster, *J. Electrochem. Soc.*, **107,** p. 1002 (1960).

[11] Rosenfeld, L., and I. K. Marshakov, *Corrosion*, **20,** p. 115*t* (1964).

[12] Slabaugh, W. H., and M. Grotheer, *Ind. Engr. Chem.*, **46,** p. 1014 (1954).

[13] Evans, U. R., *Corrosion*, **7,** p. 238 (1951).

[14] Archer, R. S., J. Z. Briggs, and C. M. Loeb, Jr., *Molybdenum Steels, Irons, Alloys,* Climax Molybdenum Company, New York, 1965, p. 131.

[15] Uhlig, H. H., *Corrosion Handbook,* John Wiley and Sons, Inc., New York, 1948, p. 569.

[16] Bates, J. F., *Ind. Engr. Chem.,* **58,** p. 19 (1966).

[17] Dix, E. H., Jr., *Trans. Inst. of Metals Div.,* AIME, **137,** p. 11 (1940).

[18] Mears, R. B., R. H. Brown, and E. H. Dix, Jr., *Symposium on Stress Corrosion Cracking of Metals,* ASTM-AIME, York, Pa., 1945, pp. 323–344.

[19] Nielsen, N. A., *Corrosion,* **20,** p. 104*t* (1964).

[20] Pickering, H. W., and P. R. Swann, *Corrosion,* **19,** p. 373*t* (1963).

[21] Priest, D. K., F. H. Beck, and M. G. Fontana, *Trans. Am. Soc. Metals,* **47,** p. 473 (1955).

[22] Pardue, W. M., F. H. Beck, and M. G. Fontana, *Trans. Am. Soc. Metals,* **54,** p. 539 (1961).

[23] Bhatt, H. J., and E. H. Phelps, *Corrosion,* **17,** p. 430*t* (1961).

[24] Fontana and Greene, *op. cit.,* Chapter 11.

[25] Leslie, W. C., and M. G. Fontana, *Trans. Am. Soc. Metals,* **41,** p. 1213 (1944).

[26] Fontana and Greene, *op. cit.,* Chapter 6.

[27] Coburn, S. K., *Materials Protection,* **6,** p. 33 (February 1967).

PROBLEMS

11.1   A 0.75×3.0 in. sample of low-carbon steel is immersed in a 10% AlCl–90% SbCl solution. HCl gas is bubbled through this solution at atmospheric pressure and 90°C for 24 hours. If the sample shows a weight loss of 1080 mg upon removal, calculate the apparent corrosion rate in mpy.   Mpy = 1357

11.2   Compute the corrosion rate in Problem 11.1 in units of in./yr, mg/cm² day, and g/cm² sec.   1.357 in./yr.   74.4 mg/cm² day 8.62×10^{-7} gm/cm² sec

11.3   Samples identical to the one described in Problem 11.1 are immersed in the same solution for differing periods of time, resulting in the following data:

SAMPLE	IMMERSION TIME (HOURS)	WEIGHT LOSS (MG)
1	24	1080
2	72	1430
3	96	1460

Using this information, compute the apparent corrosion rate in mpy for each sample. Estimate the initial corrosion rate of each steel from your results. Suggest an explanation for the decrease in the apparent corrosion rate. $R_o \approx 1900\,mpy$

11.4   A 1.0×3.0 in. sample of low-carbon steel is placed in a 2N H_2SO_4 solution where the observed activation overvoltage is $\eta_a = 0.22$ volts. Assuming that the Tafel constant is 0.08 volts, compute the corrosion rate of the Fe in mpy.
$$mpy = 7.12$$

11.5   Assuming that the steel sample in Problem 11.4 has been tested for 24 hours, compute the weight loss of the sample in mg.
$$w = 7.56\,mg$$

11.6   Compute the corrosion currents in ma/cm² for each of the samples in Problem 11.3. Assuming $i_0 = 10^{-6}$ amp/cm² and $\beta = 0.1$ volt, plot a curve of η vs time and estimate the initial value of η.
$$\eta_o \approx .86\,volts$$

✳ **11.7**   A copper plate is immersed in a solution of $CuCl_2$ in which $a_{Cu^{++}/Cu} = 10^{-1}$. A portion of the plate is exposed to a flowing stream, which reduces $a_{Cu^{++}/Cu}$ to 10^{-6}. Compute the magnitude of the cell potential that develops. Which portion of the plate is the anode? $\Delta E = .148\,volts$

11.8   Compute the free energy difference (ΔG in cal/mole) for the cell in Problem 11.7.
$$\Delta G \approx -6820\,cal/gmole$$

11.9   Compute the cell potential and ΔG in cal/mole for a standard Fe–Cu couple.
$\Delta E = -.777\,volt$ $\Delta G = 35800\,cal/gmole$

✳ **11.10**   Compute the cell potential and ΔG in cal/mole for an Fe–Zn couple.
$\Delta E = -.323\,volts$ $\Delta G = 14900\,cal/gmole$

11.11   A divalent metal electrode is operating at 25°C under a condition where $i_{corr} = \frac{1}{2}i_L$. If concentration polarization is the only operative form of polarization, compute the overvoltage for this electrode. $\eta_c = -.0089\,volts$

✳ **11.12**   Compute the overvoltage for the electrode in Problem 11.11, assuming that the concentration of metal ions in the bulk solution is doubled. (All other variables remain constant.) $\eta_c = -.00368\,volts$

11.13   A trivalent metal electrode is operating at 25°C under mixed-electrode conditions where $i_L = 20i_0$. Assuming the Tafel coefficient is 0.1, plot a polarization curve of η_{red} vs i/i_0. For what range of i/i_0 can the concentration polarization be safely neglected?

12

HIGH MOLECULAR WEIGHT POLYMERS

Organic Materials

Historically, the term "organic" is associated with living tissues as opposed to nonliving or "inorganic" minerals. To the modern chemist, however, *organic materials* are those that contain the element carbon, while inorganic materials are those that do not. Carbon dioxide, a waste by-product of many life processes, is organic; oxygen, essential to all life, is inorganic.

Since the nomenclature of organic chemistry is involved and confusing for the uninitiated, a brief introduction to it will be presented here before discussing high molecular weight organic polymers and their properties.

Saturated Compounds

PARAFFINIC HYDROCARBONS In organic compounds, the carbon atoms possess tetrahedral, molecular orbitals that show a strong tendency to form covalent bonds. Many of these compounds consist only of hydrogen and carbon covalently bonded in various configurations. The simplest of these *hydrocarbons* is methane, a compound in which a single carbon atom is tetrahedrally surrounded by four co-valently bound hydrogen atoms. Methane, a monomolecular gaseous material with a very low boiling point ($-162°C$), is composed of very symmetrical molecules possessing no net dipole moment.

Actually, methane is a member of an entire family of compounds called *paraffin hydrocarbons* having a general formula C_nH_{2n+2}. Because these compounds contain only single C–C or C–H bonds and possess no multiple, carbon-to-carbon bonds in their molecules, they are also known as *saturated hydrocarbons*. The *normal* paraffin hydrocarbon has a linear, straight-chain skeleton or carbon structure if all the hydrogen is removed; that is, the carbon atoms in the molecule are linked together in an undulating chain without side branches and are surrounded by the symmetrically distributed hydrogen atoms.

Table 12-1, which contains the names and certain thermal data for several normal paraffins, clearly shows that the melting point increases steadily with the increasing number of carbon atoms in a molecule. When displayed graphically (Fig. 12-1), this relationship also shows that the melting point (T_m) appears to approach an asymptotic limit at very high molecular weights, suggesting that the melting point varies inversely with the number of carbon atoms. The data of Fig. 12-1 are

TABLE 12–1

Thermal Data for Normal Paraffin Hydrocarbons*

FORMULA	NAME	MOLECULAR WEIGHT	BOILING POINT (°C)	MELTING POINT (°C)
CH_4	Methane	16.04	−161.49	−182.48
C_2H_6	Ethane	30.07	−88.63	−182.8
C_3H_8	Propane	44.09	−44.5	−189.9
C_4H_{10}	Butane	58.12	−0.5	−138.3
C_5H_{12}	Pentane	72.15	+36	−129.72
C_6H_{14}	Hexane	86.18	68	−95
C_7H_{16}	Heptane	100.21	98.42	−91
C_8H_{18}	Octane	114.23	125.5	−56.5
C_9H_{20}	Nonane	128.26	150.8	−51
$C_{10}H_{22}$	Decane	142.29	174.1	−29.7
$C_{15}H_{32}$	Pentadacane	212.41	270.5	+10
$C_{20}H_{42}$	Eicosane	282.54	343	36.8
$C_{26}H_{54}$	Hexacosane	366.72	262	56
$C_{30}H_{62}$	Triacontane	422.83	304	65.5
$C_{35}H_{72}$	Pantatriacontane	492.23	331	74.7

*Data from *Handbook of Chemistry and Physics, 46th ed.*, The Chemical Rubber Company, Cleveland, copyright 1966.

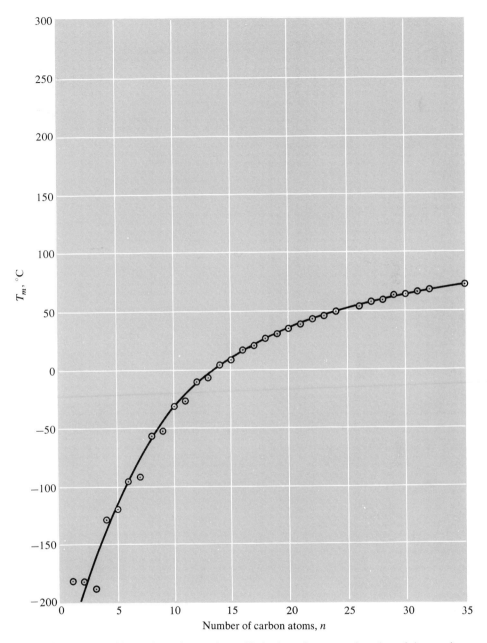

FIG. 12-1 The melting points of normal paraffin hydrocarbons as a function of the number of carbon atoms in their molecular chains.

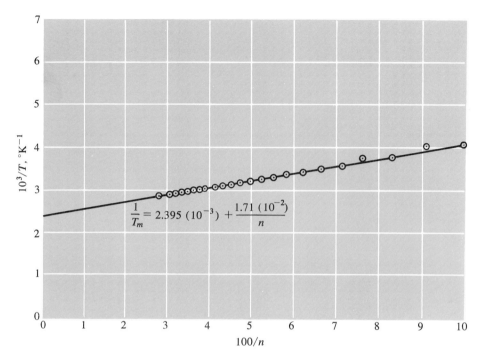

FIG. 12-2   A replotting of the data from Fig. 12-1.

replotted in Fig. 12-2 with T_m^{-1} (°K^{-1}) as ordinate and $1/n$ (n being the number of carbon atoms in the molecule) as abscissa. The data in Fig. 12-2 can be represented by the equation

$$\frac{1}{T_m} = 2.395(10^{-3}) + \frac{1.710(10^{-2})}{n} \qquad \textbf{(12.1)}$$

At room temperature, the states of aggregation in paraffin hydrocarbons vary systematically from gases to volatile liquids to heavy viscous oils and greases and, finally, to paraffin waxes containing about 30 carbon atoms per molecule. The polymer polyethylene, a relatively high melting plastic solid composed of several hundred carbon atoms, is a limiting member of this series.

In addition to the normal paraffins, there are many other types of saturated hydrocarbons. If the carbon skeleton is branched, the compounds are called *isoparaffins*; if the carbon chain closes on itself to form a saturated ring, the compounds are *cycloparaffins* or *naphthenes*. Compounds having the same chemical composition but different physical arrangements of the carbon atoms in the chain (that is, normal vs isoparaffins) are called *isomers*. The number of possible isomers increases rapidly with the increasing number of carbon atoms; the

TABLE **12–2**

Effect of Isomerism on Properties of Octane (C$_8$H$_{18}$) *

NAME OF ISOMER	STRUCTURAL FORMULA	MELTING POINT (°C)	BOILING POINT (°C)	MOLECULAR WEIGHT (g/mole)
n-octane	CH$_3$(CH$_2$)$_6$CH$_3$	−56.5	125.5	114.2
2-methylheptane	CH$_3$CH(CH$_3$)$_4$CH	−109	118	114.2
	$\vert$			
	CH$_3$			
2,2,3,3-tetramethylbutane	CH$_3$ CH$_3$	100.7	106.5	114.2
	$\vert$ $\vert$			
	CH$_3$—C——C—CH$_3$			
	$\vert$ $\vert$			
	CH$_3$ CH$_3$			

*Data from *Handbook of Chemistry and Physics, 46th ed.,* The Chemical Rubber Company, Cleveland, copyright 1966.

pronounced effect of such isomerism on the physical properties of hydrocarbons is illustrated in Table 12-2 for three of the eighteen possible isomers of octane (C$_8$H$_{18}$).

The paraffin hydrocarbons are highly combustible, yielding water and carbon monoxide or dioxide. Methane is the primary constituent of natural gas, propane and butane are common fuel gases, and hexane through octane are the primary components of the gasoline burned in internal combustion engines. Therefore, it should not be surprising that the limiting normal paraffin, polyethylene, is also combustible.

Because of their high degree of molecular symmetry, paraffins are nonpolar and therefore insoluble in polar solvents such as water. In fact, water will not even wet high molecular weight paraffin wax or polyethylene, accounting for the ease with which cubes may be freed from polyethylene ice trays.

ORGANIC HALIDES   *Organic halides* are saturated hydrocarbons with one or more hydrogen atoms replaced by halogen atoms. The fluorides and chlorides are by far the most important organic halides in the field of plastics. (Table 12-3 contains some basic data for several organic halides.)

An organic halide is less combustible than its corresponding hydrocarbon because its halogen atoms impart combustion resistance by dilution. In addition, as an organic halide burns, the halogen forms a hydrogen halogenide which further inhibits combustion by preventing complete oxidation of the hydrogen. (This is an important advantage polyvinyl chloride has over polyethylene.)

Almost every organic chloride possesses a permanent dipole mo-

TABLE **12–3**

Properties of Organic Halides*

FORMULA	NAME	MOLECULAR WEIGHT	BOILING POINT (°C)	MELTING POINT (°C)
CH_3Cl	Methyl chloride	50.49	−23.8	−97
C_2H_5Cl	Ethyl chloride	64.52	13.1	−138.7
$CH_2{=}CHCl$	Vinyl chloride	62.50	−13.9	−160.
$CH_2{=}CCl_2$	Vinylidene chloride	96.95	32	−122.1
$CH_2{=}CHF$	Vinyl fluoride	46.02	−51.	—
$CH_2{=}CF_2$	Vinylidene fluoride	64.04	< −84	—
$Cl_2C{=}CCl_2$	Tetrachloroethane	165.83	121.	−22.
$F_2C{=}CF_2$	Tetrafluoroethane	100.02	−76.3	−142.5
C_2H_5Br	Ethyl bromide	108.98	38.40	−118.9
CH_3CHCl_2	1,1-dichloroethane	98.96	57.	196.
CH_3CHF_2	Ethylidene fluoride	66.05	−24.7	—
FCH_2CH_2F	Ethylene fluoride	66.05	30.07	—

*Data from *Handbook of Chemistry and Physics, 46th ed.,* The Chemical Rubber Company, Cleveland, copyright 1966.

ment because of the six per cent partial ionic nature (computed from Eq. (2.23)) of the C–Cl bond. This dipole moment is strongly influenced by molecular symmetry. The dipole moment of methylchloride is 2.0 Debyes, for example, while the symmetrical compound carbon tetrachloride does not have any dipole moment.

ALCOHOLS Paraffin hydrocarbons with one or more hydrogen atoms replaced by —OH groups are *alcohol compounds*. These highly polar —OH groups impart considerable polarity to the alcohol molecules, making them readily soluble in water. As examples, methanol (CH_3OH) and ethanol (C_2H_5OH) are infinitely soluble in water.

Unsaturated Compounds

Unsaturated hydrocarbons possess one or more multiple, carbon-to-carbon bonds in their skeletal structures; the simplest of these compounds are ethylene ($CH_2{=}CH_2$) and acetylene ($HC{\equiv}CH$). The double or triple bonds in unsaturated hydrocarbons are extremely reactive to many electronegative substances such as chlorine or oxygen, making further reaction and the formation of more nearly saturated compounds possible. This reactivity is the basis of the polymerization reactions that result in large macromolecular polymers. Many types of unsaturated organic materials are important in the formation of polymers; some of these will be described here.

AROMATIC HYDROCARBONS   *Aromatic hydrocarbons* are compounds possessing one or more six-membered rings of carbon atoms joined by a resonating sequence of alternating single and double bonds. The simplest aromatic hydrocarbon is benzene (C_6H_6), which resonates [1] between the two Kekulé structures illustrated by

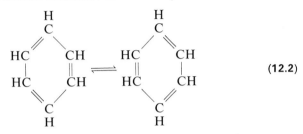

(12.2)

Even though the aromatic benzene ring is highly unsaturated, it is very stable and will not enter into a reaction except through its peripheral hydrogen atoms. This stability, derived from the conjugated resonance structure of the alternating bond sequence, is of considerable importance in the formation of the molecule polystyrene from styrene. Styrene is ethylene with one of the hydrogen atoms replaced by a benzene ring; the styrene molecule is reactive at the ethylenic double bond, but not at any of the double bonds of the benzene ring. Therefore, benzene appears as a side appendage grouping in the large polymer molecule and has a marked influence on the overall properties of polystyrene.

ALDEHYDES AND ACIDS   *Aldehydes* contain the grouping

which is reactive and important in the formation of certain types of polymeric compounds. The simplest aldehyde is formaldehyde (HCHO), an important constituent of the polymer phenolformaldehyde which is commercially known as Bakelite. Although the aldehydes are polar, hydrogen bonds are not formed because of the lack of hydrogen atoms attached to the oxygen.

Organic acids contain the grouping

$$\begin{array}{c} OH \\ | \\ -C{=}O \end{array}$$

where one oxygen atom is doubly bound to the carbon atom and the other is part of an —OH group also attached to the carbon. The —OH group is only slightly acidic and organic acids are therefore weak in comparison to strong inorganic acids such as HCl. Organic acids play an important part in the formation of many polymeric materials.

Polymers and Polymerization

Definitions

Polymers are giant organic molecules which range in molecular weight [2] from 10^4 to 10^7 and consist of one or more simple subunits repeated many times. These recurring structural units are derived from simpler organic chemicals called *monomers*. Monomers contain only one unit of the repeated chemical structure of the polymer. The chemical reaction in which the monomers are converted to large polymer molecules is called a polymerization reaction, or simply *polymerization*.

The root of both the words monomer and polymer, the simple word *mer*, literally means a unit or building block. A mer is a small grouping of atoms analogous to the unit cell in a crystalline solid; that is, when a mer unit is translated along a molecular skeletal chain by a distance equal to its own length, it reproduces the structure. Consequently, a *poly*mer is simply many mers linked together to form a giant macromolecule.

Because of the prefix mono-, meaning "one," the word monomer may lead to some confusion if it is not carefully defined. Although the chemical composition of the mer unit is the same in both a monomer and a polymer, the molecular arrangement and bond structures of the mer are markedly different in a monomer than when they occur in a polymer. For example, the monomer ethylene, with the structural formula $H_2C{=}CH_2$, has the grouping C_2H_4 as a mer unit; this *same* mer unit, C_2H_4, occurs in the polymer polyethylene as

$$[-H_2C-CH_2-]_n$$

Note that the double bond is lost here in passing from the monomer to the polymer, but the molecular weight of the mer unit, or *mer weight*, is the same in both cases. The mechanism of the loss of the double bond will be discussed in detail later in this chapter.

The term *resin*, frequently encountered in the study of polymeric materials, is very difficult to define precisely. It originates from the use in varnishes and enamels of naturally occurring resinous materials derived from trees, and it refers to a multitude of materials of varying chemical compositions. Natural resins have amorphous structures, exhibit conchoidal or shell-like fractures, are brittle, lustrous, insoluble in water, and fuse into a plastic state when heated. These substances, which may or may not be true polymeric materials, often consist of complex mixtures of low molecular weight, nonpolymeric materials. In chemical terminology, resin includes all naturally occurring or synthetic organic materials that can be dissolved in solvents and/or oils to form substances such as varnishes, enamels, lacquers, and some paints which create relatively hard, glossy coatings when applied to a surface.

In the plastics industry, the binder in molding powders—the material which softens and flows under heat and pressure and holds the entire composition together—is usually considered a resin. Thus, no unequivocal definition of this term exists.

Another common term, *plastic*,[1] is difficult to define precisely. Although many familiar items such as buttons, combs, fountain pen cases, and toys are commonly considered to be plastic, these items are made from a great variety of different materials which often are not plastic in the sense that they are easily deformable.

Plastics are solid compositions bonded by high molecular weight, polymeric, organic compounds. These materials, which are capable of being formed, or have been formed, by pressure at relatively moderate temperatures or by a chemical reaction, may be as soft and pliable as polyethylene or as hard and brittle as Bakelite.

Rubbers are important polymeric materials in our industrialized society. Natural rubber, a polymer derived from the sap of the rubber tree, is a substance that can be stretched to much more than its original length and upon release can retract rapidly and forcibly to substantially its original dimensions. Rubber must exist in a raw plastic state that can be easily molded but it must also respond to alteration by *vulcanization* to give the end product a dimensional stability. Vulcanization (discussed later in this chapter) is a cross-linking process in which the randomly coiled, unattached rubber molecules are bonded together by a vulcanizing substance such as sulfur. This process makes it possible to distinguish between rubber and otherwise similar plastic materials.

---

[1]Plastics were first used commercially in 1868 when John Hyatt compounded cellulose nitrate with camphor to produce celluloid billiard balls. In 1907, Baekeland obtained patents for various phenolic compounds which were immediately employed as insulators by the rapidly growing electrical industry.

Types of Polymers

Due to their varying molecular constitutions, some polymeric materials behave differently after they have been heated and worked. Some of these materials soften upon heating and can be molded into any desired shape; others can be molded into a given shape initially, but heating will no longer soften them thereafter. On the basis of their thermal behavior, three broad classes of polymers can be distinguished.

Thermoplastic materials soften and flow; they may be reformed repeatedly upon applications of heat and pressure. On the contrary, *thermosetting materials* lose this ability to be reformed after they are first subjected to applications of heat and pressure which set them into rigid shapes. *Elastomeric materials* are thermoplastic during molding operations, but are cured by vulcanization to form a usually highly elastic, intermediate rubbery state. If carried nearly to completion, the latter process may also produce materials that resemble thermosetting plastics.

Thermoplasts soften and become distorted under their own weights at temperatures on the order of 150 to 250°F, although some heat resistant grades have been made that can be used up to 500°F if very low stress levels exist. Increasing these stress levels will reduce the useful operating temperatures of most thermoplastic materials to room temperature.

Although most thermosets are not serviceable above 400°F where thermal degradation of the polymer usually occurs, they are harder and more heat resistant than thermoplasts.

Basic Polymer Structures

Polymers can be grouped into four structural classes: linear, branched-chain, moderately cross-linked, and highly cross-linked.

LINEAR POLYMERS   Linear polymers consist of mer units attached sequentially to form a linear or chain-like structure without any side appendages [2]. The term linear is applicable here only in the sense that this type of polymer molecule is proportionately very long in comparison with its thickness and width. Actually, linear polymer molecules rarely assume straight rod-like forms; instead they are normally coiled up and entangled with their neighbors, much like the strands of spaghetti in a bowl.

Linear polymers usually have a random, or amorphous, molecular arrangement in which only van der Waals intermolecular attractions are operative. They are usually thermoplastic and relatively soluble, although some of these polymers are relatively insoluble due to their excessive molecular lengths and the chemical natures of some constituent groups. Linear polymers are often used for forming fibers.

BRANCHED-CHAIN POLYMERS   Some materials form side branches when they polymerize; such branching may be caused by minor quantities of impurities in the reaction vessel or the existence of multiple active sites in the polymerizing molecule. The degree of branching, which varies from one sample to another, may have little effect or may be the predominant effect on the physical properties of a material. Side branches may vary in length from one carbon atom, to a chain of atoms comparable to the parent chain.

Since chain branching decreases the thermoplastic behavior and solubility of a polymer, it is frequently undesirable (particularly in elastomers) and an effort should be made to reduce it where possible.

MODERATELY CROSS-LINKED POLYMERS   In some polymers, adjacent chains, attached by primary bonds, form a *cross-linked*, three-dimensional network structure. Such a structure is usually obtained by the secondary reaction of a primarily linear unsaturated polymer, as in the vulcanization of an elastomer.

Cross-linking markedly affects the physical properties of any polymeric material in which it occurs. An increased degree of cross-linking corresponds to an enormous increase in the polymer molecular weight. Cross-linking also interferes with the complete separation of individual chains by a solvent; that is, solvents can only partially separate large cross-linked networks, causing structural swelling instead of the formation of a true solution. The volume increase due to such swelling may be ten- to 100-fold because of the great length of the coiled molecules between the cross links; thus, cross-linking will severely decrease polymer solubility.

Network extensibility of cross-linked polymers accounts for many elastomeric properties. In vulcanized rubber, the application of a stress initially uncoils and extends the molecules of the slightly cross-linked chains. When their fully extended condition is reached, further extension will require breaking the primary linking bonds, causing the elastomer to become stiff and unyielding. Upon release of the stress, these molecules will recoil and return to their original dimensions.

The rapid stretching of rubber requires that considerable energy be put into the structure, generating a momentary temperature increase; rapid release of the stress results in an energy release which cools the rubber. This is the molecular source of the thermoelastic effect described in Chapter 9.

HIGHLY CROSS-LINKED POLYMERS   Thermosetting plastics possess highly cross-linked, three dimensional structures that form amorphous rigid networks with very low mobilities. These polymers are highly insoluble, infusible, and strong; they also have very low extensibilities because the large number of cross-links make it impossible to separate the structural segments.

Polymerization Mechanism

Polymers are synthesized by two types of reactions: *addition polymerization* and *polycondensation* (or condensation polymerization).

ADDITION POLYMERIZATION   The polymer produced by this sequence of reactions has a mer unit identical with the monomer except that its double bonds have been opened to link the mers together. A typical example of this type of reaction is the formation of polyvinyl chloride

$$n\,(CH_2{=}CHCl) \rightarrow (-CH_2CH-)_n \tag{12.3}$$
$$\underset{Cl}{|}$$

The value of n, known as the *degree of polymerization* (DP), is the *average number*[2] *of mers* in the polymer molecules. Because molecular size in a given polymer varies over rather wide limits, it is necessary to determine its molecular weight distribution. (Some of the methods used for this purpose will be discussed later in this chapter.)

In order for polymerization to occur, a monomer must contain at least one active site or functional group, such as a multiple bond. A double bond is known as a *bifunctional group* because it produces two free, highly reactive, molecular bonding orbitals when activated. If a monomer is strictly bifunctional and contains no impurities, the resultant polymer should be linear. The addition of higher functional monomers that can react with those functional groups already present will result in a cross-linked structure.

Monomers that undergo addition polymerization are almost always of the bifunctional ethylenic type involving the $C{=}C$ bond. Ethylenic monomers have the general formula $CH_2{=}CHX$, where X represents the various size groups.

Addition polymerization involves four sequential reactions: initiation, propagation, transfer, and cessation. When M represents a mono-

---

[2]The use of the subscript n (as in Eq. (12.3)) to designate indefinite molecular size began when organic chemists encountered "troublesome," uncontrollable side reactions in which unidentifiable, high molecular weight "tars" were formed. Since these tars were apparently noncrystalline and no simple methods were available for determining their molecular weights, the empirical formula was followed by an n.

mer molecule, X a solvent molecule or a chain transfer agent, and the dagger an activated monomer or growing chain, these four reactions may be represented [2] as follows:

1. The initiation or activation step

$$M + \text{energy} \rightarrow M^{\dagger} \tag{12.4}$$

2. The propagation step

$$M^{\dagger} + (n - 1)M \rightarrow M_n^{\dagger} \tag{12.5}$$

3. The transfer reaction, involving one or more of the following steps:

$$M_n^{\dagger} + M \rightarrow M_n + M^{\dagger} \tag{12.6}$$

$$M_n^{\dagger} + M_m \rightarrow M_n + M_m^{\dagger} \tag{12.7}$$

$$M_n^{\dagger} + X \rightarrow M_n + X^{\dagger} \tag{12.8}$$

$$X^{\dagger} + M \rightarrow X + M^{\dagger} \tag{12.9}$$

4. The cessation reaction, which proceeds according to one of the following mechanisms:

$$M_n^{\dagger} + M_m \rightarrow M_{n+m} \tag{12.10}$$

$$M_n^{\dagger} + M_m^{\dagger} \rightarrow M_n + M_m \tag{12.11}$$

The transfer reactions can be considered cessation reactions in that they terminate the growth of a particular chain. However, in the process of terminating one chain, another is initiated; thus, the transfer reaction actually determines relative chain length.

Unsaturated bifunctional compounds differ greatly in their tendencies to polymerize. Simple olefinic hydrocarbons (low molecular weight, unsaturated hydrocarbons containing one double bond) react only in the presence of catalysts; ethylene, for example, polymerizes only if catalysts and pressure are used. Styrene, however, polymerizes while standing in its container, where spontaneous polymerization apparently results when the substituted side group polarizes and, thereby, activates the double bond.

Many catalysts have been developed that initiate addition polymerization. One group of catalysts is the so-called free-radical formers, such as organic or hydrogen peroxides. These catalysts activate the double bond to form highly reactive, intermediate free radicals[3] or carbonium ions. Each free radical, in turn, activates another monomer which attaches itself to the chain; the free radical then passes to the end of the chain to continue the propagation part of the reaction. Some catalyst residues will appear in the finished polymer. Such initiators

---

[3]Essentially a free radical is an organic ion in which a group of covalently bound atoms has a free bonding orbital which can overlap with another (such as in a double bond) to form a covalent bond.

are not *true* catalysts because they do not remain unchanged at the end of the reaction and the quantities used do affect the rate of the overall reaction.

Other catalysts, BF_3 and $AlCl_3$, for example, have incomplete outer electron shells. These compounds are highly electronegative and can polarize double bonds to form carbonium ions, which then enter into the propagation reaction. Ultraviolet light or nuclear radiation can also be used to initiate polymerization reactions [2].

POLYCONDENSATION OR CONDENSATION POLYMERIZATION   This reaction involves two or more molecules of different types and possibly different functional groups. In order for polycondensation reactions to occur, the monomers must possess at least two functional groups that will not lead to the formation of ring structures. In most polycondensation reactions, two different monomers are used to produce complicated polymer chains; for example, when dibasic acids react with dihydric alcohols, polyesters are produced.[4] In such instances, two molecules react with the splitting-out, or condensation, of a small molecule such as water or methyl alcohol—thus, the term condensation polymerization.

Condensation polymers may be thermoplastic or thermosetting, depending upon their monomers. If the monomers are exclusively bifunctional, the polymer will be linear and thermoplastic; if they are polyfunctional, thermosetting polymers will result. By varying the relative amounts of bi- and polyfunctional monomers, the properties of a polymer can also be varied over wide ranges. For example, approximately equal molal quantities of tri- and bifunctional monomers produce tight, three-dimensional networks characteristic of the thermosets. If only small amounts of the trifunctional monomer are used, a linear thermoplastic polymer will result.

Polycondensation may be illustrated by the following schematic reactions in which R and R' represent different functional groups:

$$-R-CO \boxed{OH + H} O-R'-COOH \rightarrow \qquad \textbf{(12.12)}$$
$$-R-COO-R'-COOH + H_2O$$

$$-R-COO-R'-CO \boxed{OH + H} O-R-COOH \rightarrow \qquad \textbf{(12.13)}$$
$$-R-COO-R'COO-R-COOH + H_2O$$

The product molecule reacts to extend the chain length in the same way as the monomer. The reaction that produces nylon (12.14) is a typical polycondensation reaction.

---

[4]In organic chemistry, esters correspond to organic salts—one of the products of the reaction between organic acids and alcohols. A dibasic acid contains two —COOH groups; a dihydric alcohol has two —OH groups.

$$NH_2—(CH_2)_6—NH_2 + HOOC—(CH_2)_4—COOH \rightarrow$$
(hexamethylene diamine)   (adipic acid)

$$H_2O + (NH_2)(CH_2)_6NHCO(CH_2)_4COOH \qquad \textbf{(12.14)}$$

The resultant molecule can react with either adipic acid or hexa-
methylenediamine to produce a linear polyamide which might have a
structure such as

$$[—OC(CH_2)_4CONH(CH_2)_6NHCO(CH_2)_4CONH(CH_2)_6NH—]_n$$

known as polyhexamethyleneadipamide or nylon 6, 6.

A condensation reaction producing a highly cross-linked thermosetting polymer is the two-stage reaction of phenol with formaldehyde.
In the first stage, formaldehyde reacts with phenol to form mono-,
di-, and/or trimethylol phenol:

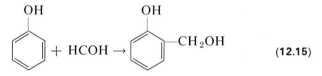

$$\qquad \textbf{(12.15)}$$

The relative amounts and types of methylol phenols formed depend
upon the ratio of phenol to formaldehyde, the catalyst used, the
temperature, and the reaction time. In the second stage, a water molecule is split-out when a reaction occurs between the methylol hydroxyl
group and the hydrogen of an adjacent benzene ring, or between two
methylol hydroxyl groups:

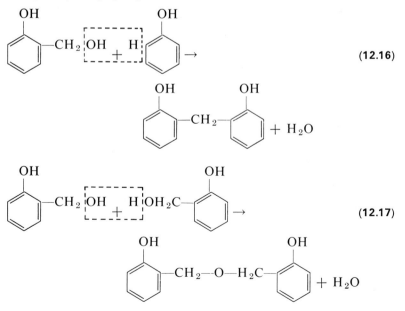

If alkaline catalysts and excess formaldehyde are used, the formation of the CH_2-O-CH_2 ether linkage predominates; smaller amounts of formaldehyde lead to the CH_2 methylene linkage.

In actual practice, the phenol-formaldehyde reaction initially reaches an only partially completed stage where it forms either a viscous resin called resol (if the structure contains ether linkages and has a low molecular weight) or resitol (a long, linear methylene-linked polymer with a slight amount of cross-linking). Resol is completely soluble in the alkaline reaction solution, while resitol is insoluble in the alkaline solution but readily soluble in organic solvents. Upon cooling, both resins become hard, brittle thermoplasts which are used separately or in combination for adhesives, castings, plastics, and laminates. When either resin is heated in the presence of curing agents, extensive cross-linking occurs which forms hard, rigid, infusible, insoluble thermosets having the structure

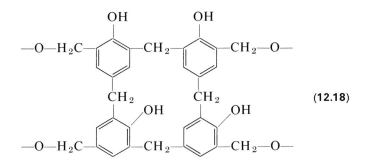

(12.18)

These phenol-formaldehyde condensation polymers are marketed under the trade name Bakelite.

COPOLYMERIZATION   In general, addition polymerization reactions occur between bifunctional monomer units of one kind. However, some addition polymerization reactions can occur between two different types of bifunctional monomers; such a reaction produces linear polymers, called *copolymers*, which have alternating types of mer units that are usually randomly distributed.

The varied ratios of monomers used in commercial copolymer manufacture account for the wide range of properties that copolymers exhibit. Thus, a polymer may be tailored to a specific property requirement, and some polymers may be fashioned from monomers that would not normally polymerize alone [2].

One example of copolymerization is the formation of GR-S rubber from styrene and butadiene by the following reaction:

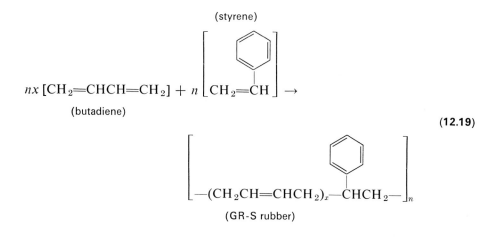

(styrene)

$$nx\,[CH_2{=}CHCH{=}CH_2] + n\,\left[CH_2{=}CH\right] \rightarrow$$

(butadiene)

(12.19)

$$\left[-(CH_2CH{=}CHCH_2)_x{-}CHCH_2{-}\right]_n$$

(GR-S rubber)

Like all elastomeric diene polymers, this linear polymer is unsaturated. Apparently, when the conjugated double bond system of the butadiene molecule is disrupted by activation, one double bond shifts to the middle of the molecule while the other enters into the polymerization reaction. The presence of the uninvolved double bond allows the rubber to react with other agents and establish cross-links by the vulcanization process.

Physical States of Polymers

POLYMER CRYSTALLIZATION   Polymerization reactions, which usually take place in the liquid phase, result in the formation of a large amorphous mass composed of molecules of varying length and sometimes varying composition. This random amorphous structure is also periodically interspersed with crystalline regions, producing a definite x-ray diffraction pattern [3], as described in Chapter 4.

The addition of *plasticizers*—nondrying vegetable oils, high boiling monomers, or low molecular weight polymers—greatly increases the flexibility of polymers by separating the larger polymer molecules and modifying their crystalline structures. Since they have low molecular weights, plasticizers may diffuse out of the polymer structure and evaporate or otherwise be lost; consequently, the material loses its plasticity and cracks with age.

ELASTOMERS   In the elastomeric (rubbery) state, the geometric arrangement of the polymer molecules permits large, reversible extensibility at room temperature. Elastomers are long, chain-like molecules which are irregularly coiled and entangled in the normal state. These coils are randomly arranged in an amorphous structure. Upon stretching, however, they partially disentangle and straighten out, orienting or crystallizing the chains, increasing interchain attraction, and stiffening the material. Thus, the introduction of cross-linking bonds greatly increases the strength of elastomeric materials.

As the temperature is lowered, polymers pass from the elastomeric condition to a glassy state, which is reached at a well-defined *glass transition temperature* characteristic of each polymer.

The glass transition of elastomers is often demonstrated in elementary physics classes by immersing a rubber ball in liquid air and reducing its temperature far below its normal glass transition temperature; the cold rubber ball shatters like glass when thrown on the floor or against a wall. Elastomeric polymer molecules are constantly in motion, with adjacent segments vibrating relative to one another. At the glass transition temperature, the molecular structure is frozen and this movement is restricted. Elastomers are polymers whose glass transition temperature is below room temperature. Since the strong intermolecular forces of crystalline polymers raise the glass transition temperature close to the softening point of the polymer, such materials are usually nonelastomeric.

VULCANIZATION   Vulcanization is a mechanism for increasing the degree of cross-linking and, therefore, the mechanical strength and hardness of unsaturated elastomers. Natural rubber is a thermoplastic elastomer that undergoes large, permanent deformation when stretched. Because the molecular chains of this polymer are only held together by van der Waals forces, the molecules are able to slide past one another when the chains are uncoiled and stretched out. Consequently, raw natural rubber is not a very useful engineering material.

In the process of vulcanization, discovered by Goodyear in 1839, rubber is heated for varying periods of time at temperatures ranging from $200-400°F$ in the presence of appropriate vulcanizing and accelerating agents. Accelerators reduce the time required for vulcanization to occur or permit the use of lower temperatures throughout the process. The most common vulcanizing agent is elemental sulfur, although sulfur-bearing materials or combinations of these with selenium or tellurium, quinoid structures, or dinitrobenzene are also used [4].

During vulcanization, the double bonds in the rubber molecules are activated, opened, and cross-linked through S–C bonds, as illustrated in (12.20) for polyisoprene (natural rubber).

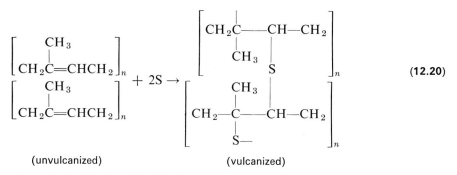

(unvulcanized) (vulcanized)

As each double bond in this isoprene unit is activated, two sulfur linkages can be formed with other polyisoprene molecules, introducing two sulfur atoms into each double bond. However, the other ends of each pair of these sulfur linkages are attached to carbon atoms in two other isoprene groups, so that only one net sulfur atom is required to completely vulcanize one double bond in the isoprene unit. This observation provides the basis for calculating the amount of sulfur necessary to vulcanize the raw rubber.

EXAMPLE 12.1

Compute the quantity of sulfur required to fully vulcanize 100 lb of raw polyisoprene and the percentage of sulfur in the final product.

Solution:

The mer weight of polyisoprene is

$$m = 5(12.01) + 8(1.008) = 68.114$$

Since 1 sulfur atom is required per double bond, we see that

$$\text{lbs S} = \frac{(100 \text{ lb isoprene})(1 \text{ mole S/mer})(32.064 \text{ lb S/mole S})}{(68.114 \text{ lb isoprene/mer})}$$

$$= 47.0 \text{ lb}$$

$$\% \text{ S} = \frac{32}{32 + 68} = 32\% \text{ S by weight}$$

In most practical applications, it is not desirable to vulcanize 100 per cent of the double bonds in the rubber. Only a certain fraction of the possible cross-linkages are formed, because the degree of vulcanization markedly affects the physical properties of the rubber. For example, the rubber ordinarily used in automobile tires is only four to six per cent vulcanized.

Vulcanization reduces the plasticity of rubber while maintaining its elasticity by the introduction of cross-linkages. The end product exhibits greater strength, less surface tackiness, and a marked increase in resistance to swelling by solvents.

Many commercially produced rubbers are vulcanized in the same manner as natural rubber, although sulfur is not always employed as the vulcanizing agent. For example, chloroprene polymers are vulcanized with MgO, where the cross-link reaction takes place between the MgO and two chlorine atoms.

One of the early steps in rubber processing is *mastication* [4] or physical breakdown. In this process, the raw rubber mass is worked between rolls that are designed to knead and mix the material until it becomes partially depolymerized and breaks down. Although such mastication produces the same effect as the addition of plasticizers, it is more advantageous because increased plasticity can be removed readily by vulcanization.

ORIENTED STRUCTURES   Since a polymer possesses different properties in its crystalline form than in its amorphous form, the properties of polymeric fibers must be considerably influenced by molecular orientation. Fibers are composed of long, linear molecules oriented more or less parallel to the long axis of the fiber; the method of *extrusion* (Fig. 12-3), where the random molecular arrangement of the melt is lost when the material is forced through a die and cast out in a long filament, is one common way to orient these fiber molecules. Because fiber-forming polymers are thermoplasts that have a high tendency to

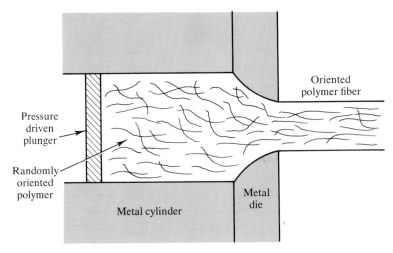

FIG. 12-3   Schematic representation of the molecular orientation process in polymer fiber extrusion.

crystallize, their molecular orientation is enhanced by the presence of strongly attracting side groups attached at regular intervals along a polymer chain. The flexibility of this type of fiber structure is due to the presence of amorphous regions interspersed throughout the oriented regions of the fiber, since these materials produce fibers with high mechanical strengths and softening points. The orientation resulting from fiber extrusion imparts considerable strength in the axial direction of the fiber, leaving it very weak in the transverse direction.

This decreased mechanical strength is unimportant because fibers are never loaded in this direction; in the formation of plastic films, however, such a decrease in transverse mechanical strength is very important. Two-way orientations can be introduced into the structure by the simultaneous drawing of the product in two directions. This is generally accomplished by blowing or inflated-tube extrusion, or by the use of sequential sets of rolls rotating at differing rates [5]. In either case, this simultaneous, "two-way" stretch orients all of the molecules in the plane of the film. Mylar film, which is amorphous, clear, and fairly strong upon extrusion, is oriented planarly to form a film equally strong in all directions in its own plane. When cast into fibers, this same polymer is known as Dacron.

CHEMICAL RESISTANCE   The chemical resistance of a polymeric material is dependent upon both its chemical nature and its molecular arrangement. The effect of chemical attack upon high molecular weight polymers is usually internal in nature and consists of softening, swelling, and loss of strength. Vulcanization of rubber increases its resistance to solvent swelling by introducing cross-links into the structure.

A chemical similarity between polymer and solvent increases the reactivity of the polymer, whereas a dissimilarity increases its chemical stability. Thus, polyvinyl alcohol is more readily soluble in water or polar solvents than polyethylene, which is susceptible to attack by gasoline and paraffinic liquids.

Increasing the molecular weight of a polymer decreases its solubility in a given solvent. High molecular weight polymers therefore produce more viscous solutions than those with low molecular weights. The crystalline form of the polymer is less soluble than the amorphous form of a given molecular weight polymer.

Because the effect of chain branching is quite similar to that of moderate cross-linking in decreasing the solubility of a given molecule, it is sometimes difficult to distinguish between branched-chain and slightly cross-linked polymers in terms of their resistance to solvent attack. Structural symmetry and bond strength are important factors in determining such polymer resistance to attack by corrosive chemicals. A good example of both factors can be seen when the structure and chemical activity of polytetrafluoroethylene, or Teflon, is examined. This material, composed entirely of carbon and fluorine, has a molecular structure exactly like polyethylene. Because of the symmetry of its molecular structure and the strength of its carbon-to-fluorine bonds, this polymer is extremely inactive. (It is only subject to attack by elemental fluorine or sodium and is completely inert to all other corrosive chemicals.) The high molecular symmetry produces close molecular packing, so that the polymer is also extremely resistant to solvents.

Solutions and Molecular Weights of Polymers

SOLVATION AND SOLUBILITY   Highly viscous solutions result when high molecular weight polymer molecules dissolve. Because they are extremely large, polymer molecules exhibit a characteristic behavior in solution that differs markedly from the normal behavior of small, low molecular weight molecules in solution.

One factor that determines the properties of polymer solutions is the relationship between the sizes of polymer and solvent molecules. A solvent molecule is ordinarily only a small fraction (say, 1/100–1/2000) of the total size of a typical polymer molecule. Therefore, it is necessary to consider the interaction of solvent molecules with parts of the polymer molecule, rather than with the entire molecule at once. In order for a solvent to dissolve a polymer mass, the individual solvent molecules must interpenetrate the polymer molecules and separate them by overcoming the attractive intermolecular forces. In thermoplastic polymers, these binding forces are van der Waals, and in some cases hydrogen bridge, forces. Since the magnitude of intermolecular forces increases with increasing molecular size, the solubility of the polymer should decrease with increasing molecular weight. In large measure, molecular crystallinity determines the fraction of the total number of possible secondary bonds that actually form.

Two polymers with similar crystallinities and the same type of attractive intermolecular forces can have different solubilities in a given solvent if the rigidity of their chains differs. A solvent dissolves a flexible polymer by breaking each intermolecular secondary bond in sequence. Completely rigid molecules cannot be as easily pried loose by a solvent; instead, all of the intermolecular bonds must be broken simultaneously, making the material less soluble.

Varying solubilities due to different degrees of packing and molecular rigidity can be illustrated with polyvinyl alcohol, starch, and cellulose. The intermolecular attractive forces in polyvinyl alcohol and cellulose are comparable, and both will exhibit the same degree of crystallinity if properly treated. Polyvinyl alcohol molecules are flexible, and the polymer is readily water soluble; however, the molecules of cellulose are rigid and, therefore, insoluble in water. By contrast, starch and cellulose molecules have comparable rigidities, but exhibit marked differences in both molecular packing and shape; thus, starch is also more soluble than cellulose.

The natures and magnitudes of intermolecular bonding forces markedly affect the solvent behavior of polymers. For example, even though a particular polymer may exhibit a relatively high value of equilibrium solubility, a very long time is frequently required to bring such a polymer into solution. The solvent molecules must interpenetrate the solid polymer and remove individual molecules from the parent mass. If the solid has relatively high degrees of crystallinity and intermolecular bonding, the interpenetration of these solvent molecules may be hindered, and considerable time may be needed to disperse the solid. When placed in benzene or another similar organic solvent, natural rubber and polystyrene actually swell before solvating.

Swelling, which usually precedes a slow, spatial dispersion of the polymer, occurs when the solvent molecules interpenetrate the polymer molecules and expand the structure before the individual polymer molecules can be removed and dissolved. The swollen solute particles become tacky and usually agglomerate, sticking to the container walls and further increasing the solvation time. In some cases, dissolution is never completed and the polymer swells until it reaches an equilibrium volume. Swelling is eventually limited by cross-links; that is, the network structure will expand to a point where elastic forces in the cross-links counterbalance the swelling forces of the solvent molecules. Swelling may be very large in some moderately cross-linked structures, while only slight swelling may occur in very highly cross-linked structures.

The properties of high molecular weight polymer solutions are considerably different from those of low molecular weight solutions, basically due to the shape and size of the polymer molecules. Consider a long chain of $-CH_2$ groups such as occurs in the polyethylene molecule. If the rotation of the groups about the C–C bonds were free and unrestricted, this long chain molecule would be completely limp and flexible. If such a molecule were then isolated in a neutral solvent medium, a broad range of molecular shapes with the same potential energy would exist. Because these various molecular segments would be in constant motion due to thermal vibrations, there would not be any fixed or static shape to the molecule at any time; instead there would be some "effective" shape to the molecule at any given instant. In dilute solutions, the effective shape assumed by a long, linear chain polymer molecule has been found to be a *random coil* or a very highly kinked and twisted chain, wound up into an approximately spherical envelope with a volume many times the actual volume of the polymer molecule.

The random coil model of the polymer molecule in solution is based on the postulate of the free rotation of mer units about the C–C bond in its chain. In many molecules, however, such free rotation is not possible because of the steric interference of multiple bonds or side groups. As the degree of stiffness of the chain increases, the amount of molecular coiling decreases, thus distorting the spherical shape of the envelope into an ellipsoid. The limiting case of this distortion occurs once the bond angles are all rigidly fixed and the molecule becomes completely stiff and rodlike. Cellulose molecules approximate this behavior in contrast to rubber molecules, which are extremely flexible and coiled.

The actual shape assumed by a polymer molecule in solution is also affected by the type of solvent and the temperature of the solution. If the solvent is inefficient for the particular polymer, the effective shape of a flexible chain is more compact and coiled than it would be in an efficient solvent. In such cases, increasing temperature causes the molecule to uncoil and elongate; when the shape of the molecule is more extended in a good solvent, the effect of temperature on its properties decreases.

The effective volume of the solvated molecule depends upon the quantity of the solvent immobilized by the individual polymer molecules. When the chain is completely extended, mutual attraction between the polar groups of the solute and solvent results in a solvent envelope that completely surrounds the polymer chain. Such a molecule is said to be completely solvated. Solvent molecules must follow the motion of the polymer molecules to which they are now attached; thus, the motion of a part of the solvent will be affected by the polymer molecules. Similarly, solvent molecules trapped in the interior of a coil, whether they have been solvated or not, must move with the polymer molecules. In addition, if the molecule is highly coiled, motion of the solvent molecules into and out of the effective volume will undoubtedly be restricted to a degree, although some diffusion of this type must occur. When the solution is induced to flow, this condition is particularly important because the polymer carries large quantities of the solvent along with it, increasing the viscosity much more than can be predicted by simply considering the absolute volume of the polymer chains. These effects are reflected primarily in an increase in the viscosity of the polymer solutions relative to that of the solvent medium, as well as in non-Newtonian viscosity. In large measure, the alteration of entrapped fluid volumes and the shapes and sizes of molecular coils by shear forces are responsible for the variable viscosities of most polymer solutions.

Viscosities of Macromolecular Solutions

Some dilute polymer solutions exhibit ordinary Newtonian viscosities that are many times greater than those observed for similar concentrations of simple, low molecular weight organic molecules.

The increase in polymer solution viscosity when compared to that of the pure solvent is often expressed as a dimensionless *specific viscosity*, which is defined as

$$\mu_{sp} = \frac{\mu - \mu_s}{\mu_s} = \mu_r - 1 \qquad (12.21)$$

where μ is the solution viscosity, μ_s is the viscosity of the pure solvent, and μ_r is the *relative viscosity*.

If a polymer solution is considered to be a suspension of rigid spheres that are perfectly wetted by a solvent, the Einstein equation [6] introduced earlier (Eq. (8.12)) can be used in the form

$$\mu_{sp} = 2.5\phi \qquad (12.22)$$

where ϕ is the volume fraction of the spheres. Equation (12.22) indicates that specific viscosity is independent of the properties of the suspended material; consequently, the amount by which μ_r exceeds unity is due to the disturbance of the basic flow field generated by the spheres. In Chapter 8, we saw that increasing particle concentrations led to greatly increased viscosities and non-Newtonian effects. Thus, physical chemists study the influences of polymer concentrations on μ_{sp} for dilute polymer solutions in order to interpret complex polymer molecular structures and interactions.

For many dilute, high polymer solutions, the effect of polymer concentration (expressed in terms of mass concentrations) can be correlated by a simple empirical relationship of the form

$$\mu_{sp} = bc + mc^2 \qquad (12.23)$$

where b and m are constants. This equation suggests that a plot of μ_{sp}/c vs c should be linear (see Fig. 12-4). The constant b, which is the intercept at zero concentration, has been named [7] the *intrinsic viscosity of the solution* and given the symbol $[\mu]$. If we let $\phi = 0.01cv$ (where v is the partial specific volume of the solute), Eq. (12.22) becomes

$$\mu_{sp} = 0.025 \, cv \qquad (12.24)$$

From this equation it is evident that Einstein's theory predicts an intrinsic viscosity of a dilute suspension of spheres to be $[\mu] = 0.025$,

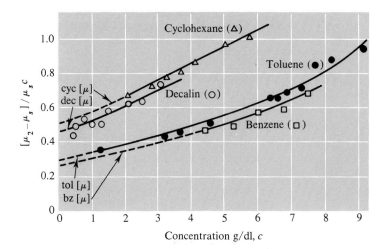

FIG. 12-4   The variation of $(\mu_\infty - \mu_s)/\mu_s c$ as a function of c (polyisobutylene concentration) in several solvents at 30°C. μ_∞ is the upper limiting Newtonian viscosity of pseudoplastic isobutylene solutions at very high rates of shearing strain. (From E. W. Merrill, *et al., J. Polymer Sci., Part A,* **1,** p. 1201 (1963). Used by permission.)

since the magnitudes of the partial specific volumes are approximately unity. Relatively few high molecular weight solutes—albumins are one example—actually possess intrinsic viscosities in this range; most high molecular weight polymers are several orders larger in magnitude.

Flory [8] suggested that $[\mu]$ is proportional to M^α, when M is the molecular weight and α is a constant varying from 0.5 to 1.0 which must be determined experimentally for a given polymer in a given solvent. The relation of the intrinsic viscosity to the molecular weight of a polymer is the basis of one of the experimental methods used in determining polymer molecular weight distributions.

Molecular Weight Distributions

Distribution Curves

Various polymerization mechanisms result in molecular chains of varying length; thus, any real polymer will exhibit some statistical distribution in molecular size and weight. It is therefore necessary to devise a method of measuring and correlating molecular weight distributions and then to define meaningful and useful averages which can be derived experimentally from such data.

Numerous methods are used to determine molecular weight distribution curves for polymers. One of these [9], *fractional precipitation*, involves adding successive portions of a liquid that will not dissolve the polymer to a polymer solution, causing various fractions of the polymer to precipitate from solution. The higher molecular weight species precipitate first, followed by successively lower weight species. In this way, it is possible to obtain a series of fractions of the original polymer solution, each having a different range of molecular weights. These fractions are then subjected to viscosity tests to determine a molecular weight from Flory's relation [8], and a distribution is obtained as shown schematically in Fig. 12-5. If the center points of the distribution bars are connected by a curve as illustrated in the figure, an approximation of the distribution curve results. An even better approximation could be obtained by using smaller fractions.

The width of the distribution peak in Fig. 12-5 is quite broad, indicating a large spread in molecular size. (This is usually found to be true in most high polymers.) Thus, the curve illustrates that the average degree of polymerization, however it is defined, does not give a complete and true picture of the polymer. Relative amounts of large and small chains will have a significant effect on the properties and behavior of the polymer.

Whenever a wide range of molecular sizes exists, the different

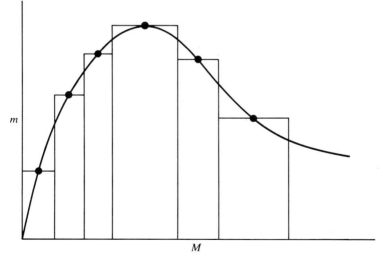

FIG. 12-5   Schematic molecular weight distribution for a polymer. *M* is the molecular weight range of the fraction, and *m* is the mass of the fraction having molecular weight *M*.

methods of determining molecular weights will yield different results, since each is based on a different method of averaging the distribution. For example, the osmotic-pressure, end-group, and freezing point techniques determine averages based solely on the *number* of molecules present in the solution; viscosity, diffusion, and light-scattering measurements, however, determine averages influenced by molecular size, shape, and flexibility, as well as the actual number of molecules present. In order to illustrate this, various types of average molecular weights as calculated from molecular size distributions will now be described.

The very commonly used *weight-average* molecular weight $\overline{M}_w$ is defined by the relationship

$$\overline{M}_w = \frac{\sum_i m_i M_i}{\sum_j m_j} = \sum_i w_i M_i \qquad (12.25)$$

where the m_i are the respective masses of individual fractions and the w_i are corresponding weight fractions. (The M_i are the center point values of the particular molecular weight range, indicated by the large dots in Fig. 12-5.) This averaging technique is significant in the analysis of properties, such as viscosity, where the weight or size of the molecules is important.

EXAMPLE 12.2

Using the data in the accompanying table, compute $\overline{M}_w$ and DP for this sample of polyvinyl acetate.

RANGE OF MOLECULAR WEIGHTS	M_i	w_i	$w_i M_i$
5,000–10,000	7,500	0.12	900
10,000–15,000	12,500	0.18	2,250
15,000–20,000	17,500	0.26	4,550
20,000–25,000	22,500	0.21	4,725
25,000–30,000	27,500	0.14	3,850
30,000–35,000	32,500	0.09	2,925
		$\sum_i = 1.00$	19,200

Solution:

Clearly, $\overline{M}_w = 19,200$ g/mole. The polyvinyl acetate mer has the nominal composition $C_4H_6O_2$, giving a mer weight of 86; thus, $DP = 19,200/86 = 224$.

Because some experimental techniques are more sensitive to the number of polymer molecules present than others, it is necessary to define a *number-average* molecular weight as

$$\overline{M}_n = \frac{\sum_i n_i M_i}{\sum_j n_j} = \sum_i N_i M_i \qquad (12.26)$$

where $n_i = w_i N_0 / M_i$, the number of molecules in the particular molecular weight fractions; $N_i = n_i / \sum n_j$, the corresponding number fractions; w_i is the weight fraction, as before; and N_0 is Avogadro's number. When the value of $\overline{M}_n$ is calculated for the sample of polyvinyl acetate in Example 12.2, $\overline{M}_n = 16{,}010$ g/mole is obtained. This illustrates the fact that whenever there is a size distribution, the number-average is less than the weight-average molecular weight due to the greater number of small molecules in the smaller weight fractions. If all the molecules were identical in size, the two average molecular weights would also be identical; such a polymer is considered *monodisperse* [10]. Most commercial polymers formed by condensation and addition have $\overline{M}_w / \overline{M}_n$ values of about 1.5–2.0, although some that are made with stereospecific catalysts may range as high as 50. Large values of this ratio indicate a large molecular weight range.

Molecular Weight Determinations

Molecular weight distributions can be determined from measurements of such phenomena as viscosity, osmotic pressure, sedimentation and diffusion, and light scattering.

VISCOSITY METHODS   One of the most widely used methods of polymer molecular weight determination involves viscosity measurements of the polymer in varying solutions. Since such determinations can be made easily and rapidly, they are a good production control technique in industrial applications.

As determined from these data, intrinsic viscosity is a function of molecular weight. Flory [8] found that viscosity data could best be correlated with molecular weights (obtained by an osmotic pressure method) by the expression

$$[\mu] = KM^{\alpha} \tag{12.27}$$

where K and α are constants characteristic of a given polymer. This relationship is now generally used to determine molecular weights from viscosity data. As indicated in Table 12-4 for different polymer-solvent combinations, the exponent α varies [11] from 0.5 to 1.0. The two parameters K and α must be obtained by absolute calibration, employing a method other than viscometry in determining molecular weights. When the value of α in Eq. (12.27) is unity, the weight-average molecular weight is obtained. For other values of the exponent, a viscosity weighted average is obtained which lies intermediate to the weight- and number-average molecular weights.

TABLE **12-4**

Dependence of Intrinsic Viscosity on Molecular Weight for Various Polymer-Solvent Systems*

POLYMER	SOLVENT	$T(°C)$	MOLECULAR WEIGHT RANGE $(\times 10^{-3})$	$10^4 K$	α
Polystyrene	Benzene	25	32–1300	1.03	0.74
Polystyrene	MEK (Methyl-ethyl ketone)	25	2.5–1700	3.9	0.58
Polyisobutylene	Diisobutylene	20	6–1300	3.6	0.64
Polyisobutylene	Cyclohexane	30	6–3150	2.6	0.70
Polyisobutylene	Benzene	24	1–3150	8.3	0.50
Polyisobutylene	Benzene	60	6–1300	2.6	0.66
Natural rubber	Toluene	25	40–1500	5.0	0.67
Polymethylmethacrylate	Benzene	20	77–7440	0.84	0.73
Polymethylmethacrylate	Benzene	25	—	0.57	0.76
Polyvinyl acetate	Acetone	25	65–1500	1.76	0.68
Poly (ϵ-caproamide)	Sulfuric acid	25	4–37	2.9	0.78

*From *Principles of Polymer Chemistry,* by P. J. Flory, copyright 1953, Cornell University Press. (Used by permission of Cornell University Press.)

OSMOTIC PRESSURE METHODS   When solutions with differing concentrations of a given solute are separated by a porous membrane through which the smaller solvent molecules can pass but the larger solute molecules cannot, a differential pressure develops across the membrane. This effect is produced by the concentration difference which causes the solvent molecules to diffuse from the dilute to the concentrated solution, increasing the pressure in the concentrated solution. The pressure difference so generated is the *osmotic pressure.*

Osmotic pressures are measured in dilute polymer solutions [12]. If such a solution is ideal, the relationship between the osmotic pressure, the molecular weight, and the concentration of the solution is expressed in terms of the van't Hoff law:

$$P_o = c \left[\frac{RT}{M} \right] \qquad (12.28)$$

in which P_o is the osmotic pressure, R is the universal gas constant, T is the absolute temperature, M is the molecular weight of the solute, and c is the mass concentration of the solute. This equation is valid for low molecular weight substances in concentrations on the order of a few per cent; it is used to compute M from measurements of P_o when temperatures and concentrations are known. Because solutions of high molecular weight polymers (even in very low concentrations) deviate significantly from the simple behavior of an ideal solution, unfortunately the van't Hoff law is not applicable. It has been found empirically that Eq. (12.28) must be corrected to the form

$$P_o = c\left[\frac{RT}{M}\right] + Bc^2 \qquad (12.29)$$

where B is an empirical constant that depends on the nature of the polymer-solvent system and expresses the interaction between the molecules of solvent and the segments of dissolved polymer. The two parameters M and B are necessary to characterize the osmotic behavior of a polymer solution, even in very dilute solutions.

The usual method of determining these two parameters is by measuring the values of P_o at a given temperature for a number of different concentrations below roughly 1.0 per cent and then plotting P_o/c vs c. This plot should be linear (see Eq. (12.29)), with intercept RT/M and slope B. In actuality, however, such plots are not always linear. Although data for relatively low molecular weight polymers do produce quite linear plots, very high molecular weight polymers characteristically exhibit nonlinear plots, indicating that the simple quadratic additive correction to the van't Hoff law does not fully account for the nonideality of polymer solutions.

Linear plot intercepts are usually independent of the nature of the solvent and directly proportional to the absolute temperature, as indicated in Eq. (12.29). The osmotic pressure method of determining polymer molecular weights is considered reliable in the 50,000–500,000 weight range; excessive errors begin when particularly large molecular weights are introduced.

In addition to those described above, many other techniques are used in measuring the molecular weights of polymers. The interested reader can find descriptions of such methods in the literature of polymer chemistry.

References Cited

[1] Pauling, L., *The Nature of the Chemical Bond*, 3rd ed., Cornell University Press, Ithaca, New York, 1960, Chapter 6.

[2] Winding, C. C., and G. D. Hiatt, *Polymeric Materials*, McGraw-Hill Book Co., Inc., New York, 1961, Chapter 1.

[3] Geil, P. H., *Chem. & Engr. News*, **43**, pp. 72–84 (Aug. 16, 1965).

[4] Winding and Hiatt, *op. cit.*, Chapter 10.

[5] *Ibid.*, Chapter 4.

[6] Einstein, A., *Ann. Physik.*, **19**, p. 289 (1906); **34**, p. 591 (1911).

[7]   Kraemer, E. O., *Ind. Engr. Chem.*, **30**, p. 1200 (1938).

[8]   Flory, P. J., *J. Am. Chem. Soc.*, **65**, p. 372 (1943).

[9]   Tamblyn, J. W., D. R. Morey, and R. H. Wagner, *Ind. Engr. Chem.*, **38**, p. 473 (1945).

[10]   DiBenedetto, A. T., *The Structure and Properties of Materials*, McGraw-Hill Book Co., Inc., New York, 1967, p. 241.

[11]   Flory, P. J., *Principles of Polymer Chemistry*, Cornell University Press, Ithaca, New York, 1953.

[12]   Doty, P. M., and H. Mark, *Ind. Engr. Chem.*, **38**, p. 682 (1946).

PROBLEMS

12.1   Predict the melting points of hexadecane ($C_{16}H_{34}$), tricosane ($C_{23}H_{48}$), and octacosane ($C_{28}H_{58}$), and compare your predictions with observed values taken from the literature. What conclusions can you draw from these data?

12.2   Two samples of polyethylene are subjected to precise melting point determinations, producing the following data:

SAMPLE	MEAN MELTING POINT ($°C$)
A	142.2
B	142.9
C	143.5

From the above, determine the average degree of polymerization for each sample. If the mean melting points listed are accurate to ± 5.0 degrees, determine the molecular size range in each sample. Is this a very precise method of determining *DP*?

12.3   The energy of a C=C bond is approximately 146 kcal/mole, while that of the C–C bond is 83 kcal/mole. Compute the net energy release per mole of ethylene when it is polymerized.

12.4   A sample of polytetrafluoroethylene (mer = C_2F_4) has an average molecular weight of 65,000. Determine its average degree of polymerization.

12.5   A copolymer of butadiene and acrylonitrile (acrylonitrile or GR-A rubber) has a general structural formula $[—(CH_2CH=CHCH_2)_x\, CH(CN)CH_2—)]_n$, where x is the number of butadiene units per mer unit and (CN) is the nitrile side group of the acrylonitrile unit. A sample of this rubber was chemically analyzed and found to contain 3.58% nitrogen by weight. A different analysis revealed its mean molecular weight to be $3.02(10^5)$ g/mole. Determine its average degree of polymerization from these data.

12.6   When fully vulcanized, a sample of butyl rubber $[—(CH_2—C(CH_3)_2—)_x\text{-}CH_2C(CH_3)=CHCH_2—]_n$ is found to contain 4.85% sulfur by weight. How much sulfur is required to vulcanize 20% of the double bonds in five tons of this butyl rubber?

12.7   If the molecular weight of the butyl rubber in Problem 12.6 is 5.5(10^5) g/mole, determine its average degree of polymerization.

12.8   How much sulfur is required to vulcanize 10% of the available double bonds in one ton of the nitrile rubber in Problem 12.5?

12.9   A certain butadiene-styrene copolymer rubber has a molecular weight of 355,300 g/mole and a *DP* of 950. Its structural formula is [—$(CH_2CH$=$CHCH_2)_x$—$CH(C_6H_5)$-$CH_2]_n$, where (C_6H_5) is the benzene ring side group of the styrene unit. Calculate the amount of sulfur required to vulcanize 12% of the available double bonds in 2.5 tons of this rubber.

12.10   What is the percentage of sulfur in the final product of Problem 12.9?

APPENDIXES

APPENDIX I

Selected Physical Constants

CONSTANT	MAGNITUDE	UNITS
Avogadro's number	$6.023(10^{23})$	molecules/g-mole
Boltzmann's constant	$1.380(10^{-23})$	joules/° K
	$1.380(10^{-16})$	ergs/° *C K molecule*
Density of water	62.4	lb_m/ft^3
	8.34	lb_m/gal
Electronic charge	$1.602(10^{-19})$	coulombs
	$4.803(10^{-10})$	esu
	$4.803(10^{-10})$	$(erg \cdot cm)^{1/2}$
Electronic mass	$9.106(10^{-28})$	g
Faraday's constant	96,493	coul/equivalent
	23,062	cal/volt-equivalent
Gas constant	1.987	cal/g-mole ° C
	1.987	Btu/lb mole ° F
	82.057	cm^3-atm/g-mole ° C
	0.082057	liter-atm/g-mole ° C
	8.3144	joules/g-mole ° K
Gravitational acceleration	980.665	cm/sec^2
Planck's constant	$6.624(10^{-27})$	erg sec
Proton mass	$1.6723(10^{-24})$	g
Speed of light in vacuum	$2.998(10^{10})$	cm/sec
Standard volume of gas at 0° C and 760 mm Hg	22.4	liter/g mole

Selected Mathematical Constants

CONSTANT	MAGNITUDE
$\sqrt{2}$	1.414 ...
$\sqrt{3}$	1.732 ...
cos 30°	0.866 ...
cos 45°	0.707 ...
cos 60°	0.5
π	3.1416 ...
e	2.718 ...
ln 10	2.30258 ...

APPENDIX II

Selected Conversion Factors

AREA

$1 \text{ m}^2 = 10^4 \text{ cm}^2 = 1550 \text{ in.}^2 = 10.76 \text{ ft}^2$

ENERGY

$1 \text{ joule} = 10^7 \text{ ergs} = 0.2389 \text{ cal}$
$1 \text{ kcal} = 10^3 \text{ cal} = 4186 \text{ joules} = 4.186 \, (10^{10}) \text{ erg}$
$1 \text{ Btu} = 778.2 \text{ ft lb}_f = 252 \text{ cal}$
$1 \text{ eV} = 1.602 \, (10^{-19}) \text{ joules} = 1.602 \, (10^{-12}) \text{ erg} = 0.386 \, (10^{-19}) \text{ cal}$
$1 \text{ kw-hr} = 3413 \text{ Btu}$

FORCE

$1 \text{ lb}_f = 4.448 \, (10^5) \text{ dynes} = 32.174 \text{ ft lb}_m/\text{sec}^2 = 32.174 \text{ poundals}$

LENGTH

$1 \text{ ft} = 12 \text{ in.} = 30.48 \text{ cm}$
$1 \text{ in.} = 2.54 \text{ cm} = 25.4 \text{ mm}$
$1 \text{ Å} = 10^{-8} \text{ cm}$
$1 \text{ micron } (\mu) = 10^{-4} \text{ cm}$
$1 \text{ m} = 39.36 \text{ in.}$
$1 \text{ mi} = 5280 \text{ ft}$

MASS

$1 \text{ lb}_m = 453.6 \text{ g} = 16 \text{ oz (Avoirdupois)}$
$1 \text{ kg} = 1000 \text{ g} = 2.2046 \text{ lb}_m$

POWER

$1 \text{ watt} = 1 \text{ joule/sec} = 3.413 \text{ Btu/hr} = 1 \text{ volt} \cdot \text{amp} = 1 \text{ amp}^2 \text{ ohm}$
$1 \text{ horsepower (hp)} = 550 \text{ ft lb}_f/\text{sec} = 746 \text{ watts}$
$1 \text{ kilowatt (kw)} = 1000 \text{ watts} = 3413 \text{ Btu/hr}$

PRESSURE

1 atm $\quad= 14.696$ psi $= 760$ mm Hg
1 dyne/cm$^2 = 1.4504\ (10^{-5})$ psi
1 atm $\quad\quad= 1.0133\ (10^6)$ dynes/cm^2

TEMPERATURE

1°C (difference) $= 1.8$ °F (difference)
°C $\quad\quad\quad\quad= (°F - 32)/1.8$
°F $\quad\quad\quad\quad= 1.8\ (°C) + 32$
°K $\quad\quad\quad\quad= °C + 273.16$
°R $\quad\quad\quad\quad= °F + 459.69 = 1.8$ °K

THERMAL CONDUCTIVITY

1 cal/sec cm °C $= 242$ Btu/hr ft °F
1 watt/cm °C $\quad= 57.8$ Btu/hr ft °F

VISCOSITY

1 poise $= 1$ g/cm sec $= 100$ cp
1 cp $\quad= 0.01$ poise $= 6.72\ (10^{-4})$ lb$_m$/ft sec $= 2.42$ lb$_m$/hr ft

VOLUME

1 m^3 $\quad= 10^6$ cm^3 $= 1000$ liters
1 ft^3 $\quad= 7.481$ gal $= 28,317$ cm^3
1 cm^3 $= 3.532\ (10^{-5})$ ft^3

APPENDIX III

Selected Equilibrium Phase Diagrams*

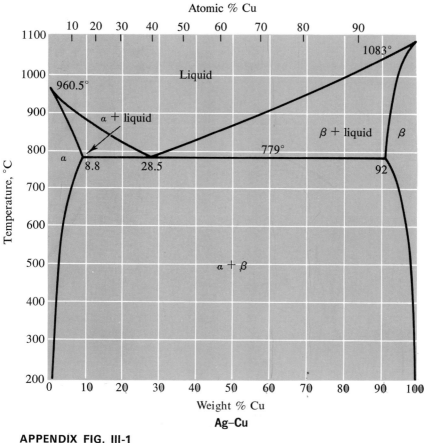

APPENDIX FIG. III-1

*The figures in Appendix III are taken from Phillips, H. W. L., *Metals Reference Book*, *4th ed.*, Volume II, ed. by C. J. Smithells, copyright 1967, Butterworth & Company, Ltd., London. Used by permission.

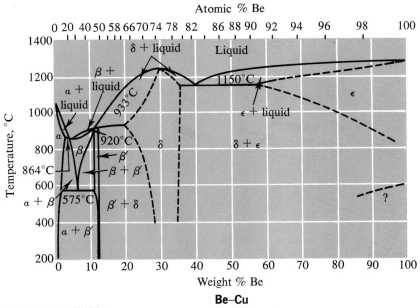

APPENDIX FIG. III-2(a)

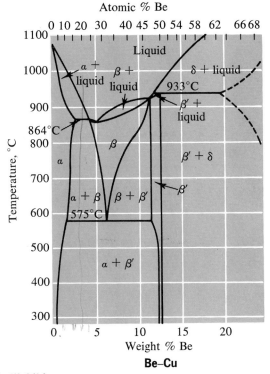

APPENDIX FIG. III-2(b)

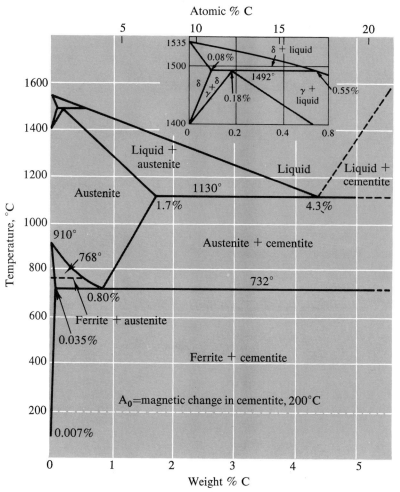

APPENDIX FIG. III-3

$$\alpha = \times \left(\frac{6.7 - 1}{6.7 - .007} \right) \qquad \text{eutectoid solid} = \times \left(\frac{6.7 - 1}{6.7 - .8} \right)$$

$$\text{cement.} = \times \left(\frac{1 - .007}{6.7 - .007} \right) \qquad \text{prim. cementite} = \times \left(\frac{1 - .8}{6.7 - .8} \right)$$

$$\text{eutectoid } \alpha = \text{eutec} \left(\frac{6.7 - .8}{6.7 - .007} \right)$$

$$\text{"} \quad \text{cement.} = \text{eutec} \left(\frac{.8 - .007}{6.7 - .007} \right)$$

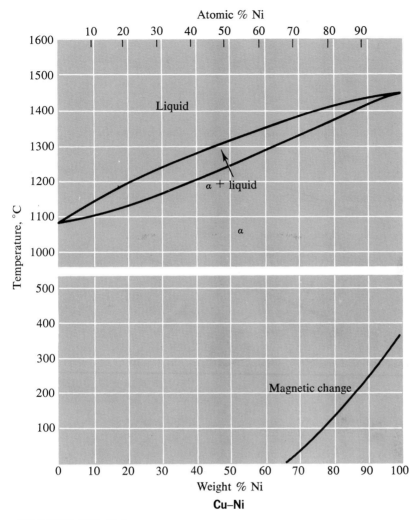

APPENDIX FIG. III-4

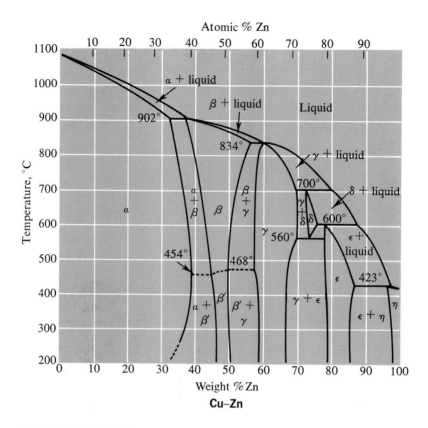

APPENDIX FIG. III-5

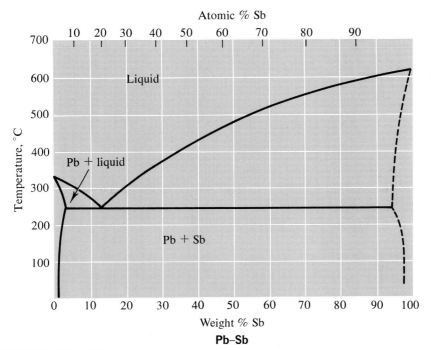

Pb–Sb

APPENDIX FIG. III-6

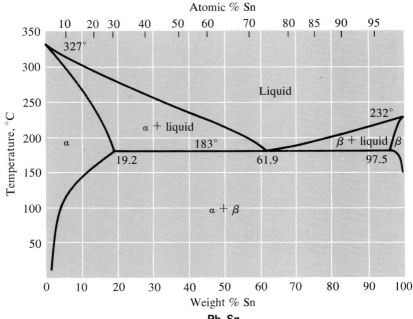

Pb–Sn

APPENDIX FIG. III-7

APPENDIX IV

Selected Properties of the Elements

ELEMENT	SYMBOL	ATOMIC NUMBER	ATOMIC WEIGHT*	MELTING POINT, °C	DENSITY OF SOLID G/CC	CRYSTAL STRUCTURE OF SOLID
Actinium	Ac	89	(227)†	1050		FCC
Aluminum	Al	13	26.982	660	2.70	FCC
Americium	Am	95	(243)		11.7	Hexag
Antimony	Sb	51	121.75	630.5	6.68	Rhomb
Argon	Ar	18	39.948	−189.4	1.67	FCC
Arsenic	As	33	74.922	817ª	5.72	Rhomb‡
Astatine	At	85	(210)	~300		
Barium	Ba	56	137.34	714	3.5	BCC‡
Berkelium	Bk	97	(247)			
Beryllium	Be	4	9.012	1277	1.85	HCP‡
Bismuth	Bi	83	208.980	271	9.80	Rhomb
Boron	B	5	10.811	2030	2.34	Tetrag‡
Bromine	Br	35	79.909	−7.2		Orthorho
Cadmium	Cd	48	112.40	321	8.65	HCP‡
Calcium	Ca	20	40.08	850	1.55	FCC‡
Californium	Cf	98	(249)			
Carbon	C	6	12.011	3727ᵇ	3.51ᶜ	DC‡
Cerium	Ce	58	140.12	804	6.77	FCC‡
Cesium	Cs	55	132.905	28.7	1.90	BCC
Chlorine	Cl	17	35.453	−101	1.9	Tetrag
Chromium	Cr	24	51.996	1875	7.19	BCC‡
Cobalt	Co	27	58.933	1495	8.85	HCP‡
Copper	Cu	29	63.54	1083	8.96	FCC
Curium	Cm	96	(247)			
Dysprosium	Dy	66	162.50	1407	8.55	HCP
Einsteinium	Es	99	(254)			
Erbium	Er	68	167.26	1497	9.15	HCP
Europium	Eu	63	151.96	826	5.24	BCC

*1961 atomic weights based on carbon-12 rounded off to a maximum of three decimal places.
†Atomic weights in parentheses are for the most stable isotopes.
‡Other crystal modifications exist.
ªMelting point at 28 atm pressure.
ᵇSublimes.
ᶜDiamond.

ELEMENT	SYMBOL	ATOMIC NUMBER	ATOMIC WEIGHT*	MELTING POINT, °C	DENSITY OF SOLID G/CC	CRYSTAL STRUCTURE OF SOLID
Fermium	Fm	100	(253)			
Fluorine	F	9	18.998	−220	1.3	
Francium	Fr	87	(223)	~27		BCC
Gadolinium	Gd	64	157.25	1312	7.86	HCP‡
Gallium	Ga	31	69.72	29.8	5.91	Orthorho
Germanium	Ge	32	72.59	937	5.32	DC
Gold	Au	79	196.967	1063	19.32	FCC
Hafnium	Hf	72	178.49	2222	13.1	HCP‡
Helium	He	2	4.003	−269.7		HCP
Holmium	Ho	67	164.930	1461	6.79	HCP
Hydrogen	H	1	1.008	−259.2		Hexag
Indium	In	49	114.82	156.2	7.31	FC Tetrag
Iodine	I	53	126.904	113.7	4.94	Orthorho
Iridium	Ir	77	192.2	2454	22.4	FCC
Iron	Fe	26	55.847	1537	7.87	BCC‡
Krypton	Kr	36	83.80	−157.3	3.0	FCC
Lanthanum	La	57	138.91	920	6.19	Hexag‡
Lawrencium	Lw	103	(257)			
Lead	Pb	82	207.19	327.3	11.34	FCC
Lithium	Li	3	6.939	180.5	0.53	BCC
Lutetium	Lu	71	174.97	1650	9.85	HCP
Magnesium	Mg	12	24.312	650	1.74	HCP
Manganese	Mn	25	54.938	1245	7.43	Cubic‡
Mendelevium	Mv	101	(256)			
Mercury	Hg	80	200.59	−38.4	14.19	Rhomb
Molybdenum	Mo	42	95.94	2610	10.2	BCC
Neodymium	Nd	60	144.24	1019	7.0	Hexag‡
Neon	Ne	10	20.183	−248.7	1.45	FCC
Neptunium	Np	93	(237)	637	19.5	Orthorho‡
Nickel	Ni	28	58.71	1453	8.9	FCC
Niobium	Nb	41	92.906	2415	8.6	BCC
Nitrogen	N	7	14.007	−210	1.03	Hexag‡
Nobelium	No	102	(253)			
Osmium	Os	76	190.2	2700	22.57	HCP
Oxygen	O	8	15.999	−218.8	1.43	Cubic‡
Palladium	Pd	46	106.4	1552	12.02	FCC
Phosphorus (white)	P	15	30.974	44.2	1.83	Cubic‡
Platinum	Pt	78	195.09	1769	21.45	FCC
Plutonium	Pu	94	(242)	640	19.3	Monoclinic‡
Polonium	Po	84	(210)	254	9.2	Monoclinic‡

(continued)

ELEMENT	SYMBOL	ATOMIC NUMBER	ATOMIC WEIGHT*	MELTING POINT, °C	DENSITY OF SOLID G/CC	CRYSTAL STRUCTURE OF SOLID
Potassium	K	19	39.102	63.7	0.86	BCC
Praseodymium	Pr	59	140.907	919	6.77	Hexag‡
Promethium	Pm	61	(147)	1027		Hexag
Protoactinium	Pa	91	(231)	1230	15.4	BC Tetrag
Radium	Ra	88	(226)	700	5.0	BC Tetrag
Radon	Rn	86	(222)	−71	4	FCC
Rhenium	Re	75	186.2	3180	21.04	HCP
Rhodium	Rh	45	102.905	1966	12.44	FCC‡
Rubidium	Rb	37	85.47	38.9	1.53	BCC
Ruthenium	Ru	44	101.07	2500	12.2	HCP‡
Samarium	Sm	62	150.35	1072	7.49	Rhomb‡
Scandium	Sc	21	44.956	1539	2.99	HCP‡
Selenium	Se	34	78.96	217	4.79	Hexag‡
Silicon	Si	14	28.086	1410	2.34	DC
Silver	Ag	47	107.870	960.8	10.5	FCC
Sodium	Na	11	22.990	97.8	0.97	BCC
Strontium	Sr	38	87.62	768	2.60	FCC‡
Sulfur (yellow)	S	16	32.064	119	2.07	Orthorho‡
Tantalum	Ta	73	180.948	2996	16.6	BCC
Technetium	Tc	43	(99)	2200	11.5	HCP
Tellurium	Te	52	127.60	450	6.24	Hexag
Terbium	Tb	65	158.924	1356	8.25	HCP
Thallium	Tl	81	204.37	303	11.85	HCP‡
Thorium	Th	90	232.038	1750	11.66	FCC‡
Thulium	Tm	69	168.934	1545	9.31	HCP
Tin	Sn	50	118.69	231.9	7.30	Tetrag‡
Titanium	Ti	22	47.90	1668	4.51	HCP‡
Tungsten	W	74	183.85	3410	19.3	BCC
Uranium	U	92	238.03	1132	19.05	Orthorho‡
Vanadium	V	23	50.942	1900	6.1	BCC
Xenon	Xe	54	131.30	−112	3.6	FCC
Ytterbium	Yb	70	173.04	824	6.96	FCC‡
Yttrium	Y	39	88.905	1509	4.47	HCP‡
Zinc	Zn	30	65.37	419.5	7.13	HCP
Zirconium	Zr	40	91.22	1852	6.49	HCP‡

*1961 atomic weights based on carbon-12 rounded off to a maximum of three decimal places.
†Atomic weights in parentheses are for the most stable isotopes.
‡Other crystal modifications exist.
ªMelting point at 28 atm pressure.
ᵇSublimes.
ᶜDiamond.

APPENDIX V

Definitions of Selected Terms Used in Materials Engineering Science

Acceptor Level An electron hole-type carrier in a *p*-type semiconductor.

Activated State An intermediate state in a rate process possessing an additional amount of activation energy.

Activation Energy The height of the energy barrier to a reaction, which must be surmounted by thermal excitation before the reaction can be activated.

Activation Free Energy The free energy barrier to a reaction, including the changes in entropy and enthalpy.

Active The condition of a metal surface when any protective or passivating films are absent and it is very susceptible to attack by corrosive agents.

Age-hardening Precipitation hardening, although the precipitate is not optically visible.

Allotropic forms Alternative arrangements of atoms in an element or compound.

Alloy A multicomponent liquid or solid in which the primary component is a metal.

Anion A negatively charged ion, resulting when an electronegative atom acquires one or more excess electrons.

Anisotropic Having properties which vary with spatial direction.

Anode The negative electrode or the metal in a cell which dissolves as ions and supplies electrons to the external circuit.

Aqueous Corrosion The solution of a metal into an electrolyte by galvanic action in the presence of oxygen and a dissimilar metal.

Arrhenius Equation An empirically based equation describing the rate of a reaction as a function of temperature and any energy barriers to the reaction.

Atactic Having side groups distributed randomly along a polymer chain.

Atomic Site A position normally occupied by an atom in a crystal lattice.

Austenite (γ) The face-centered cubic form of iron containing the highest carbon content, stable at intermediate temperatures.

Azeotrope A mixture or solution of liquids in phase equilibrium with a vapor phase of identical composition.

Bainite   A structure formed by the relatively rapid cooling of austenite, consisting of needles of cementite and highly strained ferrite. Bainite is nearly as hard as martensite, but not as brittle.

Body-centered Cubic   An atomic packing arrangement generally associated with metals, where one atom is in contact with eight atoms identical to it at the corners of an imaginary cube.

Bohr–Sommerfeld Orbits   The circular and elliptical planar orbits associated with the Bohr–Sommerfeld, solar system model of the atom.

Bond Angle   The angle between the two closest directional bonds of an atom.

Bond Length   The equilibrium center-to-center distance between atoms in the bound state.

Bond Strength   The energy which must be added to a bond in order to break it and separate its constituent atoms. Also the depth of the potential energy minimum in a plot of atomic potential energy vs relative separation distance.

Bonding Force   The force that holds two atoms together, resulting from a decrease in energy as two atoms are brought closer to one another.

Branched Polymer   A polymer having secondary chains branching from its main molecular chains.

Brittle Fracture   A mode of fracture characterized by the nucleation and rapid propagation of a crack, with little accompanying plastic deformation. Brittle fracture surfaces in crystalline materials can be identified by their shiny, granular appearance.

Brittle Transition Temperature   The temperature at which some normally ductile metals (BCC-iron, for example) undergo sharp decreases in impact strengths and become brittle.

Bulk Modulus   The proportionality constant between hydrostatic pressure and fractional decrease in volume.

Burgers Circuit   A closed path traversed mentally around a region of crystal imperfection. Its closure characteristics will reveal the nature of the particular dislocation that is causing the imperfection.

Burgers Vector   The vector by which the lattice on one side of an internal surface is displaced relative to the lattice on the other side as a dislocation moves along a surface; a property of the dislocation. Also, the closure vector of a Burgers circuit.

Cathode   The metal in a cell which accepts electrons from the external circuit; hydrogen atoms are neutralized from the solution on this cathode surface.

Cathodic Polarization   The accumulation of reaction products on the cathode surface to such an extent that the cathode is separated from the electrolyte and corrosion slows or stops.

Cation   A positively charged ion, the result of an electropositive atom losing one or more of its valence electrons.

Cation Site   The location of a symmetrically coordinated position between anions in an ionic crystal.

Cementite   The orthorhombic carbide of iron, Fe_3C.

Ceramic   A compound of metals and nonmetallic elements, usually involving ionic bonding.

Cleavage Planes   The characteristic crystallographic planes along which brittle fracture propagates. Usually the planes with the lowest surface energies.

Climb   The shifting of an edge dislocation from its original slip plane to an adjacent plane by the process of atomic diffusion.

Cold-work   Plastic deformation below the temperature at which recrystallization and extensive recovery take place.

Component   A raw material used in forming a multiphase system.

Composite Materials   Multiphase materials in which phase arrangement and distribution are controlled by mechanical instead of thermal or chemical means.

Compound   A multicomponent phase existing over a very narrow range of compositions.

Concentration Cell   A galvanic cell caused by a difference in electrolyte compositions.

Cooling Curve   A plot of time vs the temperature of a material undergoing one or more phase transformations. Such data provide a basis for equilibrium phase diagrams.

Coordination   The packing of atoms around another atom.

Coordination Polyhedron   A polyhedron resulting from connecting the centers of all of the atoms touching a central atom; each vertex represents one of the surrounding atoms.

Copolymer   The combination of two polymers into the same chain; the resulting configuration of the two may be random, alternating, block, or graft.

Corrosion   The deterioration of a material by chemical action, most commonly in the form of galvanic or electrolytic cell action.

Covalent Bond   A highly directional, primary bond arising from the energy reduction associated with the overlapping, half-filled orbitals of two atoms.

Creep   Plastic deformation which occurs as a function of time when material is subjected to a constant stress or load. Creep results from the continuous generation and movement of dislocations.

Critical Free Energy   The free energy of a particle with the critical radius and the maximum free energy of any particle of the new phase during nucleation.

Critical Radius   The minimum size which a particle of the new phase can be and still grow (that is, be a *nucleus*).

Critical Resolved Shear Stress   The resolved stress on an active slip system at which the slip is initiated. The true measure of the intrinsic strength of a material.

Cross-linking   Bonding polymer chains together at various points by means of primary bonds.

Cross-slip   Changing a slip plane by screw dislocation.

Crystal   A repetitive, three-dimensional arrangement of atoms or ions in a solid.

Crystal Structure   A mathematical representation of the relative positions of all atoms or ions in an ideal crystal.

Curie Temperature   The transition temperature between ferro- and paramagnetism.

Decarburization   Removal of carbon from the surface layers of steel by diffusion and chemical reaction at high temperatures.

Defect   Any of several types of imperfections in the crystal structure of a solid.

Degree of Freedom   The number of variables in a thermodynamic system that can be varied at will without disrupting the state of the system.

Degree of Polymerization   The number of mers in a polymer molecule of average molecular weight.

Dendrite   The crystallographic skeleton of a grain (usually in the form of plates or spines), composed of the first solid which crystallizes when a liquid or a gas is cooled.

Diffusion   The motion of matter through matter.

Diffusion Coefficient   The constant of proportionality in both of Fick's laws.

Dipole   Two charges with opposite signs coupled together, as in a molecule with a small, but definite, separation of positive and negative charge centers.

Dipole Moment   The product of the two charges of a dipole and the distance between their centers.

Dislocation   A line imperfection that can be visualized as the boundary between a region of an internal surface over which slip has occurred and another region over which no slip has occurred. The dislocation is called *edge* when the Burgers vector is perpendicular to the line vector of the dislocation; it is called *screw* when the Burgers vector is parallel to the line vector. When neither of these cases is true, the dislocation is called *mixed*.

Domain (Magnetic)   A small crystalline region of a magnetic material in which the individual atomic magnetic moments are aligned parallel to one another.

Donor Levels   Energy levels which result when dopant atoms with excess electrons create *n*-type semiconductors.

Dopant   An element added in very small quantities (usually ppm) to ultrapure silicon, germanium, or other similar host materials in order to create extrinsic semiconductors.

Ductile Fracture   A mode of fracture characterized by slow crack propagation which usually follows a zigzag path along planes where a maximum resolved shear stress has occurred. A surface experiencing ductile fracture generally has a dull, fibrous appearance.

Ductility   Strain at fracture.

Elastic Deformation   Change in the shape of a stressed body, recoverable when the stress is released.

Elastic Limit   The stress at which a material deviates from linear elastic behavior.

Elastic Strain   A dimensional change that is caused by a stress and is completely recoverable when that stress is removed.

Elastomer   A noncrystalline polymer which can be stretched to more than twice its original length, but contracts quickly when the load is released.

Electrode Potential   Voltage that develops at an electrode when it is placed in an electrolyte and compared with a standard electrode.

Electrolyte   An ionic solute.

Electrolytic Corrosion   Chemical deterioration of a metal by galvanic or electrolytic cell action.

Electromotive Force Series   An arrangement of metallic elements according to the voltage or electrochemical potentials they exhibit in a 1N ionic solution.

Electron Hole   The absence of an electron charge carrier in a semiconductor crystal, creating an effective positive charge carrier. A hole in a conduction energy band.

Electron Shell   A group of electrons having the same principal quantum number n (often designated by capital letters: K shell for $n = 1$, L shell for $n = 2$, etc.).

Electronegativity   A numerical system describing the relative tendencies of atoms to acquire electrons.

Endurance Limit   The stress level below which a material can be cyclically stressed for an infinite number of times without experiencing failure.

Energy Band   A "smeared-out" group of allowed energy states with similar magnitudes in a metal.

Energy Gap   The energy difference between the conduction and valence energy bands in a semiconductor.

Engineering Stress   The load on a sample divided by the original cross-sectional area.

Engineering Strain   The change in length of a sample divided by its original length.

Equicohesive Temperature   The temperature at which the effect of the grain boundary area on the strength of a material ceases to hinder and begins to assist dislocation movement by jog or climb.

Equilibrium   A state in which no changes in existing variables take place with time.

Equilibrium Diagram   A graphical representation of the pressures, temperatures, and compositions for which various phases are stable at equilibrium.

Equilibrium Distance   The interatomic distance at which the force of attraction equals the force of repulsion between two atoms.

Eutectic   An equilibrium phase transformation in which all the liquid phase transforms to two solid phases simultaneously on cooling.

Eutectoid Reaction $S \rightarrow S_A + S_B$ (all solid phases) on decreasing temperature.

Extrusions and Intrusions   Eruptions and depressions caused by dislocation concentrations on the surfaces of fatigued samples.

Fatigue Strength   The maximum cyclic stress a material can withstand for a given number of cycles before failure occurs.

Ferrite (α)   The body-centered cubic form of iron, stable at low temperatures.

Filler   A second phase added to a polymer to increase its mechanical strength.

Frenkel Imperfection   A point imperfection in which a cation vacancy is associated with an interstitial cation in an ionic crystal.

Fluidity   A measure of the ease with which a substance flows under shear stress. The reciprocal of viscosity.

Galvanic Cell   Two dissimilar metals in electrical contact in the presence of an aqueous electrolyte.

Galvanic Protection   To protect a metal by connecting it to a sacrificial anode or by impressing a d-c voltage and making it a cathode.

Galvanic Series   An arrangement of metallic elements and alloys according to their electrochemical potential in sea water.

Gel   A solid framework of colloidal particles linked together and containing a fluid in its interstices; the gel may be either rigid or elastic, depending on the nature of the linking and the geometry of the interstices.

Gibbs Phase Rule   The statement that at equilibrium the number of phases plus the degrees of freedom must equal the number of components plus two.

Glass   Any noncrystalline solid; applied more commonly to noncrystalline inorganic oxides than to noncrystalline polymers.

Glass Former   An oxide which forms a glass easily. Also an oxide which contributes to the network of silica glass when added to it.

Glass Transition Temperature   The center of the temperature range in which a non-crystalline solid changes from glass brittle to viscous or elastomeric.

Grain   A single crystal in a polycrystalline material.

Grain Boundary   A surface imperfection which separates crystals of the same crystal structure, but different orientations, in a polycrystalline aggregate.

Grain Growth   An increase in the average size of the grains in a polycrystalline material.

Graphitization   The transformation of cementite (Fe_3C) to iron and graphite in steel or cast iron.

Ground State   The lowest energy quantum state.

Growth   Increase in the size of the nucleus.

Hardenability   The ease with which martensite forms in a given material.

Hardness   A measure of the resistance of a material surface to penetration.

Heterogeneous   A mixture of two or more distinct phases.

Heterogeneous Nucleation   Nucleation occurring at surfaces, imperfections, severely deformed regions, or other structural features which lower the critical free energy.

Hole   A vacancy in a crystal structure caused by the absence of an atom.

Homogeneous   Uniform in structure and composition.

Homogeneous Nucleation   Nucleation occurring in a perfectly homogeneous material, such as in a pure liquid.

Homogenizing Eliminating all concentration gradients in each phase, usually by elevated temperature heat treatment.

Hot-work Deformation above the temperature at which recrystallization and extensive recovery take place.

Hybridization A rearrangement of orbitals which corresponds to replacing one set of solutions to the time-independent Schrödinger equation by an equivalent, but different, set; the tetrahedral molecular orbitals of carbon are a classical example of hybridization.

Hydrogen Bond A secondary bond arising from dipole attractions in which hydrogen is the positive end of the dipole.

Hypereutectic A composition to the right of the eutectic transformation point in a binary phase diagram.

Hypoeutectic A composition to the left of the eutectic transformation point in a binary phase diagram.

Hysteresis The noncoincidence of elastic loading and unloading curves under cyclic stress. Also the similar separation of curves observed in magnetization experiments and in the stress cycling of thixotropic, pseudoplastic non-Newtonian fluids.

Ideal Solution A solution involving zero enthalpy of mixing.

Inclusion A foreign or impurity phase in a solid.

Inhibitors Compounds added to an electrolyte to coat the anode or cathode and stop corrosion.

Intergranular Corrosion The preferential attack of grain boundary regions.

Intermediate An oxide which may be either a glass former or modifier, depending on the composition of the glass involved.

Intermediate Phase A phase which occurs in a range of compositions between those of the terminal phases.

Interstitial A point imperfection in which a foreign atom fits in an interstice between matrix atoms.

Interstitial Solid Solution A solid solution in which the atoms of one component are dissolved in the interstices between the atoms of the other component.

Invariant Reactions Equilibrium phase transformations involving zero degrees of freedom.

Ion Any atom or molecular group which has become charged.

Ion Concentration Cell The galvanic cell formed when two pieces of the same metal are electrically connected, but are in solutions of different ionic concentrations.

Ion Core An atom without its valence electron or electrons.

Ionic Bond A primary bond arising from the electrostatic attraction between two or more oppositely charged ions.

Ionic Crystal A crystal in which the predominant form of bonding is ionic.

Ionization Potential   The energy which must be added to an atom to remove an electron and create a positive ion.

Isomers   Compounds of the same composition, but different molecular or atomic configurations.

Isotactic   Having side groups all on the same side of a vinyl polymer chain.

Isotherm   A constant temperature line on a phase diagram.

Isothermal   At a constant temperature.

Isotropic   Having uniform properties in all directions.

Jog   An increase in the length of a dislocation line created by an intersection of dislocations. Often a geometric step in this line.

Jominy End Quench Test   A specialized test to measure the hardenability of a steel.

Lattice   A geometrical array of lines or points in which atoms are considered spheres, representing a certain type of crystal structure.

Lattice Point   One point in a lattice array, all the points of which have identical surroundings.

Lattice Translation   A vector connecting any two lattice points in the same array.

Ledeburite   The eutectic solid formed from austenite and cementite.

Lever Rule   Equations used for calculating the relative amount of each phase in an equilibrium, two-phase microstructure.

Linear Molecule   A long chain molecule with no side branches.

Linear Thermal Expansion Coefficient   A coefficient giving the proportionality between a linear dimensional change and the temperature change causing it.

Low-angle Boundary   A surface imperfection separating two misoriented regions of a crystal. The angle of misorientation is small (a few degrees or less).

Magnetic Saturation   The value of magnetic induction or the condition when all magnetic domains are aligned with their external magnetic field.

Magnetization Curve   An experimental plot of induced magnetization (B, Gauss) as a function of the applied magnetic field (H, oersteds), showing hysteresis properties.

Martensite   The structure formed when a diffusionless transformation of austenite is cooled rapidly enough to avoid the extensive formation of pearlite or bainite.

Mer   The repeating unit of a polymer molecule.

Mer Weight   The molecular weight of one mer.

Metallic Bond   A primary bond arising from the increased spatial extension of the valence electron wave functions when an aggregate of metal atoms is brought close together.

Metastable   Unstable, but having a very long lifetime.

Molecular Crystal   A crystal in which the subunits associated with each lattice point are molecules.

Molecular Orbital   An energy state for electrons, characteristic of molecules as opposed to free atoms.

Monomer   A substance possessing one or more chemical reaction sites from which a polymer can be made.

Modifier   An oxide which promotes crystallization when added to silica glass by breaking up its silica network.

Necking   The concentration of plastic deformation in a localized region of a sample under tension.

Negative Deviation   Negative enthalpy of mixing.

Network Structure   An atomic or molecular arrangement in which primary bonds form a three-dimensional network.

Newtonian Fluid   A fluid for which the applied shear stress is directly proportional to the resultant shear strain rate.

Node   A surface along which the wave function is zero and, therefore, the probability of finding an electron is zero.

Non-Newtonian Fluid   Any fluid whose shear stress-shear strain rate relation is other than a simple proportionality.

Nucleation   The beginning of a phase transformation, marked by the appearance of tiny regions of the new phase, called *nuclei*, which grow until the transformation is complete.

Nucleus   A small crystalline region formed in a melt from which larger crystals may grow.

Orbital   The quantum mechanical description of the state of an electron, including its energy and its time-averaged spatial distribution.

Overaging   Age-hardening carried to the point of precipitate agglomeration, so that dislocation pileup is reduced and softening occurs.

Oxidation   The increase in the positive valence of a metallic element during a chemical reaction. Principally used when referring to reactions with oxygen.

Oxygen Concentration Cell   The galvanic cell formed when two pieces of the same metal are electrically connected, but are in solutions of different oxygen concentrations.

Packing Factor   The ratio of the volume of atoms present in a crystal to the theoretical volume of the unit cell.

Passivation   The formation of a film of atoms or molecules on the anode surface to the extent that it is effectively separated from the electrolyte and corrosion slows or stops.

Pauli Exclusion Principle   The statement that no more than two electrons can occupy the same orbital, and that these electrons must spin in opposite directions.

Pearlite   The lamellar or layered eutectoid structure of ferrite and cementite, formed from austenite on cooling.

Peritectic Reaction $\alpha + L \rightarrow \sigma$ on decreasing temperature.

Permanent Dipole Bond   A secondary bond arising from dipole attractions in which the oppositely charged ends of the dipoles are electronegative and electropositive atoms.

Phase   A physically distinct region of matter with characteristic atomic structure and properties which changes continuously with temperature, composition, and any other thermodynamic variable. In principle, the various phases of a system are mechanically separable.

Phase Diagram   A graphical representation of the pressures, temperatures, and compositions for which phases are predicted under specified conditions (not necessarily of equilibrium).

Pilling–Bedworth Ratio   The ratio of the volume of oxide formed to the volume of metal consumed during oxidation.

Pit Corrosion   Rapid, local attack resulting from the formation of small anodic regions at a break in a passive film or corrosion deposit.

Plastic Deformation   The change in the shape of a stressed body which is not recoverable upon release of the stress.

Plasticizer   A lower molecular weight material added to a polymer to separate its molecular chains and prevent crystallization.

Plane of Reflection   A symmetry plane across which a lattice is completely and identically reproduced, as if it were reflected from the opposite side (a {100} plane in a cubic crystal, for example).

Poisson's Ratio   The ratio of transverse to axial strain caused by axial stress.

Polymer   A solid composed of long molecular chains. Also referred to as a plastic or a resin.

Precipitation Hardening   Strengthening due to the nonequilibrium precipitation of a second phase.

Preferred Orientation   The preferential alignment of either crystals or molecular chains, producing a similar orientation in every part of the solid.

Primary Creep   A transient component of creep for which the creep rate diminishes with time.

Primary Phase   A solid phase which forms in a microstructure at a temperature above that of an invariant reaction, and is still present after the invariant transformation has occurred.

Protective Oxide   An oxide which forms readily and covers a metal surface to such an extent that it essentially stops further reaction.

Pseudoplastic Fluid   A fluid whose apparent viscosity decreases with increasing shear stress, but which possesses no yield stress.

Quantum Jump   A change in energy from one *allowed* value to another.

Quantum Mechanics   A branch of physics in which the systems studied only possess discrete values of energy that are separated by forbidden regions. Quantum mechanics should be distinguished from continuum mechanics in which a continuum of energies is assumed possible.

Quantum Numbers   A set of integers (n, l, and m) representing the discrete solutions to the time-independent Schrödinger wave equation. A quantum number S may be included to indicate the spin of the electron ($\pm\frac{1}{2}$).

Rate Laws   Relationships between oxide thickness and time of oxidation.

Recovery   Relief of internal stress by thermally activated dislocation motion.

Recrystallization   The nucleation and growth of unstrained regions in a plastically deformed material.

Recrystallization Temperature   The temperature at which a material recrystallizes in one hour, or some other definitive period.

Relaxation Time   The time required for the time-independent component of strain to reach $1/e$ of its final value.

Resistivity   The reciprocal of conductivity.

Schmid's Law   The statement that critical resolved shear stress is a material constant at constant temperature and constant strain rates.

Schottky Imperfection   A point imperfection in an ionic crystal which occurs when a cation vacancy is associated with an anion vacancy.

Secondary Creep   The constant, strain-rate component of creep.

Selective Oxidation   Oxygen preferentially attacking one of the components or phases in an alloy.

Sessile Dislocation   A dislocation in which the active slip system does not contain both the Burgers vector and the dislocation line. An immobile dislocation.

Shear Modulus   The proportionality constant between elastic shear strain and shear stress.

Sintering   Bonding powders together by solid state diffusion, resulting in the absence of a separate bonding phase.

Slip   The sliding displacement of one part of a crystal relative to another part by the motion of a dislocation or dislocations. The movement of a dislocation in one of its slip planes.

Slip Plane   Any crystallographic plane which contains both the Burgers vector and the line vector of a dislocation.

Space Lattice   A three-dimensional array of points, called *lattice points*, each of which has identical surroundings. Also known as a Bravais lattice.

Spherulite A grain of a crystalline polymer in which the orientations of molecular chains vary, giving the molecular structure of the grain an approximately spherical symmetry.

Stacking Fault A surface imperfection which results from the stacking of one atomic plane on another out of sequence; the lattices on both sides of the fault have the same orientation, but are translated with respect to one another by less than a lattice translation.

Stacking Sequence The manner in which close-packed planes are stacked on top of one another in a particular structure. Planes stacked so that their atoms are directly above one another are represented by the same letter; for example, *ABABA* . . . represents HCP stacking, while *ABCABCAB* . . . represents FCC stacking.

Steric Hindrance The interference caused when two or more atomic groups try to occupy the same space.

Stress Concentration The magnification of the level of an applied stress in the region of a notch, void, or inclusion.

Stress Corrosion The preferential attack of stressed areas in corrosive environments which, alone, would not cause corrosion.

Substitutional Impurity Atom A point imperfection in which a foreign atom occupies a site normally filled by a matrix atom in a perfect crystal.

Substitutional Solid Solution A solid solution in which some of the atomic sites normally occupied by atoms of the matrix component are filled by atoms of another component.

Syndiotactic Possessing side groups distributed in a regularly alternating manner along a vinyl polymer chain.

Thermoplast A polymer which softens and melts upon heating.

Thermoset A polymer which undergoes thermal degradation of its structure upon heating, instead of softening and melting.

Thermoelastic Effect The temperature change caused by a change in the stress state.

Tie Line The locus of all points in a two-phase region of an equilibrium diagram which represents equilibrium between the same two phase compositions at the same termperature.

Tilt Boundary A low-angle boundary in which the misorientation is a rotation about an axis lying within the boundary.

Toughness Modulus A measure of the energy absorbed by a material in breaking it. $\hat{T}$ is the area under the stress-strain curve.

Transistor A semiconductor device composed of two or three segments separated by *p-n* junctions. A transistor performs the same operations that some electron vacuum tubes do.

True Strain The natural logarithm of the ratio of the instantaneous length of a plastically deformed sample to its original length. Also the natural logarithm of the ratio of the original cross-sectional area of such a sample to its instantaneous cross-sectional area.

True Stress   The load on a sample divided by its instantaneous cross-sectional area.

Twinning   A homogeneous shear which reorients a deformed lattice into a mirror image of its parent lattice across the twinning plane.

Ultimate Tensile Strength   The maximum engineering stress a material can withstand.

Unit Cell   A convenient repeating lattice unit with lattice translations as its edges. A crystal structure unit cell is a unit cell which indicates atom positions as well as lattice points.

Viscoelastic   Simultaneously possessing one of the numerous combinations of the characteristics of viscous flow and elastic strain.

Viscosity   The proportionality constant between shear stress and rate of shearing strain in viscous flow. Apparent viscosity is the ratio of stress to strain rate.

Work-hardening (Strain-hardening)   An increase in hardness and flow stress which occurs with increasing plastic deformation.

Yield Strength   The stress needed to produce a specified amount of plastic deformation (usually a 0.2 per cent change in length).

Young's Modulus (E)   The proportionality constant between elastic strain and uniaxial stress.

INDEX

5,7 a) AMT. OF LEDEBURITE
   IS AMT. OF LIQUID

PRIMARY $\gamma = 75\left(\frac{4.3-3}{4.3-2.06}\right) = 43.5$

$L = 75\left(\frac{3-2.06}{4.3-2.06}\right) = 31.5$ gm LEDEBURITE

γ IN LEDEBURITE $= 31.5\left(\frac{6.67-4.3}{6.67-2.06}\right) = 16.2$ gms EUTECTIC γ

gms $Fe_3C = 31.5 - 16.2 = 15.3$

$31.5\left(\frac{4.3-2.06}{6.67-2.06}\right) = 15.3$

TOT. $\gamma = 75\left(\frac{6.67-3}{6.67-2.06}\right) = 59.7 = 43.5 + 16.2$

b)
PRIMARY $Fe_3C = 75\left(\frac{3-.8}{6.67-.8}\right) = 28.1$ g

$\gamma = 75\left(\frac{6.67-3}{6.67-.8}\right) = 46.9$ g

γ changes to pearlite, so
46.9 g OF PEARLITE

c) α IN PEARLITE $= 46.9\left(\frac{6.67-.8}{6.67-.02}\right) = 41.3$ g OF α

Fe_3C IN PEARLITE $= 46.9\left(\frac{.8-.02}{6.67-.02}\right) = 5.6$ g OF Fe_3C

TOT. $Fe_3C = 28.1 + 5.6 = 33.7$ g

PRIMARY Fe_3C

$75\left(\frac{3-.02}{6.67-.02}\right) = 33.7$ g .

5,8 a) $55\left(\frac{12-3.5}{12-.5}\right) = 40.7$ g

b) HOW MUCH EUTECTOID SOLID (β)

$\beta = 55\left(\frac{3.5-1.4}{6-1.4}\right) = 25.1$ g

α IN EUTECTOID $= 25.1\left(\frac{11.4-6}{11.4-1.4}\right) = 13.56$ @ 575°C

α IN EUTECTOID @ R.T. $= 25.1\left(\frac{12-6}{12-.5}\right) = 13.1$

@ R.T. β' IN EUTECTOID SOLID $= 25.1 - 13.1 = 12$ g

$25.1\left(\frac{6-.5}{12-.5}\right) = 12$

TOT β' @ R.T. $= 55\left(\frac{3.5-.5}{12-.5}\right) = 14.3$

β' FROM $\alpha = 14.3 - 12 = 2.3$ g

$20g$ · B , $50gm$ - A , $30g$ - H_2O

@ $30°C$

a) HOW MANY PHASES PRESENT

b) WHAT IS THE COMP. OF EACH PHASE

c) HOW MUCH OF EACH PHASE IS PRESENT

20% B , 50% A , 30% W

a) 2 PHASES

b)

A. .015 B   B. .345 B

     .4 A   .59 A

     .585 W   .065 W

WATER - RICH   BENZENE - RICH

c) A - WATER RICH PHASE

$\dfrac{100g}{100} \left(\dfrac{.345 - .20}{.345 - .015} \right) = 43.9$

$43.9 g$ OF W RICH PHASE , SO, $55 g$ OF BENZENE RICH PHASE

7.14 $\epsilon_y = \dfrac{\sigma_y}{E} = \dfrac{2 \times 10^4 \frac{\#}{IN.^2}}{30 \times 10^6 psi} = .000667$

$\nu = .31$ $\epsilon_x = \epsilon_z = -\nu \epsilon_y = -(.31)(.000667) = -.000207$

$\epsilon_x = \epsilon_y = -\nu \epsilon_z = -(.31)(.000333) = -.000103$

$\epsilon_y = \epsilon_t = -\nu \epsilon_x = -(.31)(.000667) = .000207$

$\delta_v = \dfrac{\Delta V}{V_o} = \dfrac{(x + \epsilon_x)(y + \epsilon_y)(z + \epsilon_z) - xyz}{xyz}$

11. $mpy = \dfrac{534\ W}{\rho A T}$

$\dfrac{mils}{YEAR} = \dfrac{mg}{g}\Bigg|\dfrac{cm^3}{IN.^2}\Bigg|\dfrac{g}{hns.}\Bigg|\dfrac{IN.^3}{1000mg(2.54)^3 cm^3}\Bigg|\dfrac{1000\ mils}{IN.}\Bigg|\dfrac{24(365)\ hns}{YEAR}$

$= 534$

11.4 = $\eta = 22\ VOLTS$ $\qquad 1'' \times 3''\ SAMPLE$ $\qquad \eta_A = \beta\ ln\ {}^i/_{i_0}$

TAFEL CONST. = .08 $= \beta$

$i_0 = CURRENT\ DENSITY = 10^{-6}\dfrac{AMP}{cm^2}$

$i = i_0 e^{\eta_A/\beta} = 10^{-6}\ e^{22/.08} = 10^{-6}\ e^{2.75} = 10^{-6}(15.6) = 1.56\times10^{-5}\dfrac{AMP}{cm^2}$

$\dfrac{1.56\times10^{-5}\ AMP}{cm^2}\Bigg|\dfrac{1\ coul}{AMP\ SEC}\Bigg|\dfrac{gm\ equi\ wt.}{96,500\ coul.}\Bigg|\dfrac{55,847\ g}{2\ equi\ wts.} = 4.51\times10^{-9}\dfrac{g}{cm^2 sec}$

$\dfrac{4.51\times10^{-9}\ g}{cm^2\ sec}\Bigg|\dfrac{}{7.87}\Bigg|\dfrac{}{2.54}\Bigg|\dfrac{1000\ mils}{IN.}\Bigg|\dfrac{(3600)(24)(365)\ sec}{YEAR}$

$\boxed{1.56\times10^{-5}\ A/cm^2 = 7.12\ mpy}$

$\underline{mpy = 2.19\times10^{-3}\ mg/cm^2}$

11.7 $\quad \Delta E = \dfrac{RT}{NF}\ ln\ {}^{a_1}/_{a_2}$

$= \dfrac{8.32\ joules}{mole\ °K}\Bigg|298°K\Bigg|\dfrac{mole}{2\ equi}\Bigg|\dfrac{equi}{96,500\ coul.}\Bigg|\dfrac{volt\ coulomb}{joule}\ ln\ {}^1/_{10^{-6}}$

$= .0286 \cancel{(10} .0286\ ln(10^5) = .0286\ (11.5) = .148\ volts$

11.8 $\quad G = -nFe$

$= -2equi\Bigg|\dfrac{96,500\ coul.}{equi}\Bigg|\dfrac{.148\ volts}{}\Bigg|\dfrac{1\ joule}{volt\ coul.}\Bigg|\dfrac{kcal}{4186\ joule}$

$= -6.820\ kcal/mole$

11.9 $\quad$ Fe - Cu $\quad$ couple

$Fe = Fe^{2+} + 2e^-$ $\quad \sim t_0 = -.44\ v$ $\quad$ anode

$Cu = Cu^{2+} + 2e^-$ $\quad \sim E_0 = .337v$ $\quad$ cathode

$\quad E_{0_{Fe}} - E_{0_{Cu}} = -.44 - .337 = -.777$